Water-Soluble Synthetic Polymers: Properties and Behavior

Volume I

Author

Philip Molyneux

Department of Pharmacy
Chelsea College
University of London
England

CRC Press
Taylor & Francis Group
Boca Raton London New York

CRC Press is an imprint of the
Taylor & Francis Group, an **informa** business

First published 1983 by CRC Press
Taylor & Francis Group
6000 Broken Sound Parkway NW, Suite 300
Boca Raton, FL 33487-2742

Reissued 2018 by CRC Press

Library of Congress Cataloging in Publication Data

Molyneux, Philip.
Water-soluble synthetic polymers.

Bibliography: p.
Includes indexes.
1. Water-soluble polymers. I. Title.
QD382.W3M64 1983 547.7 82-22649
ISBN 0-8493-6135-4 (v. 1)
ISBN 0-8493-6136-2 (v. 2)

A Library of Congress record exists under LC control number: 82022649

ISBN 13: 978-1-315-89857-5 (hbk)
ISBN 13: 978-1-351-07767-5 (ebk)

Visit the Taylor & Francis Web site at http://www.taylorandfrancis.com and the
CRC Press Web site at http://www.crcpress.com

PREFACE

Water-soluble synthetic polymers are a family of materials that have been developed commercially and studied scientifically at an accelerating pace in recent years. Partly, this is a reflection of the increasing diversity in the applications of water-soluble polymers as a group — in mineral-processing, water-treatment, oil-recovery and many other areas. Partly, also, it reflects an increasing awareness of the remarkable features of water as a solvent towards polymers and other substances, and of the need for research on simple polymers of this type to help us understand the way in which the more complex biopolymers function in their native aqueous environment of the living cell.

Although several monographs and reviews have appeared on individual polymers of this type, and their applications and other technical aspects have also been discussed, this is apparently the first book to deal with the physical chemistry of water-soluble synthetic polymers as a group. This collective survey enables their properties and behavior to be compared, and to be correlated with their molecular structures for predictive purposes. However, this has made it necessary to critically re-appraise much of the earlier fundamental work, so that current discussions of more recent work can be put on a proper basis. Thus, of the 1800 or so references cited, the middle two-thirds relate to the twenty-year period centered on about 1968. Nevertheless, sufficient key recent references have also been included so that the existing "state of the art" is delineated.

To this same end, postscripts have been added to Chapters 2, 3, and 4 of Volume I, and to Chapters 1 and 2 of Volume II, listing some significant work that came to light after the main texts of these had been completed.

Clarification of the chapter format, nomenclature, deka-deci notation, notes on the tables, and abbreviations and symbols, will be found in Appendixes I to V, placed at the end of these volumes.

Many of the proper names used for commercial materials (polymers and others) mentioned in the text are registered trademarks, and these are accordingly put in italics to distinguish them from those not so-protected (e.g. common names for dyes, official names for pharmaceuticals). Such a distinction is uncertain in several cases because of the looseness of usage of such names, particularly in the scientific literature.

It is a pleasure to express my thanks to my colleagues on the staff of Chelsea College, particularly those in the Library and in the Audio-Visual Service Unit, for their unfailing help given in these increasingly difficult times.

Philip Molyneux
July 1982

THE AUTHOR

Philip Molyneux graduated with a B.Sc. (Honours Chemistry) degree from the University of Birmingham in 1955, and obtained his Ph.D. degree in Polymer Chemistry from the same University in 1958. After research work at Illinois Institute of Technology, Chicago, and at the University of Keele, Staffordshire, he joined Chelsea College, University of London, in 1963 as a Lecturer in Pharmaceutical Chemistry.

Dr. Molyneux was elected a Fellow of the Royal Institute of Chemistry (now, Royal Society of Chemistry) of Great Britain in 1970; he is also a Chartered Chemist.

Dr. Molyneux has presented invited lectures to various international meetings, contributed to several books, and published numerous papers on the properties and behavior of water-soluble polymers, as well as on polymer characterization, chemical kinetics, micelle behavior, and adsorption. His current major research interests include the study of the interactions between polymers and other solutes in aqueous solution, and the theory of permeation through polymer films.

To my wife, Aree, for her patience and encouragement

"And a thousand thousand slimy things"
S. T. Coleridge; ***The Rime of the Ancient Mariner***

VOLUME OUTLINES

Volume I

Volume II

TABLE OF CONTENTS

Volume I

Chapter 1

INTRODUCTION

"In my beginning is my end . . . "
T. S. Eliot: *East Coker*

I. PREAMBLE

The aims of this book are to survey, summarize, and discuss the present state of our knowledge of the physics and chemistry of water-soluble synthetic polymers; the intention is to provide, in this way, a basis from which ultimately the observed properties and behavior of these materials may be correlated with their molecular structures and with the interactions that take place at the molecular level. Although these and closely related materials have many diverse commercial and technical applications, it is not intended to deal with these aspects in any detail, particularly because they have been covered in the various texts and reviews to be listed subsequently. However, in many cases the published scientific studies that are to be considered here have clearly been influenced by the commercial availability and industrial importance of these materials, while any theories and concepts that are developed to explain their properties and behavior will also be valuable in interpreting and guiding their practical applications.

This introductory chapter is designed to put the subject into its wider context and to deal with some of the background to this area, before proceeding to consider the properties and behavior of the specific polymers.

II. WATER-SOLUBLE POLYMERS

The great majority of polymers or polymer-based materials that we encounter are water insoluble — and in many cases indeed they are required to be water resistant — for evident reasons concerned with their applications and end uses; we may cite as examples materials of natural origin such as rubber, leather, wood, silk, cotton, and paper, as well as the plastics, elastomers, fibers, textiles, and similar materials that have been made synthetically.*

However, there is a relatively small but important group of polymers that are water soluble; these materials are often referred to collectively as "water-soluble gums" or "water-soluble resins", particularly where their solubility is an important feature of their applications. As with polymers in general, it is convenient to divide these into three groups according to their origin, i.e., natural, modified natural (semisynthetic), and synthetic. The natural water-soluble polymers include many important examples of the so-called biopolymers,[1] although this latter term is often restricted to materials which are actively involved in the functioning of biological systems, whereas many technically important materials of this group (e.g., bacterial polysaccharides, vegetable gums, and other exudates) are seemingly more or less accidental by-products of the system.

It is only within the last 20 years or so that water-soluble polymers have been recognized as a group in its own right. This recognition was signaled by the publication in 1962 of a text edited by Davidson and Sittig concerned with the sources, properties, and applications

* The word "polymer" is used throughout these volumes as a general term for any macromolecular substance having chains composed of large numbers of identical or at least closely similar units. On this basis the term is used to cover even complex biopolymers such as, for example, the proteins, which are then viewed as highly specific copolymers of the twenty or so naturally occurring amino acids.

of these materials, and further confirmed by the appearance of a second edition 6 years later;[2] the contents of this latter book indicate the range and types of materials of main interest at that time.[4-13] More recently, a similar but more extensive handbook, edited again by Davidson,[14] has been published which in many respects is an expanded and updated version of the previous text; the individual chapters indicate again the types of these materials that are currently of interest and their applications.[15-38]

Various conferences and symposia have been held on this subject and the proceedings of some of these have been published.[39-41] The general applications of water-soluble polymers in drugs, cosmetics, and foods have been reviewed by Idson;[42] their properties and applications in synthetic-resin emulsions (latexes) and the associated paints have been discussed by Warson,[43] where one important aspect is their interactions with small-molecule components such as preservatives.[43-45] The polysaccharide gums have been reviewed in a volume edited by Whistler and BeMiller.[46] Many of the water-soluble polymers have medical and pharmaceutical applications, and this area has been cataloged, reviewed, and discussed.[47-53]

III. WATER-SOLUBLE SYNTHETIC POLYMERS

A. General Features

The synthetic type forms of course only a subgroup of the water-soluble polymers as a whole, and at first sight might seem to have very restricted subject matter.* However, as the contents of the subsequent chapters show, this is far from the case since in common with other such areas the study of the scientific aspects of these materials has greatly expanded over the past few years. The literature in certain areas, particularly that of polyelectrolytes (see Chapter 1 of Volume II) and of the interfacial activity of polymers (see Chapter 4 of Volume II), has become so extensive that in many cases it is necessary to limit the present discussion to a representative selection of key references from the published work. Furthermore, the development of more sophisticated — but at the same time more highly specialized — techniques for studying aqueous solutions of polymers (e.g., laser light scattering, nuclear magnetic relaxation) has made it more difficult to discuss critically the whole range of available data.

The principal objective has therefore been to construct a sound framework in which information on the properties and behavior of these materials can be presented in summarized form, with a bibliography sufficient to form the basis for anyone wishing to become involved in the infinite regress of computer searching.

Certain of the relevant information is indeed available from standard handbooks,[54,55] textbooks,[56-60] and encyclopedias,[61] but in the main such sources are concerned more with the behavior of water-insoluble synthetic polymers because of their greater technical importance, and correspondingly with the behavior of polymers in nonaqueous solvents — media which are often preferred even for water-soluble polymers because of doubts about the interpretation of aqueous solution behavior. Only Morawetz, in his monograph[62] on the behavior of polymers in solution, has dealt specifically with the aqueous solution behavior of synthetic and other polymers.

The only monograph that has appeared so far on the physical chemistry of water-soluble synthetic polymers is a recently published Russian one.[63] The present author has also previously discussed the physiochemical behavior of these polymers vis-à-vis water, alongside

* There is, of course, an ambiguity in the use of the term "synthetic" in this context, since great advances are being made in the synthesis in vitro of "natural" polymers such as proteins, nucleic acids, and carbohydrates. However, in this book the term has its common connotation, i.e., for carbon chain polymers of the vinyl and acrylic types, as well as heterochains of the polyoxide and polyimine types, which are currently of the greatest interest from the commercial and scientific viewpoints.

that of the insoluble ones.[65] Reviews and discussions of specific polymers are given in the general references already cited, and are naturally also listed in the corresponding sections of subsequent chapters of the book. The general applications of individual polymers are also listed in these same sources. Amongst more specific applications of these polymers as a group, we may note their use in affinity chromatography (in cross-linked form),[66] as supports (again, in cross-linked forms) in systems for insolubilized enzymes,[67] and in dental cements.[68] However, all applications of these materials need to take into account possible carcinogenicity and other hazards to health not only from the polymer itself but also from any residual monomer.[69]

Emphasis has already been placed upon the need to correlate the properties and behavior of these materials with the molecular structures of the components. To some extent, this is achieved by studies on the same polymer in a number of different solvent media (including water); correspondingly, for this purpose any types of third component present (small-molecule cosolute; second polymer solute; interface) need to be varied in a regular logical fashion. However, to obtain information on the correlation with the structure of the polymer, it is necessary to study different polymers (preferably, closely related ones) under comparable conditions in aqueous solution; among the few cases of published work of this kind we may cite as exemplars the studies by Silberberg et al. on the light scattering and viscosity of poly(acrylic acid), poly(methacrylic acid), poly(acrylamide), and polymethacrylamide;[70] by Andersson on partial specific volume and its pressure dependence of poly(ethylene oxide), poly(methacrylic acid) (nonionized and ionized), poly(vinyl alcohol), and polyvinylpyrrolidone (as well as a variety of natural and modified natural polymers in aqueous solution, and of synthetic polymers in nonaqueous solution);[71] and by Roy-Chowdhury and Kale on density and ultrasonic velocity measurements (giving partial specific volume and adiabatic compressibility) of poly(acrylic acid) (nonionized and ionized) and polyacrylamide,[72] poly(methacrylic acid) (nonionized and ionized),[73] polymethacrylamide,[74] poly(maleic acid-*co*-styrene) (nonionized and ionized),[75] and poly(4-vinyl-N-butylpyridinium bromide).[76]

In selecting the synthetic polymers for discussion in these volumes, usually the examples are restricted to those which are soluble in water at normal temperatures. This has meant, in particular, excluding certain important water-sensitive (i.e., marginally water-insoluble) materials which even in the linear (uncross-linked) form are only highly swollen into "hydrogels" by water under these conditions; amongst such polymers excluded in this way from further discussion is poly(hydroxyethyl methacrylate) [poly(ethylene glycol monomethacrylate); polyHEMA; PHEMA]:

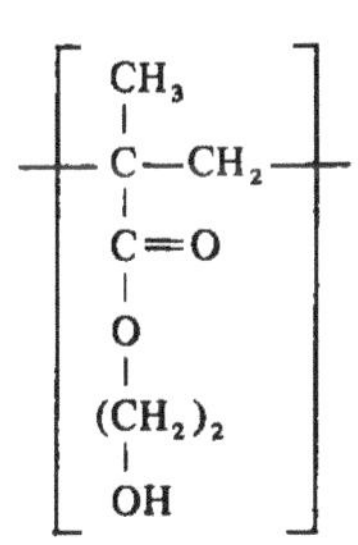

which is of current interest both from the scientific and medical viewpoints.[77-80] The propyl homolog of this polymer has also been studied.[81]

However, the water-insoluble high polymers of propylene oxide are considered along with the water-soluble oligomers in Chapter 2, and the vinyl acetate-vinyl alcohol copolymers with high vinyl acetate content (and hence water insoluble) are considered with the water-soluble types of low vinyl acetate content in Chapter 4.

B. Comparisons with Other Water-Soluble Polymers

In this section we look at some useful comparisons between the features of the synthetic water-soluble polymers and those of the natural and the semisynthetic types.

From the fundamental standpoint, despite the extensive work that has been carried out on the behavior in aqueous solution of natural polymers such as the polysaccharides,[82] proteins and polypeptides,[83,84] and polynucleotides,[83] much remains to be learned about the interactions taking place at the molecular level in these systems. With the proteins, much helpful information on these interactions can be obtained from the study of the synthetic homopolypeptides;[85] these are indeed also interesting in the present context, since they have structural similarities to the N-substituted polyamides (see Chapter 3), while N-substituted polypeptides such as polysarcosine:[86]

$$\text{-}[\text{N(CH}_3\text{)}-\text{CH}_2-\text{CO}]\text{-}$$

have similarities both to N,N-disubstituted polyamides (see Chapter 3) and to polyvinylpyrrolidone (see Chapter 4). Synthetic polypeptides with side chains containing N-(hydroxyalkyl)-substituted amide groups have been studied for the relative ease of their formation of the α-helix,[87] and for their binding ability towards sodium dodecyl sulfate;[88] such types of polymers are also interesting clinically as blood plasma expanders.[89]

The modified natural polymers comprise mainly the diverse derivatives of cellulose. However, from the fundamental standpoint, the interpretation of studies on these is generally complicated by the wide molecular weight distribution of common commercial samples, and also by uncertainties in the manner in which the substituent groups introduced by the chemical modification are distributed on each monosaccharide unit and between different monosaccharide units.

In these respects, synthetic polymers have the advantage that they are generally much simpler in their chain structure than most natural or modified natural polymers, while with synthetic copolymers a variety of degrees of complexity can be introduced in a controlled fashion. This makes these water-soluble synthetic polymers almost ideal model substances for developing techniques for studying aqueous solutions of polymers, and for testing theories of the behavior of such systems.

Synthetic water-soluble polymers have indeed been suggested from time to time as models of the natural polymers; this type of suggestion dates back more than 40 years, to Kern's proposal of poly(acrylic acid) as a model for albumin.[90] Even if their chain structures supported this type of analogy the differences in chain conformation would not, since water-soluble synthetic polymers mainly have the random coil type of conformation which is shared only by certain polysaccharides and their modified forms, but not by the proteins or the nucleic acids in their native states. These differences are bridged to some extent by the fact that many of the globular proteins can be unfolded from their native compact conformation into an essentially random coil state,[91] while on the other hand certain synthetic polymers such as alkylated poly(ethylene imine) ("synzymes") (Chapter 2), poly(methacrylic acid) (Chapter 3), maleic acid copolymers (Chapter 1 of Volume II), and "polysoaps" such as N-alkylated poly(4-vinylpyridine)[92] (Chapter 1 of Volume II) can all have relatively compact conformations in aqueous solution.

However, although in general they can seldom be considered as good, close analogs of the biopolymers, nevertheless the study of the relatively simple-structured synthetic polymers should help in evaluating the strengths of interactions amongst their component groupings and the water molecules in aqueous solution. Basic information of this kind should then be applicable to interpreting the behavior of the more complex-structured biopolymers.

One special feature of the synthetic polymers, which has its uses and its drawbacks, is that the presence of many linked chain units has an amplifying effect on interactions which would be weak or even undetectable for wholly small-molecule solutes. For example, considering the conformation of an isolated flexible-chain polymer molecule, it only requires two distantly connected units on a chain of, say, 1000 units to "complex" together (forming a transient cross-link) for there to be a marked effect on the volume of its domain; significant interactions between the two polymer molecules can also result from such an apparently minor effect occurring intermolecularly. Similarly, in polymer/small-molecule cosolute interactions (see Chapter 2 of Volume II), the cosolute may be bound by a chelating or enveloping effect from several consecutive monomer units; also, the bound cosolute molecule may form transient cross-links by binding to two distantly connected units on the same chain, or to units on different polymer chains. This amplifying effect shows itself even more markedly in the interactions between different polymers (Chapter 3 of Volume II), and in the interactions between polymers and interfaces (Chapter 4 of Volume II).

Furthermore, nonionic water-soluble synthetic polymers form a very useful halfway house between two important and widely studied groups, i.e., polymers in nonaqueous solution and polyelectrolytes (which include many biopolymers) in aqueous solution, since potentially they enable a separation to be made between the effects arising from the change of the solvent (nonaqueous → water), and those arising from the introduction of the charged groups. What is particularly useful in the present context is that certain synthetic polymers are soluble in nonaqueous as well as in aqueous media [this applies to poly(ethylene oxide), poly(propylene oxide), poly(vinyl methyl ether), poly(ethylene imine), poly(acrylic acid), poly(methacrylic acid), and polyvinylpyrrolidone], and some can be ionized to controlled extents [i.e., weak acids such as poly(acrylic acid) and poly(methacrylic acid) and weak bases such as poly(ethylene imine)].

Comparing the practical applications of these different types of water-soluble polymers, the synthetic type is often used in competition with others of the same type or with the natural or modified natural polymer. The different types may be used together, as in a complex formulation such as a latex (emulsion) paint where the various polymers are each assumed to have a definite and distinct function;[43] in other applications an active but relatively expensive polymer such as poly(vinyl alcohol) is "bulked up" by a cheaper but less active natural material such as starch.[8,34]

Of course, in choosing a polymeric material the designer or technologist is not greatly concerned about its origin or its molecular structure as long as it is efficient in the effect or function desired and also fulfills certain other requirements such as price, availability, and physical and chemical stability. With natural polymers such as the gum exudates there are sometimes climatic and political factors leading to limitations in availability of material with the required properties; also, the natural and the modified natural polymers are in many cases easily hydrolyzed and oxidized, and they are also biologically labile. By contrast, synthetic polymers are generally not limited in their availability (although they are normally more expensive than the other two types) while they are also in most cases much more chemically and biologically stable.

C. Water and Aqueous Solutions

In studying the behavior of synthetic polymers in aqueous solution, two questions have to be considered (and if possible, answered). First, to what extent does this behavior differ from that of polymers in nonaqueous solution, and indeed to what extent must water be viewed as a special (if not unique) solvent? Second, to what extent does this behavior differ from that of small-molecule analogs? (This is, of course, a perennial problem in polymer science, but in this case there may be specific differences that arise when it is water that is the solvent medium.)

This immediately brings to the forefront the peculiarities of water as a liquid and as a solvent. Not so long ago, even with aqueous solutions the solvent was often treated as an essentially structureless medium in which the solute molecules reacted and interacted — a viewpoint which with water was seemingly a case of "familiarity breeds contempt". Over the last 20 years, however, and particularly over the last 10 years, there has been an expansion of interest in water and a greater appreciation of its idiosyncrasies. This is reflected in the publication of a multivolume treatise on this subject, edited by Franks.[64,93-97] There have been numerous other reviews and published discussions on this subject, amongst which are particularly useful articles by Frank[98] and by Symons.[99]

It would not be appropriate to enter here into a detailed discussion of water and aqueous solutions, neither is the author competent to do so, but a brief account is necessary to put aqueous solutions of synthetic polymers into their context, and to provide background for later discussion.

It is generally agreed that water is a highly hydrogen-bonded liquid, and that this bonding has many features (especially, an essentially tetrahedral coordination) in common with those in ice. However, it is immediately beyond this point that the differences of opinion arise, the protagonists being the proponents of the "uniformist" school and those of the "mixture" school. The "uniformist" school views the liquid as a somewhat broken down form of ice, with the extent of the breakdown increasing steadily with rise in temperature from 0 to 100°C (and indeed, beyond that up to the critical point). By contrast, the "mixture" school views the liquid as a solution of a bulky ("ice-like"?) component in a compact component (monomeric water molecules?). This second model, although it can be critized on several grounds, has the merit that it takes into account the known highly cooperative character of hydrogen bonding, i.e., that the formation of a donor bond from a water molecule to a second one increases the activity of the latter as a donor (but by the same token decreases its activity as an acceptor). This means, e.g., that of the three possible forms of the linear hydrogen-bonded trimer:

```
     H        H        H
I    O—H···O—H···O—H

     H        H        H
II   O—H···O···H—O
              H

        H               H
III  H—O···H—O—H···O—H
```

the "head-to-tail" form I, because of the cooperative character of its bonding, is markedly more stable than the "introverted" form II or the "extraverted" form III.[100] Higher multimers of the cooperative form I acquire even more stability by linking up the ends to form rings, e.g., to produce the cyclic "head-to-tail" hexamer:[100]

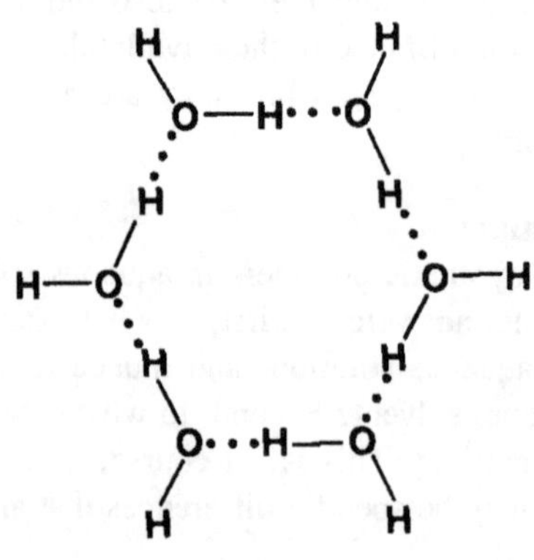

It is possible that the "ice-like" or "cluster" regions within the liquid contain rings of the type shown here, and that these are also involved in the "hydrophobic hydration" formed around nonpolar solutes and solute groupings (see below).

Solutes and solute groupings introduced into water disrupt the local structure, but then lead to new forms of structure in the regions under their influence. Hydrophilic nonionic solutes (e.g., glycols, polyols, and sugars) form hydrogen bonds to and from the water molecules, more or less compensating for the water-water bonds broken and hence leading to approximately ideal behavior. Hydrophobic (i.e., nonpolar) solutes (noble gases, alkanes, arenes, etc.), as far as their limited solubility allows, form local regions of "hydrophobic hydration" where the water molecules are even more ordered and even more highly hydrogen bonded than in the bulk, possibly with structures similar to those formed in the clathrate hydrates, or with cooperative ring structures of the type already illustrated above. Simple ionic solutes (Na^+, Cl^-, Ca^{2+}, SO_4^{2-}, etc.) orient the local water molecules by ion-dipole forces, the H_2O being turned with its oxygen towards the solute molecule in the case of cations and with the hydrogens towards it (and possibly one of them hydrogen bonded to it) in the case of anions.

In general, molecules will contain more than one of these types of groups, and each will therefore have the appropriate form of "hydration", although it is a debatable point to what extent there is interference between the different forms of hydration on neighboring groups on the molecule. Thus, in amphiphilic molecules there will be the corresponding forms of hydration around the hydrophilic and the hydrophobic parts, but there must be some conflict between these regions at their junction.

These forms of hydration then determine the types of interactions between solute molecules in aqueous solution. Thus, the hydration around hydrophilic nonionic molecules greatly weakens the net strength of any solute-solute hydrogen bonds, and similarly that around ions greatly weakens the cation/anion attractions (although these are still sufficient to lead to effects treated by Debye-Hückel theory and the like).

By contrast, since the formation of "hydrophobic hydration" is disfavorable, hydrophobic groups in aqueous solution are attracted to one another by the "hydrophobic effect" (also variously termed the "hydrophobic bond" or "hydrophobic interaction") which thereby allows some of the highly structured water to be released back into the bulk when the two groups come close to one another.[101,102] Since the formation of hydrophobic hydration is exothermic (representing even more efficient hydrogen bonding than in bulk water), then the hydrophobic effect is endothermic, i.e., it becomes stronger with rise in temperature; this is a characteristic feature, although not a necessarily diagnostic one, of this type of interaction.

D. Sources of Polymer Samples

Many of the polymers dealt with here are produced commercially, and hence are available in certain cases directly from the manufacturers; they may also be obtained from general chemical suppliers, or sometimes only from specialist suppliers[103] — although in these cases of indirect supply there may be doubts about the characteristics of the sample because of the uncertainty of its provenance.

However, it is also necessary to bear in mind certain limitations and complicating factors in making laboratory studies on these commercial materials (even when direct from the manufacturer) and in interpreting the results. In the first place, there is often much uncertainty about the manner by which the polymer was produced, which raises corresponding uncertainties regarding such important features as the average molecular weight, the molecular weight distribution and the tacticity of the polymer, and the comonomer composition and distribution for copolymers. The main source of information on these materials is necessarily the technical literature of the manufacturer concerned, where for commercial reasons the data are generally limited, often ambiguous, and naturally not necessarily as trustworthy as those from published work in scientific journals. Furthermore, examination of the specification for a particular grade often shows that an actual sample supplied can differ widely in, for example, its molecular weight, from the nominal value while still remaining within the limits of the specification; such variations are acceptable in industrial applications of the material (which may have been tailored to take much more into consideration, including such factors as ease and reproducibility of manufacture, and price), but they would be of more serious consequences in basic scientific studies. Again, many reviews of specific polymers in the scientific literature emanate from the research and development groups of these same firms, which leads to a certain narrowing of the range of the types of samples discussed.

One apparent advantage in studying the commercial grades of these polymers is that the results should presumably be directly applicable to the actual use of the material. However, even for this purpose it is wise to extend the studies to samples with, say, somewhat different molecular weights, or compositions for copolymers, to show up any differences that will arise from variations between the different samples of the same grade in actual practice.

All these limitations are an inevitable consequence of the commercial origin of these materials, whose macromolecular nature introduces special features not encountered with small-molecule materials. Nevertheless, it indicates that in addition to subjecting the sample supplied to standard methods of chemical analysis (IR, UV, NMR, etc.), it is always wise to characterize it at least, say, by measurement of its intrinsic viscosity in a suitably defined solvent medium (preferably, that in which the main studies are being performed), which may also allow its (viscosity-average) molecular weight to be estimated (see Section III.G). Similarly, where the material is a copolymer (or even has only a minor content of another monomer unit which may nevertheless affect its properties), not only should the composition be determined but also the distribution of the comonomer units (random, blocky, etc.).

It is presumably with these considerations in mind, and particularly for the sake of closer control over the characteristics of their polymer samples, that many workers prefer to synthesize these for themselves. It is mainly for this reason that the sections in later chapters on specific polymers contain details and references to the methods of polymerization that have been so used, and which indeed may even be those used in commercial practice. Even here, one should be mindful of the fact that the structural details of the polymer are often sensitive to the synthesis method used; furthermore, side reactions are more serious in their implications than with small-molecule products, since with polymer products any such formed "impurities" may be fixed essentially inextricably in the polymer chain.

E. Analysis of Water Content

Many of these polymers are hygroscopic, and the determination of the water content of the bulk polymer is therefore a perennial problem. Commercially supplied samples must be expected to have an appreciable water content; similarly, when a sample of a polymer solution has been evaporated down or freeze-dried to determine the polymer (or total solids) content of the solution, the solid obtained is often found to have several percent of water retained. In either case the water content of the polymer needs to be determined if the true concentration of a solution is to be known.

The methods available for the determination of water in polymers have been reviewed by Mitchell,[104] and by Urbański and colleagues,[105] while the methods used in the present context are considered specifically with the individual polymers. However, the simplest method is to heat the polymer further in vacuum and observe the weight loss; preferably this needs to be done just above the glass transition temperature for an amorphous polymer, or just above the melting point for a highly crystalline one, to increase the mobility of water molecules while minimizing possible decompósition or degradation reactions. Alternatively, there may be a water-miscible volatile organic liquid (e.g., methanol, ethanol, or acetone) which is a marginal nonsolvent for the polymer, and which therefore when brought into contact with the polymer swells it sufficiently for the water to be extracted, when the residual organic liquid can then be removed by decantation followed by mild vacuum heating. Another possibility is to use a reactive dehydrating agent such as thionyl chloride ($SOCl_2$), so long as this does not react with the polymer itself.

The absence of water in the treated polymer may be verified by infrared spectroscopy and comparison with reference spectra — except that many published reference spectra[106] contain bands quite evidently due to water in the polymer sample that was used!

F. Solubility and Precipitation Behavior*

Many of these polymers, although they are indeed water soluble at normal temperatures, precipitate out either when their solutions are cooled (e.g., PAA, PMAAm) or heated (e.g., PEO, PPO, PVME, PMAA), or on the addition of various cosolutes (e.g., salts, organic liquids, phenols, etc.; see Chapter 2 of Volume II), or when subjected to both of these. Furthermore, certain related synthetic polymers may be water insoluble at normal temperatures, but dissolve at lower temperatures, e.g., PPO oligomers with $M \approx$ 2D3, and poly(vinyl acetate-*co*-vinyl alcohol) (PVAC-VAL) with 30% VAC, or they may dissolve on the addition of ionic amphiphiles such as detergents (see Chapter 2 of Volume II). The investigation of these effects is therefore an important part of the study of the properties and behavior of these materials.

It should be noted, firstly, that solubility or insolubility with polymers is much more of an "all-or-none" phenomenon than it is with small-molecule substances; with the latter, the equilibrium solubility changes monotonically with change in temperature or with concentration of additive, whereas a polymer is often indefinitely soluble up to a certain point (temperature, concentration of cosolute) and essentially insoluble beyond that point.

With a highly crystalline polymer below its melting point, the solubility equilibrium may be that between the polymer itself and its solution; as with other crystalline solids, the extent of solubility is likely to be critically dependent on the lattice structure of the material. On the other hand, with amorphous polymers it is a liquid/liquid phase separation that is involved, with the "precipitate" actually a relatively concentrated polymer solution or at most polymer more or less swollen with the solvent.

So long as the polymer concentration is relatively low, when phase separation takes place the precipitated polymer remains colloidally suspended in the system; this point is therefore often referred to as the "cloud point". Thus, for temperature-induced precipitation, this point may be symbolized by t_p (for the Celsius temperature) or T_p (for the absolute, i.e., thermodynamic temperature). However, it is also necessary, for self-evident reasons, to specify the *direction* of temperature change which leads to precipitation, so that t_{p-} is used to symbolize the cloud point for systems where the polymer precipitates on cooling, and t_{p+} correspondingly for those where it precipitates on heating. These two contrasting forms of behavior are often referred to as "normal" and "inverse" solubility-temperature behavior, since these terms are indeed descriptive of the behavior of nonaqueous polymer solutions; however, with aqueous systems the "inverse" type of behavior is if anything that more commonly encountered.**

The value of t_p will depend upon the molecular weight M and the concentration c_2 of the polymer (as well as the nature of the polymer and the natures and concentrations of any other solutes present). For any particular amorphous polymer, the variation of t_{p-} (i.e., "normal" solubility behavior) with polymer concentration has the form of an approximately inverted U-shaped plot, with its maximum at the critical precipitation temperature t_{c-} (and critical polymer concentration c_{2c}) for that molecular weight. Extension of this to a range

* Solubility, precipitation, and phase separation in polymer systems have been treated in the standard textbooks,[56,59,62] and also (with some specific discussion of aqueous and other hydrogen-bonded systems) by Small[107] and by Molyneux.[65]

** Viewing phase separation as an undesired occurrence, then the "normal" type of behavior may be referred to more graphically as *thermophilic* and the "inverse" type as *thermophobic*.

of polymer samples (fractions) of different molecular weights shows that t_{c-} rises (or, if it is less than 0°C, becomes less negative) with increasing $\overline{M}$, but that it approaches a limiting value for $M \rightarrow \infty$ (where also $c_{2c} \rightarrow 0$) which is the ideal (or Flory, or theta temperature) Θ_-, for the system.

Conversely, "inverse" solubility-temperature behavior gives rise to approximately U-shaped plots of t_{p+} against c_{-2}, with the maximum at t_{c+}, and with Θ_+ as the corresponding limiting value of these critical precipitation temperatures for the ideal state.

In certain cases, polymer solutions may show both forms of behavior in different temperature ranges; these may merge into one another, in which case with $t_{c-} > t_{c+}$ this gives rise to a closed curve for the phase diagram (e.g., oligomers of PEO — see Chapter 2), while with $t_{c-} < t_{c+}$ an "hourglass"-shaped phase diagram may result.

The theta temperature for the system has a wider significance, since it is the point at which the second virial coefficient A_2 (relating to deviations from ideal behavior) becomes zero, and the Flory-Huggins interaction parameter (reduced residual chemical potential) χ has the value of 0.5 exactly.

It is not intended to enter here into a detailed discussion of the thermodynamics of these systems, since this is a subject which has been treated at length in several standard texts and reviews,[56,59,62] and also covered more briefly in a previous treatment of synthetic polymers vis-à-vis water by the present author.[65] However, it is pertinent to try to relate solubility, precipitation, and other observed effects in these systems with the types of molecular interactions involved, particularly because this can lead to useful correlations between macroscopic behavior and the molecular structures of the components.

The thermodynamic behavior of the polymer solution depends upon the characteristics of the exchange equilibrium:

$$M \sim M + S \sim S \rightleftharpoons M \sim S + M \sim S \qquad (1)$$

where M represents a monomer unit (or some other appropriate unit of the chain) and S represents a solvent molecule or, more exactly, a group of solvent molecules with the same volume as the unit M; the symbol "$\sim$" then represents the noncovalent interactions between the species in contact, with M $\sim$ M in particular representing the interaction either between two distantly connected units on the same chain (i.e., the ambiguously named "long-range interaction") or between units of different molecules, in each case equivalent to a transient and reversible cross-link between the two. The equilibrium (Equation 1) is, for example, forced over to the right when a polymer solution (sufficiently concentrated for the molecules to be overlapping) is diluted, or when a more or less isolated polymer molecule increases the volume of its encompassed domain.

With nonionic solutes in aqueous solution these interactions will involve van der Waals forces, hydrogen bonds, and hydrophobic effects as appropriate for the particular species; with electrolytes present (small-molecule salts, or polyelectrolytes, or both) there is the added possibility of ion/ion, ion/dipole, and ion/induced-dipole forces. The position of the equilibrium depends upon the balance (often a delicate one) between the three types of interaction (M $\sim$ M, S $\sim$ S, and M $\sim$ S), which will themselves be complex, and which will be affected in different ways by change of temperature or the presence in their neighborhood of different types of molecules (i.e., polymer units, solvents, cosolutes, etc).

Thus if the process as written is exothermic (as may be shown directly by measuring the heat of dilution) then it will be disfavored by rise in temperature, which will give rise to a greater number of M $\sim$ M and S $\sim$ S contacts, a more compact conformation for a loosely coiled polymer molecule, and ultimately possibly to phase separation (i.e., a t_{p+} for the particular mixture and Θ_+ for the system in general), since all these processes lead to the desired enhancement of M $\sim$ M and S $\sim$ S contacts. The converse will apply if the equilibrium

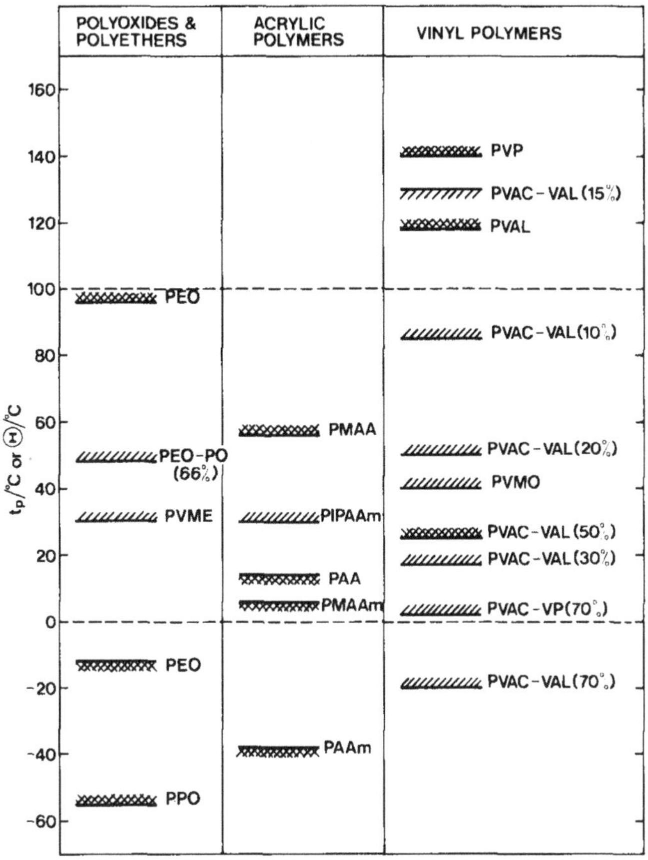

FIGURE 1. Phase separation and solubility behavior of synthetic polymers in aqueous solution; cloud points, t_p (single hatching) and ideal temperatures, Θ (double hatching), where the placement of the shading shows the direction of temperature change for phase separation. See text (Section III.F) for data sources and discussion. The copolymers listed are random, the percent value quoted is the molar content of the first monomer.

is endothermic, where the changes described will take place on the reduction of the temperature, with the possibility of a Θ_- for the system.

The application of these general concepts to the present systems is shown in Figure 1, which summarizes the phase separation and solubility behavior of these and some related polymers dealt with in more detail later in the book. The t_p values (single hatching) relate in general to relatively high molecular weight samples studied at relatively low concentrations (i.e., near the Θ-point), but in view of their sensitivity to the values of these two variables they should be regarded as only illustrating the solubility of these polymers; even the Θ values (double hatching) are in many cases estimated by rather long extrapolations — especially, of course, these below 0°C or above 100°C. The polymers listed in Figure 1 are dealt with in the following locations: PEO, Chapter 2 (Θ_+, Section II.G.2; Θ_-, Section II.F.1); PPO, Chapter 2 (Section III.F) PEO-PO (random copolymer), Chapter 2 (Section V); PVME, Chapter 2 (Section VI.F); PAA, Chapter 3 (Section II.F.2— especially Reference 1); PAAm, Chapter 3 (Section III — especially Reference 1); PMAA, Chapter 3 (Section IV.F.2 — especially Reference 1); PMAAm, chapter 3 (Section V.F. — especially Reference 1, but see also Postscript); PIPAAm, Chapter 3 (Section VI.B); PVAL, Chapter 4 [Section II.H.2 — the value of Θ_+ = 117°C plotted in Figure 1 is the mean of those from the temperature dependence of intrinsic viscosity (97°C)[68] and light-scattering second virial coefficient (137°C)[143] over the range 0 to 80°C, but Reference 143 shows that this does not correspond to an actual precipitation point since the linearly falling plot of A_2 vs. temperature deviates from linearity above 80°C, the value of A_2 reaching a minimum at about 100°C and then rising steadily beyond this); PVAL-VAC, Chapter 4 (Section II.H.2 — especially References 67, 80, 89, 118, 155, and 157); PVMO, Chapter 4 (Section IV); PVAC-VP, Chapter 4 (Reference 157); and PVP, Chapter 4 (Section III.H.4).

G. Solution Viscosity Behavior

From the practical viewpoint, one of the most useful features of polymers is their ability to increase the viscosity of a liquid in which they are dissolved, even at low concentrations. On the other hand, when it is turbulent rather than streamline flow that is involved, then, paradoxically, the presence of a dissolved polymer at low concentrations can lead to an increase in the rate of flow; this behavior, sometimes referred to as the Toms effect, is shown not only in nonaqueous systems but also in aqueous solutions of polymers such as PEO and PAA-AAm (i.e., partially hydrolyzed polyacrylamide).[108] In the present context, however, the discussion will be confined to phenomena in streamline flow.

From the scientific viewpoint, the viscosity behavior of polymers in dilute solution (and in streamline flow) can give valuable information on the conformational behavior of the polymer in the solution, as well as potentially enabling the molecular weight to be estimated.

When the viscosity of a polymer solution, η, and that of the pure solvent, η_1, have been measured, then thisin turn gives the *specific viscosity* of the solution, η_{sp}, defined by;

$$\eta_{sp} = (\eta - \eta_1)/\eta_1 \quad (2)$$

Division of this quantity by the polymer concentration, c_2, gives the *viscosity number* or *viscosity index*, η_c, i.e.:

$$\eta_c = \eta_{sp}/c_2 \quad (3)$$

Measurement of η_c at different polymer concentrations generally reveals that, for nonionic polymers at least,* it is a linear-rising function of c_2 according to the Huggins equation:

$$\eta_c = [\eta] + kc_2 \quad (4)$$

where $[\eta]$ is the *intrinsic viscosity* (also termed the limiting viscosity number, or Staudinger index) with the units of reciprocal concentration. Division of the slope k by $[\eta]^2$ gives the dimensionless *Huggins parameter* (viscosity slope parameter) k':

$$k' = k/[\eta]^2 \quad (5)$$

which for many of the present systems has a value between 0.3 and 0.4; although the theoretical interpretation of the value of this parameter is disputed, it is evidently a measure of the strength of the interactions between polymer molecules, somewhat analogous to the second virial coefficient A_2.

These measurements are customarily carried out using capillary viscometers, conveniently of the Ubbelohde (suspended-level dilution) type. The flow times for these instruments are proportional to the viscosity as long as the kinetic energy corrections are negligible, which is the case when the flow time is fairly long, so that these times may be substituted directly in Equation 2 for the calculation of η_{sp}. Certain experimental precautions have to be taken

* The viscosity behavior of polyelectrolytes is discussed in Chapter 1 of Volume II.

for results of the highest precision and accuracy — notably, ensuring that the viscometer is vertical and that all solutions are free of particles that may obstruct the capillary. The temperature also needs to be closely controlled, not so much for the benefit of the final values of $[\eta]$ which are generally only somewhat sensitive to temperature (see below), but for the individual viscosity values which are much more sensitive. With certain systems the viscosity depends on the shear rate (i.e., the flow is not Newtonian) so that it may be necessary to make measurements over a range of shear rates and extrapolate to zero rate.

It is also important that the concentrations of any additive are carefully controlled. With ionizable polymers such as weak acids or weak bases the pH should be adjusted so that the ionization is either fully suppressed or fully complete; in the latter case an adequately high ionic strength should be used to prevent marked changes in conformational size during the dilution of the polymer solution in this viscometer (see Chapter 1 of Volume II).

The value of $[\eta]$ is a direct measure of the hydrodynamic volume of the isolated polymer molecule in that environment; application of the Einstein equation relating η_{sp} for very dilute solutions to the volume fraction of particles shows that $[\eta]$ is inversely proportional to the $\bar{\rho}_2$, the average density of polymeric material within the volume encompassed by the polymer molecule,

$$[\eta] = 2.5/\bar{\rho}_2 \tag{6}$$

The value of $[\eta]$ is also useful in providing an orientation when considering ranges of polymer concentration in solution studies; from Equation 6 it is evidently inversely related to the critical overlap concentration, c_2^*, so that we can put:

$$c_2^* \approx 1/[\eta] \tag{7}$$

Thus, at $c_2 < 1/[\eta]$ the molecules are moving more or less independently (although still interacting with one another); at $c_2 > 1/[\eta]$ the molecules are overlapping to a greater or lesser extent, and at concentrations appreciably higher than this the system will be almost uniformly pervaded by polymer chains.

For a given solvent medium and temperature, the intrinsic viscosity is usually related to the molecular weight of the polymer by the customarily named Mark-Houwink-Sakurada (MHS) equation:

$$[\eta] = KM^a \tag{8}$$

where the MHS parameters a and K are characteristic of that polymer under those particular conditions. Thus, if the values of these parameters have been obtained, preferably using closely fractionated polymer samples covering a wide range of molecular weights, then they may be used to estimate the molecular weight of the given polymer sample under study from its intrinsic viscosity; with heterodisperse samples this gives, by definition, the viscosity-average value, $\bar{M}_v$, where the exact type of average depends upon the value of a, but is generally close to the weight average value, $\bar{M}_w$.

One complication that arises with these aqueous systems is that with poly(ethylene oxide) (PEO) and with poly(propylene oxide) (PPO) the oligomers are also important — indeed, with PPO it is only the oligomers which are water soluble (see Chapter 2). For such short chains the MHS equation does not apply, except over narrow ranges of molecular weight. However, the equation may be modified to extend its applicability to these low values of molecular weight. One such modification, due to Rempp,[109,110] involves the inclusion of an additive constant, $[\eta]_o$, to give what is conveniently termed the MHSR equation:

$$[\eta] = [\eta]_o + KM^a \tag{9}$$

This relation has been found to apply to oligomeric PEO in aqueous and nonaqueous media,[109,110] and likewise to PPO (see Chapter 2, Tables 5 and 6). For sufficiently high molecular weights the constant $[\eta]_0$ becomes negligible compared with $[\eta]$ so that the simple MHS relation is regained.

Another approach is that of Müh,[111] whose modified form of the MHS equation, i.e., the MHSM equation, is

$$[\eta] = f(M, \mu)\ KM^a \tag{10}$$

where $f(M,\mu)$ is a particular function of M and of the so-called "steric hindrance parameter" ("Behinderungsparameter") μ. This function is unity for $\mu = 0$, and greater than unity with positive value of μ and low values of M, as with PEO oligomers in several solvent systems studied by Müh;[111] however, it approaches unity (regardless of the value of μ) for sufficiently high values of M, so that again the MHS equation is regained.

Comparing these two methods, the method of Müh does have a theoretical background but this is rather obscure, while the method for calculating the function $f(M,\mu)$ is also not very clear; the method of Rempp, although quite empirical, is simple and seems to work well, and it is therefore the preferable one.

In addition, the values of the MHS parameters, more particularly that of a, give information on the conformation of the polymer in solution and the solvent power of the water (or other solvent) towards the polymer. In the present context, nonionic synthetic polymers generally have quite flexible chains so that they take up a random-coil type of conformation, with the value of a lying between 0.5 (corresponding to a true random coil, with the system in the Θ state) and a maximum value of about 0.8 (for a very good solvent system). With flexible chain polyelectrolytes the coulombic repulsions between the charges on the chain give a more expanded conformation with correspondingly higher values of a, which may approach the value of 2 to be expected for a fully extended chain; however, addition of salts, which shield these charges and also exert a general "salting-out" action, can reduce the conformational size and bring the value back down again, to 0.5 or even below (see Chapters 1 and 2 of Volume II).

Although the MHS index a is thus a useful parameter which is fairly readily determined, experience shows that there are commonly marked discrepancies between the values obtained by different, independent groups of workers, which can often be traced to the individual polymer samples used not being sufficiently narrow in their molecular weight distribution, or to their collectively not covering a sufficiently wide range of molecular weights (see Chapters 2 to 4 of Volume I and Chapter 1 of Volume II). Such discrepancies are serious, since the maximum range of a is so relatively narrow, i.e., from 0.5 for an ideal (i.e., relatively poor) solvent up to 0.8 for a nonionic polymer in the best solvent medium; this needs to be borne in mind when making deductions about the system on the basis of published values of this parameter.

A further aspect to be considered is the variation of $[\eta]$ with temperature. It is found that this is closely correlated with the thermicity of the exchange equilibrium (Equation 1); thus if this process is exothermic, then $[\eta]$ is found to fall with rise in temperature, which is evidently the result of the associated increase in the number of $M \sim M$ contacts and hence a corresponding shrinkage of the polymer coil. The converse applies to endothermic dilution, the value of $[\eta]$ then rising with rise in temperature. Thus, the temperature dependence of $[\eta]$ can give information upon the relative energies of the interactions involved; furthermore, a fall in the value of $[\eta]$ with change in temperature may presage the approach of polymer precipitation when the change is continued further.

REFERENCES

1. **Walton, A. G. and Blackwell, J.,** *Biopolymers,* Academic Press, New York, 1973.
2. **Davidson, R. L. and Sittig, M., Eds.,** *Water-Soluble Resins,* 2nd ed., Van Nostrand Reinhold, New York, 1968.
3. **Anderson, A. T.,** Introduction, in *Water-Soluble Resins,* 2nd ed., Davidson, R. L. and Sittig, M., Eds., Van Nostrand Reinhold, New York, 1968, chap. 1.
4. **Rutenberg, M. W.,** Modified starches, in *Water-Soluble Resins,* 2nd ed., Davidson, R. L. and Sittig, M., Eds., Van Nostrand Reinhold, New York, 1968, chap. 2.
5. **Scheffel, K. G.,** Methyl- and hydroxypropyl methyl-cellulose derivatives, in *Water-Soluble Resins,* 2nd., ed., Davidson, R. L. and Sittig, M., Eds., Van Nostrand Reinhold, New York, 1968, chap. 3.
6. **Eastman, N. C. and Rose, J. K.,** Hydroxyethylcellulose, in *Water-Soluble Resins,* 2nd ed., Davidson, R. L. and Sittig, M., Eds., Van Nostrand Reinhold, New York, 1968, chap. 4.
7. **Baird, G. S. and Speicher, J. K.,** Carboxymethylcellulose, in *Water-Soluble Resins,* 2nd ed., Davidson, R. L. and Sittig, M., Eds., Van Nostrand Reinhold, New York, 1968, chap. 5.
8. **Argana, C. P. and Czerwin, E. P.,** Polyvinyl alcohol, in *Water-Soluble Resins,* 2nd ed., Davidson, R. L. and Sittig, M., Eds., Van Nostrand Reinhold, New York, 1968, chap. 6.
9. **Azorlosa, J. L. and Martinelli, A. J.,** Polyvinylpyrrolidone, in *Water-Soluble Resins,* 2nd ed., Davidson, R. L. and Sittig, M., Eds., Van Nostrand Reinhold, New York, 1968, chap. 7.
10. **Glavis, F. J.,** Poly(acrylic acid) and its homologs, in *Water-Soluble Resins,* 2nd ed., Davidson, R. L. and Sittig, M., Eds., Van Nostrand Reinhold, New York, 1968, chap. 8.
11. **Montgomery, W. H.,** Polyacrylamide, in *Water-Soluble Resins,* 2nd ed., Davidson, R. L. and Sittig, M., Eds., Van Nostrand Reinhold, New York, 1968, chap. 9.
12. **Harding, R. H. and Rose, J. K.,** Ethylene oxide polymers, in *Water-Soluble Resins,* 2nd ed., Davidson, R. L. and Sittig, M., Eds., Van Nostrand Reinhold, New York, 1968, chap. 10.
13. **Davis, L. E.,** Polyethylenimine, in *Water-Soluble Resins,* 2nd ed., Davidson, R. L. and Sittig, M., Eds., Van Nostrand Reinhold, New York, 1968, chap. 11.
14. **Davidson, R. L., Ed.,** *Handbook of Water-Soluble Gums and Resins,* McGraw-Hill, New York, 1980.
15. **Durso, D. F.,** Introduction, in *Handbook of Water-Soluble Gums and Resins,* Davidson, R. L., Ed., McGraw-Hill, New York, 1980, chap. 1.
16. **Cottrell, I. W. and Kovacs, P.,** Alginates, in *Handbook of Water-Soluble Gums and Resins,* Davidson, R. L., Ed., McGraw-Hill, New York, 1980, chap. 2.
17. **Greminger, G. K. and Krumel, K. L.,** Alkyl and hydroxyalkylcellulose, in *Handbook of Water-Soluble Gums and Resins,* Davidson, R. L., Ed., McGraw-Hill, New York, 1980, chap. 3.
18. **Stelzer, G. I. and Klug, E. D.,** Carboxymethylcellulose, in *Handbook of Water-Soluble Gums and Resins,* Davidson, R. L., Ed., McGraw-Hill, New York, 1980, chap. 4.
19. **Guiseley, K. B., Stanley, N. F., and Whitehouse, P. A.,** Carrageenan, in *Handbook of Water-Soluble Gums and Resins,* Davidson, R. L., Ed., McGraw-Hill, New York, 1980, chap. 5.
20. **Seaman, J. K.,** Guar gum, in *Handbook of Water-Soluble Gums and Resins,* Davidson, R. L., Ed., McGraw-Hill, New York, 1980, chap. 6.
21. **Meer, W.,** Agar, in *Handbook of Water-Soluble Gums and Resins,* Davidson, R. L., Ed., McGraw-Hill, New York, 1980, chap. 7.
22. **Meer, W.,** Gum arabic, in *Handbook of Water-Soluble Gums and Resins,* Davidson, R. L., Ed., McGraw-Hill, New York, 1980, chap. 8.
23. **Meer, G.,** Gum ghatti, in *Handbook of Water-Soluble Gums and Resins,* Davidson, R. L., Ed., McGraw-Hill, New York, 1980, chap. 9.
24. **Meer, W.,** Gum karaya, in *Handbook of Water-Soluble Gums and Resins,* Davidson, R. L., Ed., McGraw-Hill, New York, 1980, chap. 10.
25. **Stauffer, K. R.,** Gum tragacanth, in *Handbook of Water-Soluble Gums and Resins,* Davidson, R. L., Ed., McGraw-Hill, New York, 1980, chap. 11.
26. **Powell, G. M.,** Hydroxyethylcellulose, in *Handbook of Water-Soluble Gums and Resins,* Davidson, R. L., Ed., McGraw-Hill, New York, 1980, chap. 12.
27. **Butler, R. W. and Klug, E. D.,** Hydroxypropylcellulose, in *Handbook of Water-Soluble Gums and Resins,* Davidson, R. L., Ed., McGraw-Hill, New York, 1980, chap. 13.
28. **Seaman, J. K.,** Locust bean gum, in *Handbook of Water-Soluble Gums and Resins,* Davidson, R. L., Ed., McGraw-Hill, New York, 1980, chap. 14.
29. **Pedersen, J. K.,** Pectins, in *Handbook of Water-Soluble Gums and Resins,* Davidson, R. L., Ed., McGraw-Hill, New York, 1980, chap. 15.
30. **Volk, H. and Friedrich, R. E.,** Polyacrylamide, in *Handbook of Water-Soluble Gums and Resins,* Davidson, R. L., Ed., McGraw-Hill, New York, 1980, chap. 16.
31. **Greenwald, H. L. and Luskin, L. S.,** Poly(acrylic acid) and its homologs, in *Handbook of Water-Soluble Gums and Resins,* Davidson, R. L., Ed., McGraw-Hill, New York, 1980, chap. 17.

32. **Powell, G. M.,** Polyethylene glycol, in *Handbook of Water-Soluble Gums and Resins,* Davidson, R. L., Ed., McGraw-Hill, New York, 1980, chap. 18.
33. **Braun, D. B.,** Poly(ethylene oxide), in*Handbook of Water-Soluble Gums and Resins,* Davidson, R. L., Ed., McGraw-Hill, New York, 1980, chap. 19.
34. **Modi, T. W.,** Polyvinyl alcohol, in *Handbook of Water-Soluble Gums and Resins,* Davidson, R. L., Ed., McGraw-Hill, New York, 1980, chap. 20.
35. **Blecher, L., Lorenz, D. H., Lowd, H. L., Wood, A. S., and Wyman, D. P.,** Polyvinylpyrrolidone, in *Handbook of Water-Soluble Gums and Resins,* Davidson, R. L., Ed., McGraw-Hill, New York, 1980, chap. 21.
36. **Rutenberg, M. W.,** Starch and its modifications, in *Handbook of Water-Soluble Gums and Resins,* Davidson, R. L., Ed., McGraw-Hill, New York, 1980, chap. 22.
37. **Gerard, T.,** Tamarind gum, in *Handbook of Water-Soluble Gums and Resins,* Davidson, R. L., Ed., McGraw- Hill, New York, 1980, chap. 23.
38. **Cottrell, I. W., Kang, K. S., and Kovacs, P.,** Xanthan gum, in *Handbook of Water-Soluble Gums and Resins,* Davidson, R. L., Ed., McGraw-Hill, New York, 1980, chap. 24.
39. **Stainsby, G., Ed.,** *The Chemistry and Rheology of Water Soluble Gums and Colloids,* S. C. I. Monograph No. 24, Society of Chemical Industry, London, 1966.
40. **Bikales, N. M., Ed.,** *Water-Soluble Polymers (Polymer Science and Technology,* Vol. 2), Plenum Press, New York, 1973.
41. **Finch, C. A., Ed.,** *Chemistry and Technology of Water-Soluble Polymers,* Plenum Press, New York, 1983.
42. **Idson, B.,** Gums in drugs, cosmetics, foods, *Drug Cosmet. Ind.,* 113(5), 42; (6)48, 1973.
43. **Warson, H.,** *The Applications of Synthetic Resin Emulsions,* Ernest Benn, London, 1972.
44. **Warson, H.,** Emulsion paint performance: influence of non-polymer constituents, *Paint Oil Colour J.,* 142, 214, 1962.
45. **Smith, A. L.,** *Guide to Preservatives for Water-Borne Paints: A Summary of Chemical and Toxicological Data on 130 Commercial Products and a Critical Review of Current Literature,* Spec. Publ. No. 14, Paint Research Association, Teddington, Middlesex, 1980.
46. **Whistler, R. L. and BeMiller, J. N., Eds.,** *Industrial Gums: Polysaccharides and Their Derivatives,* 2nd ed., Academic Press, New York, 1973.
47. **Atik, M.,** Dextrans, their use in surgery and medicine, with emphasis on the low molecular weight fractions, *Anesthesiology,* 27, 425, 1966.
48. **Kliman, A.,** Presently useful plasma volume expanders, *Anesthesiology,* 27, 417, 1966.
49. **Florence, A. T.,** Pharmaceutical uses of polymers: some recent developments, *Pharm. J.,* 213, 36, 1974.
50. **Alexander, B., Odake, K., Lawlor, D., and Swanger, M.,** Coagulation, hemostasis and plasma expanders: a quarter century enigma, *Fed. Proc. Fed. Am. Soc. Exp. Biol.,* 34, 1429, 1975.
51. **Labrude, P.,** Les macromolécules synthètiques d'intérêt thérapeutique, *Lyon Pharm.,* 29, 467, 1978.
52. **Reynolds, J. E. F., and Prasad, A. B.,** *Martindale: The Extra Pharmacopoeia,* 28th ed., The Pharmaceutical Press, London, 1982.
53. **Juniper, R.,** *Pharmaceutical Applications of Polymers: A Selected Bibliography,* Rubber and Plastics Research Association of Great Britain, Shawbury, Shropshire, 1981.
54. **Roff, W. J. and Scott, J. R.,** *Fibres, Films, Plastics and Rubbers: A Handbook of Common Polymers,* Butterworths, London, 1971.
55. **Brandrup, J. and Immergut, E. H., Eds.,** *Polymer Handbook,* 2nd ed., Wiley-Interscience, New York, 1975.
56. **Flory, P. J.,** *Principles of Polymer Chemistry,* Cornell University Press, Ithaca, N.Y., 1953.
57. **Tanford, C.,** *Physical Chemistry of Macromolecules,* John Wiley & Sons, New York, 1961.
58. **Rodriguez, F.,** *Principles of Polymer Systems,* McGraw-Hill, New York, 1970.
59. **Elias, H. -G.,** *Macromolecules,* Vol. 1, Plenum Press, New York, 1977.
60. **Elias, H. -G.,** *Macromolecules,* Vol. 2, Plenum Press, New York, 1977.
61. **Mark, H. M., Gaylord, N. G., and Bikales, N. M., Eds.,** *Encyclopedia of Polymer Science and Technology: Plastics, Resins, Rubbers, Fibers,* Vols. 1 to 16, Wiley-Interscience, New York, 1964 to 1972.
62. **Morawetz, H.,** *Macromolecules in Solution,* 2nd ed., Wiley-Interscience, New York, 1975.
63. **Bekturov, E. A. and Bakayova, Z. Ch.,***Sinteticheskie Vodorastvorimie Polymeri v Rastvorach,* [Synthetic Water-Soluble Polymers in Solution], Kazakh SSR Academy of Sciences, Alma Ata, 1981.
64. **Franks, F., Ed.,** *Water: A Comprehensive Treatise,* Vol. 4, Plenum Press, New York, 1975.
65. **Molyneux, P.,** Synthetic polymers, in *Water: A Comprehensive Treatise,* Vol. 4, Franks, F., Ed., Plenum Press, New York, 1975, chap. 7.
66. **Guilford, H.,** Chemical aspects of affinity chromatography, *Chem. Soc. Rev.,* 2, 249, 1973.
67. **Melrose, G. J. H.,** Insolubilized enzymes; biochemical applications of synthetic polymers, *Rev. Pure Appl. Chem.,* 21, 83, 1971.
68. **Wilson, A. D.,** The chemistry of dental cements, *Chem. Soc. Rev.,* 7, 265, 1978.

69. **Anon.,** *IARC Monographs on the Evaluation of the Carcinogenic Risk of Chemicals to Humans,* Vol. 19, World Health Organization/International Agency for Research on Cancer, Lyon, France, 1979.
70. **Silberberg, A., Eliassaf, J., and Katchalsky, A.,** Temperature-dependence of light scattering and intrinsic viscosity of hydrogen bonding polymers, *J. Polym. Sci.,* 23, 259, 1957.
71. **Andersson, G. R.,** A study of the pressure dependence of the partial specific volume of macromolecules in solution by compression measurements in the range 1-8000 atm, *Arkiv Kemi,* 20, 513, 1963.
72. **Roy-Chowdhury, P. and Kale, K. M.,** The adiabatic compressibility of poly(acrylic acid) and polyacrylamide in aqueous solution, *J. Appl. Polym. Sci.,* 14, 2937, 1970.
73. **Roy-Chowdhury, P.,** Adiabatic compressibility of polyelectrolytes in aqueous solution: poly(metha crylic acid), *J. Appl. Polym. Sci.,* 12, 751, 1968.
74. **Roy-Chowdhury, P.,** Ultrasonic velocity adiabatic compressibility of polymethacrylamide in aqueous solution at 25°C, *Indian J. Chem.,* 7, 692, 1969.
75. **Roy-Chowdhury, P. and Kale, K. M.,** The adiabatic compressibility of a poly(dibasic acid) in aqueous solution: copolymer of styrene and maleic acid, *Makromol. Chem.,* 171, 145, 1973.
76. **Roy-Chowdhury, P.,** Adiabatic compressibility of polyelectrolytes in aqueous solutions. II. Poly(4-vinyl-*N-n*-butylpyridinium bromide), *J. Polym. Sci. Part A-2,* 7, 1451, 1969.
77. **Tuzar, Z and Kratochvil, P.,** Light scattering. XVIII. The behaviour of polymers in mixed solvents, *Coll. Czech. Chem. Commun.,* 32, 3358, 1967.
78. **Fort, R. J. and Polyzoidis, T. M.,** The unperturbed dimensions of poly(2-hydroxyethyl methacrylate), *Makromol. Chem.,* 178, 3229, 1977.
79. **Mackeen, D. L. and Green, K.,** Chlorhexidine kinetics of hydrophilic contact lenses, *J. Pharm. Pharmacol.,* 30, 678, 1978.
80. **Richardson, N. E., Davies, D. J. G., Meakin, B. J., and Norton, D. A.,** The interaction of preservatives with polyhydroxyethylmethacrylate (polyHEMA), *J. Pharm. Pharmacol.,* 30, 469, 1978.
81. **Choi, S., Jhon, M. S., and Andrade, J. D.,** Nature of water in synthetic hydrogels. III. Dilatometry, specific conductivity, and dielectric relaxation of poly(2,3-dihydroxypropyl methacrylate), *J. Colloid Interface Sci.,* 61, 1, 1977.
82. **Suggett, A.,** Polysaccharides, in *Water: A Comprehensive Treatise,* Vol. 4, Franks, F., Ed., Plenum Press, New York, 1975, chap. 6.
83. **Eagland, D.,** Nucleic acids, peptides and proteins, in *Water: A Comprehensive Treatise,* Vol. 4, Franks, F., Ed., Plenum Press, New York, 1975, chap. 5.
84. **Franks, F. and Eagland, D.,** The role of solvent interactions in protein conformation, *Crit. Rev. Biochem.,* 3, 165, 1975.
85. **Bamford, C. H., Elliott, A., and Hanby, W. E.,** *Synthetic Polypeptides,* Academic Press, New York, 1956.
86. **Fessler, J. H. and Ogston, A. G.,** Studies of the sedimentation, diffusion and viscosity of some sarcosine polymers in aqueous solution, *Trans. Faraday Soc.,* 47, 667, 1951.
87. **Lotan , N., Yaron, A., and Berger, A.,** The stabilization of the α-helix in aqueous solution by hydrophobic side-chain interaction, *Biopolymers,* 4, 365, 1966.
88. **Murai, N., Makino, S., and Sugai, S.,** Interaction of surfactants with nonionizable water-soluble peptides, *J. Colloid Interface Sci.,* 41, 399, 1972.
89. **Antoni, G., Neri, P., Pedersen, T. G., and Ottesen, M.,** Hydrodynamic properties of a new plasma expander: polyhydroxyethylaspartamide, *Biopolymers,* 13, 1721, 1974.
90. **Kern, W.,** Über heteropolare Molekülkolloide. I. Die Polyacrylsäure, ein Modell des Eiweisses. 185. Mitteilung über hochpolymere Verbindungen, *Z. Phys. Chem. (Leipzig) Abt.A,* 181, 249, 1938.
91. **Tanford, C.,** Properties of protein polypeptide chains as random coils, in *Solution Properties of Natural Polymers,* Spec. Publ. No. 23, Greenwood, C. T., Ed., The Chemical Society, London, 1968, 1.
92. **Strauss, U. P. and Williams, B. L.,** The transition from typical polyelectrolyte to polysoap. III. Light scattering and viscosity studies of poly-4-vinylpyridine derivatives, *J. Phys. Chem.,* 65, 1390, 1961.
93. **Franks, F., Ed.,** *Water: A Comprehensive Treatise,* Vol. 1, Plenum Press, New York, 1973.
94. **Franks, F., Ed.,** *Water: A Comprehensive Treatise,* Vol. 2, Plenum Press, New York, 1973.
95. **Franks, F., Ed.,** *Water: A Comprehensive Treatise,* Vol. 3, Plenum Press, New York, 1973.
96. **Franks, F., Ed.,** *Water: A Comprehensive Treatise,* Vol. 5, Plenum Press, New York, 1975.
97. **Franks, F., Ed.,** *Water: A Comprehensive Treatise,* Vol. 6, Plenum Press, New York, 1979.
98. **Frank, H. S.,** Restrictions for an acceptable model for water structure, in *Structure of Water and Aqueous Solutions,* Luck, W. A. P., Ed., Verlag Chemie & Physik Verlag, Weinheim, 1974, 9.
99. **Symons, M. C. R.,** Water structure and hydration, *Phil. Trans. R. Soc. Ser. B.,* 272, 13, 1975.
100. **Del Bene, J. E. and Pople, J. A.,** Theory of molecular interactions. III. A comparison of studies of H_2O polymers using different molecular-orbital basis sets, *J. Chem. Phys.,* 58, 3605, 1973 (and previous papers in this series).
101. **Franks, F.,** The hydrophobic interaction, in *Water: A Comprehensive Treatise,* Vol. 4, Franks, F., Ed., Plenum Press, New York, 1975, chap. 1.

102. **Tanford, C.,** *The Hydrophobic Effect: Formation of Micelles and Biological Membranes,* 2nd ed., John Wiley & Sons, New York, 1980.
103. **Anon.,** *Polysciences Catalog 1982-1983,* Polysciences Inc., Warrington, Pa., 1982.
104. **Mitchell, J.,** Methods for the determination of water in polymers, *Anal. Chim. Acta,* 81, 231, 1976.
105. **Urbański, J., Czerwiński, W., Janicka, K., Majewska, F., and Zowall, H.,** *Handbook of Analysis of Synthetic Polymers and Plastics,* Ellis Horwood, Chichester, England, 1977.
106. **Nyquist, R. A.,** *Infrared Spectra of Plastics (Polymers) and Resins,* 2nd ed., The Dow Chemical Co., Midland, Mich., 1961.
107. **Small, P. A.,** Some factors affecting the solubility of polymers, *J. Appl. Chem.,* 3, 71, 1953.
108. **Virk, P. S.,** Conformational effects in drag reduction by polymers, *Nature (London),* 262, 46, 1976.
109. **Rempp, P.,** Contribution à l'étude des solutions de molécules en chaîne à squelette oxygéné. II. Viscosités intrinsèques et coefficients de diffusion de translation des polyoxyéthylèneglycols, *J. Chim. Phys.,* 54, 432 and 466, 1957.
110. **Sadron, C. and Rempp, P.,** Viscosités intrinsèques des solutions de chaînes courtes, *J. Polym. Sci.,* 29, 127, 1958.
111. **Müh, G.,** Über den Zusammenhang zwischen dem Molekulargewicht und der Viskositätszahl von Lösungen kurzkettiger Polyäthylenglykole. II. Mitteilung über Polyäthylenglykole, *Kolloid-Z. Z. Polym.,* 196, 140, 1964.

Chapter 2

NONIONIC POLYMERS — POLYOXIDES, POLYETHERS, AND POLY(ETHYLENE IMINE)

''Like a patient etherized upon a table''
T. S. Eliot: *Love Song of J. Alfred Prufrock*

I. INTRODUCTION

This chapter is concerned with the general properties and behavior (including water solubility, and behavior alone in aqueous and nonaqueous solution) of the group of nonionic synthetic polymers comprised by the organic polyoxides and polyethers. In this context, the term ''polyoxide'' is used to denote polymers with oxygen atoms in the main chain, particularly those with monomer units of the type $[O\text{–}(CH_2)_n]$, together with their derivatives (this group has also been termed ''polyepoxides''); the term ''polyether'' is then used to denote polymers with oxygen atoms in the side chains, particularly those with monomer units of the type $[CH(OR)\text{–}CH_2]$ where R is an alkyl group (the term ''polyether'' has also been used elsewhere to denote both of these two groups together). It is useful to consider these polymers first of all because of the relative simplicity of their structures; however, as Table 1 shows, even with this simplicity of structure there is a fairly complex pattern of water solubility, since poly(ethylene oxide) and poly(vinyl methyl ether) are the only two members for which the true high polymers are water soluble, while with poly(propylene oxide) it is only the oligomers which are appreciably water soluble. It is also convenient to include poly(ethylene imine) along with these oxygen-containing polymers in the chapter, since structually it is similar to poly(ethylene oxide), at least in the nonionized (nonprotonated) form considered at this stage.

II. POLY(ETHYLENE OXIDE) — PEO

$$\text{-}[O\text{—}CH_2\text{—}CH_2]\text{-}$$

$$M_0(C_2H_4O) = 44.05$$

Other chemical names, abbreviations, and trade names are polyethylene glycol; poly(ethylene glycol); polyethyleneglycol; polyethylene oxide; polyethyleneoxide; poly(oxirane); polyoxyethylene; polyoxyethylene glycol; PEG; POE; POEG; *Carbowax* (Union Carbide); Macrogol (B.P.); *Poly-G* (Olin Matheson); *Polyglycol E* (Dow); *Polyox WSR* [Water soluble resin] (Union Carbide); *Pluracol E* (Wyandotte).

A. General Features

PEO is the simplest structure of the water-soluble polymers. This fact, coupled with the wide range of uses to which PEO has been put, has made it the most extensively and intensively studied water-soluble polymer from the scientific viewpoint.

However, one of the complicating features of this polymer is that up to quite recently only relatively short chain polymers were available, many of which were really oligomers. More recent developments have led to the situation where the whole range of chain lengths from the precursor ethylene glycol up to very high polymers ($M >$ D6) is available. Rather than either excluding from consideration those materials with M below a certain arbitrary value, or alternatively including these materials but under a distinctive name such as polyethylene glycol (which is the name commonly applied to shorter-chain samples), it was

Table 1
AQUEOUS SOLUBILITIES OF POLYOXIDES AND POLYETHERS OF MONOMER UNIT FORMULA: $C_mH_{2m}O$

Polymer	Monomer unit	*m* (C:O ratio)	Solubility[a]
Poly(oxymethylene)	$—O—CH_2—$	1	I
Poly(ethylene oxide)	$—O—(CH_2)_2—$	2	S[b]
Poly(acetaldehyde)	$—O—CH(CH_3)—$	2	I
Poly(trimethylene oxide)	$—O—(CH_2)_3—$	3	I
Poly(propylene oxide)	$—O—CH(CH_3)—CH_2$	3	O[c]
Poly(vinyl methyl ether)	$—CH(OCH_3)—CH_2—$	3	S
Poly(tetramethylene oxide)	$—O—(CH_2)_4—$	4	I
Poly(but-1-ene oxide)	$—O—CH(CH_2CH_3)—CH_2—$	4	I
Poly(vinyl ethyl ether)	$—CH(OCH_2CH_3)—CH_2—$	4	I

[a] Aqueous solubility at ambient temperatures (t = 20 to 30°C): I, insoluble: O, only oligomers soluble: S, soluble.
[b] Inverse solubility – temperature behavior: $\Theta_+ = 96 \pm 3$°C (see Section II.G.2).
[c] Soluble for $M \leqslant 1500$; inverse solubility-temperature behavior: $\Theta_+ \approx -55$°C (see Section III.F).
[d] Inverse solubility – temperature behavior: $\Theta_+ \approx 30$°C (see Section VI.F).

thought preferable to include the whole range of these substances under the same general name of poly(ethylene oxide); this then frequently necessitates an accompanying statement of the average molecular weight (or degree of polymerization) of the sample considered.

A further consequence of this situation is that much scientific work has been done on technically important materials which because of their low molecular weights show behavior atypical for true polymers, in that, e.g., their melting points vary markedly with molecular weight (whereas normally for true high polymers the dependence observed is relatively small), while with properties known to vary with molecular weight (e.g., intrinsic viscosity in solution) the dependence is not of the normal form for true polymers.

Another consequence is that there is a greater prominence of the end groups in their effects both on the physical properties and the chemical behavior of the material. Names such as "polyethylene glycol" imply the presence of a hydroxyl group at each end, but as discussed below there is strong evidence for some terminal unsaturation in many instances.

The polymer has been the subject of a number of reviews[1-6] and also of a recent monograph.[7] Many of these publications come from associates of the Union Carbide Corporation, which has been responsible for the development of the high molecular weight PEO (*Polyox WSR*) described in a series of articles.[8-10] The properties of the commercially available grades of low and medium molecular weight are described in the technical literature.[11,11a,12] The pharmaceutically important grades of the polymer have also been described.[13]

B. Synthesis and Purification

The main commercial methods used to produce PEO from ethylene oxide are described in the general reviews already listed.[1-10] Because commercial materials of a wide range of average molecular weights are readily available, it is seldom necessary to undertake the synthesis of a sample of PEO, except perhaps where a special form such as the perdeuteropolymer is required.[14] From the historical viewpoint, it is interesting to note the early work by Fordyce and colleagues on the synthesis of several monodisperse forms with degrees of polymerization up to 186,[15,16] and also the early work by Staudinger and Lohmann[17] on heterogeneous catalysis of the polymerization of ethylene oxide which was the starting point for the much more recent development of methods for the synthesis of very high molecular weight PEO.[8]

However, the synthetic methods which are now used may not necessarily give the two hydroxyl end groups per molecule widely assumed to be the case with these polymers. Thus, even in the production of relatively low molecular weight polymers by conventional alkali-catalyzed polycondensation there is the possibility of a termination step which leads to a terminal double bond. Furthermore, in methods using catalysts based on aluminium alkoxides, one end group is a double bond and the other an alkoxy group derived from the catalysts.[18] These possibilities need to be considered when applying hydroxyl end-group estimations to the determination of $\overline{M}_n$ values of PEO samples.

The polymer may be purified, i.e., freed of low molecular weight contaminants, by precipitation in a variety of solvent/nonsolvent systems; amongst suitable systems quoted in the literature are benzene/heptane; benzene + ethanol/hexane; carbon tetrachloride (hot)/hexane; and chloroform/diethyl ether. Such procedures must be expected to lead to the loss of the shortest chain oligomers.

C. Fractionation and Characterization

For relatively low and medium molecular weight PEO, because of the mechanism of its polymerization the distribution is generally of the Poisson type so that the range will be relatively narrow and the material seldom require or repay fractionation; determinations of the $\overline{M}$ values for commercial samples are given in a number of publications.[19,20] For high molecular weight PEO (i.e., $\overline{M} \approx$ D6), however, the distribution seems to be very wide ($\overline{M}_w/\overline{M}_n \approx 20$), and these materials therefore require fractionation if narrow-range samples are required. The principal difficulty in this fractionation is that the bulk polymer is highly crystalline at normal temperatures ($t_m \approx 66°C$), so that conventional fractional precipitation procedures at these temperatures are generally not successful. Work in this area has been described by Booth and colleagues,[21-23] and also by Ring and colleagues;[24] the latter have used as one successful method the distribution between water and benzene + chloroform mixtures which was originally developed by Schulz and Nordt.[25]

The molecular weight distribution may be estimated in the conventional manner by measurements of both $\overline{M}_n$ and $\overline{M}_w$, or more fully by gel chromatography;[20,22] for oligomers, gas-liquid chromatography (GLC) may be used.[2] The average molecular weights can be determined by the usual methods, examples of which will be found in the literature studies for aqueous and nonaqueous solutions tabulated later in the section. For short-chain materials, the $\overline{M}_n$ value is frequently estimated by determination of the hydroxyl end-group content;

the uncertainties in this method arising from the presence of other types of end-groups have already been discussed in Section II.B. Hydroxyl content is generally determined by an acetylation procedure;[12] spectrophotometric methods have also been developed, using either complex formation with Ce(IV),[24] or reaction with phenylazobenzoic anhydride.[26]

D. Bulk Polymer

Since the materials subsumed under the general name poly(ethylene oxide) range all the way from the shortest oligomer, ethylene glycol (M = 62), up to extremely high molecular-weight samples, there is therefore a correspondingly wide span in the physical properties of these materials, which range in appearance at normal temperatures from viscous liquids through semisolids to hard crystalline solids. The dependence of melting point upon molecular weight is available from various sources, including data for true monodisperse samples with degree of polymerization P = 6, 18, 42, 90, and 186,[15,16] and some precise measurements on commercial (but still effectively monodisperse) samples;[19,27] the specification values for the melting points of commercial materials are also available from the technical literature.[2,11,12] At high molecular weights the melting point tends towards a limiting value between 66 and 68°C. However, in these materials there will be some defects, involving folding of the polymer chains, in the crystal lattice. The melting point for the perfectly crystalline material with fully linear (nonfolded) chains has been estimated by Afifi-Effat and Hay[19] to be 75 ± 3°C; this value was obtained by using an extrapolation procedure in which experimental data for the higher molecular weights were omitted. The corresponding enthalpy and entropy changes of fusion were found to be 9.5 ± 0.1 kJ basemole^{-1} and 27.7 ± 3.0 J basemole^{-1} K^{-1}, where the basemole is the ethylene oxide unit.[19]

The enthalpy (heat content) for various samples of PEO has been measured by Beaumont and co-workers[27] and by Braun and co-workers.[28] Other thermal properties, including the thermal pressure coefficient, internal pressure, and isothermal compressibility, have been measured and discussed by Allen and co-workers.[29,30]

The glass transition behavior of PEO (i.e., for the residual amorphous regions in the mainly crystalline state) has been discussed by Bailey and Koleske.[7] For well-vitrified samples which have not been allowed to crystallize the value of t_g rises monotonically from about −65°C for M = 1D3, up to a limiting value for extremely high M of about −55°C. In practice, the presence of the crystalline component leads to effectively higher values of t_g, which is found to rise to a maximum of about −20°C for M = 6D3. Nevertheless, for normal temperatures any amorphous component must still be above the t_g value, in the rubbery rather than the glassy state.

Data for the density of PEO samples around normal temperatures and at rather diverse scattered molecular weights are available from various sources, e.g., at 15°C for M = 1D3—3D6,[28] at 20°C for P = 1, 2, 4, and 6 (M = 62—282: monodisperse samples)[31] and for M = 2D2—2D3 (commercial samples),[12] and at 25°C for M = 2D2—2D4 (commercial samples).[11] By extrapolation of the molten state densities of high molecular weight samples to room temperature, Braun and colleagues obtained at value of 1.126 g cm^{-3} for the fully amorphous state;[28] this is therefore also the value to be expected for the "effective density" of PEO in solution in an indifferent solvent, so that the corresponding value for the partial specific volume $\bar{v}_2^\circ$ should be the reciprocal of this, i.e., 0.888 cm^3 g^{-1}.

Measurements from X-ray diffraction studies give the density of perfectly crystalline PEO at room temperature to be 1.23 ± 0.03 g cm^{-3} (Reference 28). With the above value for the fully amorphous state and assuming a linear relation between density and degree of crystallinity, this enables the degree of crystallinity of a sample to be estimated from its density; by such methods Braun and co-workers have obtained degrees of crystallinity lying between 76 and 90% for samples varying in M from 1D3 to 3D6.[28]

Data for the refractive indexes of PEO samples are also available both for monodisperse samples, e.g., at 20°C for $P = 1, 2, 4$, and 6,[31] and at 60°C for $P = 6, 18, 42, 90$, and 186,[15,16] and also for commercial samples at 20°C for $M = 200$—600.[11,12]

The infrared spectrum of PEO has been given at the moderate resolution suitable for identification purposes,[32] and it has also been studied at higher resolution for both the solid and the liquid (molten) state to obtain more detailed structural information on chain conformation and folding.[14,33-37] Raman spectra of PEO have also been studied for the latter purpose.[37]

The pure polymer has also been examined by nuclear magnetic resonance and relaxation techniques, including measurements of the chemical shift values for liquid (short-chain) polymers,[38] of proton spin-lattice relaxation times,[39,40] and of proton side-band satellite spectra.[41,42] As well as giving more information on the molecular structure and conformation of the polymer, these NMR investigations also enable the degree of crystallinity for solid samples to be estimated; this method indicates degrees of crystallinity up to 95%.[1,2]

Studies by X-ray diffraction, along with the IR and NMR studies already mentioned, have shown that in the crystal lattice the polymer molecules are present as 7_2 helices, i.e., with seven turns for each two monomer units, in the fiber identity period of 19.3 Å.[14,43]

In this conformation, the methylene groups are on the periphery of the helix and the oxygen atoms clustered inside when viewed along the axis. This shape results from the chain being essentially in the conformational sequence . . . t g t . . . about the bond sequence . . . –O–CH_2–CH_2–O– . . . ; the conformations are not exactly *trans* (t) and *gauche* (g), however, the actual internal rotation angles are 188 and 65°, compared with the nominal values of 180 and 60°. (In fact, the crystal lattice would be expected to consist of equal numbers of chains of opposite chirality, i.e., with conformational sequences . . . t g t . . . and . . . t g′ t . . . , since if one form or the other were preferred the sample would be optically active in the bulk.) The interpretation of the factors leading to the production of this structure, in relation to the rotational isomeric theory of chain conformation, has been examined by Mark and Flory,[44] while Mark[45] has also discussed the conformational partition function and entropy for PEO in relation to that of other polyoxide chains. The structures of PEO and of the other crystalline polyoxides (polyethers) have been compared by Tadokoro.[46]

E. Crystalline Complexes with Small-Molecule Substances

Poly(ethylene oxide) is not only itself highly crystalline but it also forms highly crystalline complexes with a wide variety of small-molecule substances. The structures of these complexes have been reviewed by Tadokoro,[46] and their formation and uses have been discussed in the technical literature.[47] The complexes of this type which are known include those with bromine;[47] cadmium chloride;[48] calcium chloride;[49] guandinium chloride;[50] iodine;[47] mercuric chloride and other mercuric halides;[48,51] phenobarbital;[50] thiourea;[52] and urea.[52-56] These complexes are interesting not only intrinsically and for their applications, but also for the light that examination of their structures may cast on the possible interactions in aqueous solution between PEO and the solvent or other solute molecules.

F. Solution Behavior

Poly(ethylene oxide) is notable for its ready solubility both in water and in a variety of organic liquids. It is convenient to consider the behavior in aqueous solution separately from that in nonaqueous liquids; comparisons of the two are necessary to show whether there are any special features in water as a solvent.

It is not the present aim to consider the results of these extensive solution studies in any great detail or depth, and the general reviews already listed together with the more specific discussions noted in Tables 2 and 3 should be consulted for this purpose. Only three aspects will be dealt with briefly here in each case, i.e., partial specific volumes specific refractive index increment, and the relations between intrinsic viscosity and molecular weight.

Table 2
STUDIES ON PEO IN AQUEOUS SOLUTION

Entry	Year	Polymer samples No.	Range of $\overline{M}$	Techniques used	Temperature (t/°C)	Polymer concentration	Comments	Ref.
1	1936	3	2D4—9D4	OP	27	0.1—2%	→ $\overline{M}_n(A_2, \chi)$	57
2	1942	Evaluation of interaction parameter χ from OP data of Schulz[57] (entry 1)						58
3	1951	1	750	VP	25	w_2 = 0.83—1.0	2 = Methoxypolyglycol 750; water vapor uptake	59
4	1957	3	3D2—5D3	VP	30—65		→ TP	60
5	1957	3	3D2—5D3	HS	80	w_2 = 0.21—0.95	→ ΔH	60
6	1957	12	62—1D4	RI	20?		436, 546 nm: → $\bar{n}$ = 0.142 (M = 1D4), 0.139 (M = 6D3) $cm^3\ g^{-1}$	31
7	1957	11	62—1D4	TD	20	1—20%	→ $D°$	61
8	1957	11	62—1D4	VI	20	1—20%	→ [η]; k'; MHSR (0.72; 3.3d2; 2.0)	61,62
9	1958	8	3D4—1D7	LS			→ $\overline{M}_w$	63
10	1958	6	3D4—1D7	SR	20,30		→ $s°$ ($\overline{M}_{sD}$, D)	63
11	1958	?	"High"	SV	30		→ $\bar{v}_2°$ = 0.8392 $cm^3\ g^{-1}$	63
12	1958	6	3D4—1D7	VI	20,30		→ [η]; MHS (30°: 0.78; 1.25d2)	63
13	1958		General review of solution properties					9,10
14	1959	1	8D3	LS		1—20%	436 nm; + NaCl (0,0.2M): → $\overline{M}_w$; osmotic coefficients	64
15	1959	5	3D4—4D6	VI	35,45		→ [η]; MHS (35°: 0.82; 6.40d3) (45°: 0.81; 6.9d3)[a]	65
16	1959	1	4D6	VI	20—35		+ K_2SO_4(0.45M): → [η]; Θ_+	66
17	1959	2	6D3, 1D4	SV	25?		→ $\bar{v}_2°$ = 0.820 $cm^3\ g^{-1}$	67
18	1960		General review of solution properties					1
19	1960	4	400—4D3	VI	35	3—22%	→ [η]; k'; MHS (0.82; 1.66d2)	68
20	1961	5	1D4—8D5	SE	25	~0.04%?	→ A_2	69
21	1961	5	1D4—8D5	TD	25	~0.04%	→ $D°$	69
22	1961	3	4D3—2D4	OP	25		Permeable membranes: → A_2	70,71
23	1961	3	4D3—2D4	TD	25	0.1—0.4%	→ $D°$	70
24	1961	3	4D3—2D4	VI	25		→ [η]	70
25	1963	2	240, 5D5	SV	22	0—3%	e = 1—4556 atm: → $\bar{v}_2°$ = 0.846, 0.834 $cm^3\ g^{-1}$ (1 atm)	72
26	1964		Reevaluation of VI data of Rempp[61] (entry 8): → MHSM (0.58; 0.106; 0.81)					73

27	1964	4	2D5—5D6	VI	30	2—8%	→ Rheology	74
28	1964	8	194—7770	VI	25		→ [η]; MHS (M ≤ 1090: 0.50, 1.56)	75
29	1965	1	6D3	NMR	30—90		60 MHz; ^{13}C-H satellite side-band spectra: → J	41
30	1966	3	6D3—3D4	RI	25		546 nm: → $\bar{n}$ = 0.135 cm^3 g^{-1}	76
31	1966	7	4D3—4D4	SE	25		→ $\overline{M}_w$	76
32	1966	7	62—1D3	VPO	25		→ $\overline{M}_n$; A_2; no aggregation	76
33	1966	(1)	(P = 2)	USA			2 = Dioxane (cyclic dimeric monomer unit analog)	77
34	1966	2	8D3, 2D4	USA	10,25	5, 9%	10—185 MHz	78
35	1966	2	8D3, 2D4	VI	10, 25		→ [η]	78
36	1966	5	335—4150[b]	HS	27	ϕ_2 = 0.01—0.5	→ Δ H	79[b]
37	1966	Review of EDM values						81
38	1966	4	5D3—3D5	LS	27		→ $\overline{M}_w$	24
39	1966	6?	5D3—3D5?	RI	27?		436,546 nm: → $\bar{n}$ = 0.134, 0.132 cm^3 g^{-1}	24
40	1966	6	5D3—3D5	VI	25		→ [η]; MHSR (0.73; 2.4d2; 2.0)[c]	24
41	1966	4	P = 1,3; M = 1D3, 2D4	VI	10—90	w_2 = 0—1.0	→ η	82
42	1967	General reviews of solution properties						2,83
43	1967	?	62—4D4	SV			→ $\bar{v}_2^{\circ}$ = 0.832 cm^3 g^{-1} (M → ∞)	84
44	1967	?	62—4D4	VI	25		→ [η]; K_o = 0[d]	84
45	1967	5	1D3—2D4	VI	10		→ [η]; MHS (0.66; 5.8d2)	85
46	1967	6	600—2D4	USA	10	4—18%	10—175 MHz	85
47	1967	7	200—6D5	HD	25	ϕ_2 = 0.03—0.6	→ Δ H	86
48	1967	2	200, 600	HS	25	ϕ_2 = 0.06—0.7	→ Δ H	86
49	1967	1	8D3	RI			546 nm: → $\bar{n}$ = 0.137 cm^3 g^{-1}	87
50	1967	1	8D3	SE	23	0.4—10%	→ A_2; A_3	87
51	1967	1	8D3	SV	25	2%	→ $\bar{v}_2^{\circ}$ = 0.837 cm^3 g^{-1}	87
52	1968	2	8D3, 2D4	OP			→ $\overline{M}_n$; A_2; χ	88
53	1968	2	8D3, 2D4	RI			λ = ?: → $\bar{n}$ = 0.137, 0.136 cm^3 g^{-1}	88
54	1968	2	8D3, 2D4	SE			→ $\overline{M}_w$; A_2; χ	88
55	1968	2	8D3, 2D4	SV	25		→ $\bar{v}_2^{\circ}$ = 0.837 cm^3 g^{-1}	88
56	1968	7	1D3—2D4	USA/V	10	9—17%	10—165 MHz	89
57	1968	7	1D3—2D4	VI	10		→ [η]; MHS (0.63, 7.5d2)[c]	89
58	1968	4	335—4D3	SV	25, 35	w_2 = 0.01—0.5	→ $\bar{v}_1$	90
59	1968	4	335—4D3	VP	25,35	w_2 = 0.01—0.7	→ Thermodynamic parameters	90
60	1968	11	62—4D6	NMR	2—82	w_2 = 0.1—0.5	H_2O and D_2O: → T_1,(^{1}H); T_2(^{1}H)	91
61	1968	12	150—4D5	NMR	35?	5%	D_2O: → δ(^{1}H)	38
62	1968	3	5D5—4D6	VI	25,35	0.1—0.3%	→ [η]	92

Table 2 (continued)
STUDIES ON PEO IN AQUEOUS SOLUTION

Entry	Year	Polymer samples No.	Range of $\overline{M}$	Techniques used	Temperature (t /°C)	Polymer concentration	Comments	Ref.
63	1968	3	5D5—4D6	LS	25		→ $\overline{M}_w$; A_2; MHS(0.76; 1.19d2)	93
64	1969	7	1580—1D6	VI	25		+ K_2SO_4 (0.45M): → [η]; MHS (0.46; 0.28)	21
65	1969	4	200—1450	TD	15—35	2.5—6.5%	→ D°_2	94
66	1969	1	2D4	USA	20		10—200 MHz	95
67	1969	1	600	IR		ϕ_2 = 0.2—0.8	D_2O; 2.5—4.0 μm, 6.5—12.5 μm	96
68	1969	1	400	NMR	4—80	ϕ_2 = 0.05—1.0	D_2O: → δ(^{1}H)	96
69	1970	Discussion of US studies (especially from Reference 95)						97
70	1970	Discussion of US studies (especially from References 95 and 97)						98
71	1970	6	150—2D4	USA	4—32	w_2 = 0.2	1.5—163 MHz	99
72	1970	1?	6D3?	RA		10%	$\Delta\nu$ = 750—1550 cm^{-1}	37
73	1970	1	4D3	NMR	35	10%	D_2O: → T_1(^{1}H)	100
74	1970	1	4D4	HS	10—40	w_2 = 0.003—0.01	Also for small-molecule analogs: → Δ H	101
75	1970	1?	1D4?	VI	30		→ [η]; TP	102, 103
76	1971	1	4D4	OP	20—30	w_2 = 0.02—0.07	→ $\overline{M}_n$; A_2; A_3; A_4 (=0); TP	104
77	1971	1	4D4	SV	25	w_2 = 0.02—0.07	→ $\bar{v}_2^\circ$ = 0.830 cm^3 g^{-1}	104
78	1971	1	4D4	VI	25?		→ [η]	104
79	1972	1	6D3	HD	30	ϕ_2 = 0.016—0.07	→ Δ H	105
80	1972	1	6D3	HS	30	ϕ_2 = 0.005—0.06	→ Δ H	105
81	1973	8	62—3D4	VI	25		→ [η]; MHSR (0.72; 2.9d2; 1.5)	106
82	1974	2	3D4, 7D5	VI	35		+ K_2SO_4(0.45M) (Θ system): → [η]^f	107
83	1974	1	2D4	NMR	10,25	9%	D_2O: → δ; T_1; T_2	108
84	1974	1	4D3	NMR	25—80	30%	D_2O; 100 MHz; ^{13}C-H satellite side-band spectra: → J	42
85	1975		Review of solution properties					109
86	1975		Review of US studies					110
87	1976	2	4D3, 6D3	SS	25		→ s°	111
88	1976	4	1D3—6D3	TD	8—35		→ D°	111
89	1976	4	1D3—6D3	VI	8—35	0.5, 1%	→ [η]	111
90	1976	2	6D5, 4D6	...		10—1000 ppm	Effect on pulsation noise of small air bubbles	112
91	1976	1	8D5	VI			+ K_2SO_4(0-0.6M); drag reduction in turbulent flow	113

92	1977	5	600—8D3	VPO	35—70	1—20%	→ A_2; χ; ($\overline{M}_n$)	114
93	1978	1	2D4	OP	25		→ $\overline{M}_n$; A_2	115
94	1978	6	1D4—2D4	VI	30		→ [η]; k'	115
95	1978	1	2D4	ESR	5—45	w_2 = 0.04	End groups spin labeled: → τ_R[g]	116
96	1979	4	P = 1—4	EP		0—100%	→ ϵ	117
97	1979	4	P = 1—4	VI		0—100%	→ η	117

[a] Also MHS parameters for the two Θ systems: 35°C, 0.45M K_2SO_4 (0.50; 0.13); 45°C, 0.39M $MgSO_4$(0.50; 0.10).

[b] See Reference 80 for revision of the molecular weight values.

[c] MHSR parameters evaluated using the combined data from References 24, 61, 67, and 73.

[d] K_a is the association constant for aggregation of the polymer molecules in solution.

[e] MHS parameters evaluated using the combined data from References 85 and 89.

[f] See also general discussion of solution behavior of PEO in Reference 107.

[g] See also general discussion and review of segmental motion of PEO in solution in Reference 116.

Table 3
STUDIES ON PEO IN NONAQUEOUS AND MIXED SOLVENTS

Entry	Year	Polymer samples No.	Polymer samples Range of $\overline{M}$	Techniques used	Temperature (t/°C)	Polymer concentration	Comments	Ref.
					$H_2O + CH_4O$: Water and Methanol			
1	1970	1	4D3	NMR	35	10%	D_2O ϕ = 0—1 + CD_4O: → $T_1(^1H)$	100
2	1970	1	4D3	VI	35	10%	D_2O ϕ = 0—1 + CD_4O: → η	100
					CCl_4: Carbon Tetrachloride			
3	1939	5	P = 6—186	VI	20—60	4.4%	Monodisperse polymers: → [η]	118
4	1957	9	200—1D4	VI	20		→ [η]; MHSR (0.61; 6.9d2; 0)	61,62
5	1964	Reevaluation of the VI data of Rempp[61] (entry 4): → MHSM (0.62; 6.02d2; 0)						73
6	1965	5	335—4150[a]	HS	27	w_2 = 0.003—0.91	→ ΔH	119
7	1966	2	600, 4D4	VI	25		→ [η]; k'	76
8	1966	3	400—2500	VPO	25		→ $\overline{M}_n$; A_2; aggregation	76
9	1967	5	600—4D4	SV	25		→ $\bar{v}_2^{\circ}$ = 0.868 cm^3 g^{-1}	84
10	1967	5	600—4D4	VI	25		→ [η]; K_o = 11.7M^{-1} (footnote b)	84
11	1968	5	194—1D3	NMR	35,75	5%	→ $\delta(^1H)$	38
12	1969	8	1580—1D6	VI	25		→ [η]; MHS (0.64; 6.2d2)	21
13	1970	1	1.4D4	RI	25		546 nm: → $\tilde{n}$ = 0.028 cm^3 g^{-1}	120
14	1970	1	1.4D4	VI	25		→ [η]	120
15	1972	4	P = 1—4	LS		ϕ_2 = 0.2—0.5	Methoxy end groups: depolarized Rayleigh scattering	121
16	1973	4	P = 1—4	LS	25		Depolarized Rayleigh scattering	122
17	1979	1	9D3	ESR	23—35	1%	End groups spin labeled: → τ_R	123
18	1979	1	9DR	NMR	30	10%	→ $T_1(^{13}C)$; τ_R	123
					$CCl_4 + CH_4O$: Carbon Tetrachloride + Methanol			
19	1970	1	1.4D4	LS	25		546 nm; $\phi(CCl_4)$ = 0.1—0.65: → $\overline{M}_w$; A_2; PS	120
20	1970	1	1.4D4	RI	25		546, 584 nm; $\phi(CCl_4)$ = 0.1—0.85: → $\tilde{n}$	120
21	1970	1	1.4D4	VI	25		$\phi(CCl_4)$ = 0.1—0.96	120
22	1974	1	3D4	ED	r.t		$\phi(CCl_4)$ = 0.20—0.835: → PS	124
23	1974	1	3D4	RI	25		$\phi(CCl_4)$ = 0.20—0.835: → $\tilde{n}$	124

$CHBr_3$: Bromoform

24	1979	1	9D3	ESR	23—35	1%	End groups spin labeled: $\rightarrow \tau_R$	123
25	1979	1	9D3	NMR	30	10%	^{13}C spectra: $\rightarrow T_1$	123

$CHCl_3$: Chloroform

26	1964	12	200—3D4	VI	20		$\rightarrow [\eta]$; MHSM (0.74; 3.75d2; 0.60)	73
27	1964	10	200—3D3	VI	20		$\rightarrow k'$	125
28	1964	6	238—1570	VI	25		$\rightarrow [\eta]$; MHS ($M \leq 732$: 0.5; 0.206)	75
29	1965	1	6D3	NMR	10—120		60 MHz; ^{13}C-H satellite side-band spectra: $\rightarrow J$	41
30	1968	12	150—4D5	NMR	35?	5%	$CDCl_3$: $\rightarrow T_1(^1H)$	38
31	1969	1	400	NMR	4—51	ϕ_2 = 0.05—1.0	$\rightarrow \delta(^1H)$	96
32	1970	1	6D3	RA			$\Delta\nu$ = 150—1500 cm^{-1}	37
33	1970	1	4D3	NMR	34.5	10%	$CDCl_3$: $\rightarrow T_1(^1H)$	100
34	1972	1	6D3	HD	30	ϕ_2 = 0.01—0.29	$\rightarrow \Delta H$	105
35	1972	1	6D3	HS	30	ϕ_2 = 0.002—0.05	$\rightarrow \Delta H$	105
36	1974	1	1D5	LS	30		436 nm: $\rightarrow \overline{M}_w$; aggregation	126
37	1974	1	ID5?	RI	30		436 nm: $\rightarrow \tilde{n}$ = 0.054 cm^3 g^{-1}	126
38	1974	1	4D3	NMR	25—60	30%	100 MHz; ^{13}C-H satellite side-band spectra: $\rightarrow J$	42
39	1979	1	9D3	ESR	23—35	1%	End groups spin labeled: $\rightarrow \tau_R$	123
40	1979	1	9D3	NMR	30	10%	$\rightarrow T_1(^{13}C)$; τ_R	123

$CHCl_3$ + C_6H_{14}: Chloroform + Hexane

41	1961	?	?	RI	20	0.5—3%	$\phi(CHCl_3)$ = 0.32(Θ system): $\rightarrow \tilde{n}$ = 0.092 cm^3 g^{-1}	20
42	1961	7	600—4D4	VI	20		$\phi(CHCl_3)$ = 0.32(Θ system): $\rightarrow [\eta]$; k'; MHS (0.50, 0.23)	20
43	1964	7	200—3D4	VI	20		$x(CHCl_3)$ = 0.595(Θ system): $\rightarrow [\eta]$; MHSM (0.50; 0.185; 0.0)	73

$CHCl_3$ + C_7H_{16}: Chloroform + Heptane

44	1964	12	200—3D4	VI	20		$x(CHCl_3)$ = 0.456—0.883: $\rightarrow [\eta]$; MHSM (0.50—0.71; 0.175—4.55d2; 0—0.68)	73
45	1964	9	200—3D4	VI	20		$x(CHCl_3)$ = 0.50—0.66: $\rightarrow k'$	125

Table 3 (continued)
STUDIES ON PEO IN NONAQUEOUS AND MIXED SOLVENTS

Entry	Year	Polymer samples No.	Polymer samples Range of $\overline{M}$	Techniques used	Temperature (t/°C)	Polymer concentration	Comments	Ref.
					CH_2Br_2: Dibromomethane (Methylene Bromide)			
46	1979	1	9D3	ESR	23	1%	End groups spin labeled: $\rightarrow \tau_R$	123
					CH_2Cl_2: Dichloromethane (Methylene Chloride)			
47	1970	1	4D3	NMR	34	10%	$\rightarrow T_1(^1H)$	100
48	1972	1	6D3	HD	30	ϕ_2 = 0.02—0.24	$\rightarrow \Delta H$	105
49	1972	1	6D3	HS	30	ϕ_2 = 0.02—0.06	$\rightarrow \Delta H$	105
50	1979	1	9D3	ESR	23	1%	End groups spin labeled: $\rightarrow \tau_R$	123
51	1979	1	9D3	NMR	30	10%	$\rightarrow T_1(^{13}C)$; τ_R	123
					CH_2I_2: Diiodomethane (Methylene Iodide)			
52	1979	1	9D3	ESR	23	1%	End groups spin labeled: $\rightarrow \tau_R$	123
53	1979	1	9D3	NMR	30	10%	$\rightarrow T_1(^{13}C)$;τ_R	123
					CH_3NO: Formamide			
54	1961	1	4D3	OP	25		Permeable membranes	70
55	1961	1	4D3	TD	25	1—2%	$\rightarrow D°$	70
					CH_3NO_2: Nitromethane			
56	1974	1	4D3	NMR	25—94	30%	100 MHz; ^{13}C-H satellite side-band spectra: $\rightarrow J$	42
					CH_4O: Methanol (See Also: $H_2O + CH_4O$; $CCl_4 + CH_4O$)			
57	1957	9	300—1D4	LS	20?		546 nm: $\rightarrow \overline{M}_w$; A_2	31
58	1957	5	62—1D4	RI	20?	1—4%	546 nm: $\rightarrow \bar{n}$ = 0.149 cm^3 g^{-1} ($M \rightarrow \infty$)	31

59	1957	17	62—2D4	VI	20		→ [η]; MHSR (0.76; 1.61d2; 2.0)	61,62
60	1959	11	1D3—3D4	LS	25		546 nm: → $\overline{M}_w$; A_2	67
61	1961	3	4D3-2D4	OP	25		Permeable membranes	70
62	1961	2	4D3, 2D4	TD	25	0.5—2%	→ $D°$	70
63	1961	3	4D3—2D4	VI	25	1.5—4%	→ [η]; (k′)	70
64	1961	8	600—4D4	LS	20		546 nm: → $\overline{M}_w$	20
65	1961	?	?	RI	25	0.5—3%	546 nm: → $\tilde{n}$ = 0.143 cm^3 g^{-1}	20
66	1961	?	?	SV	25	3%	→ $\bar{v}°$ = 0.785 cm^3 g^{-1}	20
67	1961	?	?	VI	25		→ MHS (0.57; 8.52d2)	127
68	1964	12	240—3.5D4	LS			436, 546 nm: → $\overline{M}_w$	128
69	1964	Reevaluation of VI data of Rempp[61] (entry 59): → MHSM (0.58; 0.103;?) (20°)						73
70	1966	22	62—4D4	LS	25		546 nm: → $\overline{M}_w$; A_2	76
71	1966	11	62—3D4	RI	25	≤1%	546 nm: → $\tilde{n}$ = 0.149—0.150 cm^3 g^{-1}	76
72	1966	7	200—6D3	VPO	25	1—30%	→ $\overline{M}_n$; A_2; no aggregation	76
73	1966	5	335—4150[a]	HS	27	w_2 = 0.03 —0.96	→ Δ H	129
74	1967	9	8D4—5D6	LS	45		546 nm: → $\overline{M}_w$; A_2; $\hat{R}$	130
75	1967	?	?	RI	45		546 nm: → ñ = 0.152 cm^3 g^{-1}	130
76	1967	8	62—4D4	SV	25		→ $\bar{v}°_2$ = 0.846 cm^3 g^{-1} ($M \to \infty$)	84
77	1967	11	62—4D4	VI	25		→ [η]; K_o = 0 (footnote b)	84
78	1968	10	90—6D3	NMR	34.5	1%	CD_4O: → δ (^1H)	131
79	1969	2	6D3, 3D5	NMR	34.5	<8%	CD_4O: → T_1 (^1H)	132
80	1970	1	1.4D4	LS	25		546 nm: → $\overline{M}_w$; A_2	120
81	1970	1	1.4D4	RI	25		546 nm: → $\tilde{n}$ = 0.150 cm^3 g^{-1}	120
82	1970	1	1.4D4	VI	25		→ [η]	120
83	1971	5	62—600	HD	30	ϕ_2 = 0.25—0.7	→ Δ H	133
84	1971	5	62—600	HS	30	ϕ_2 = 0.2—0.9	→ Δ H	133
85	1971	?	?	VI	25		→ TP	134
86	1974	3	1D5—1D6	LS	30, 45		→ $\overline{M}_w$ (?); A_2; no aggregation	126
87	1974	?	?	RI	30, 45		436 nm: → $\tilde{n}$ = 0.145, 0.150 cm^3 g^{-1}	126
88	1974	?	?	RI	30		546 nm: → $\tilde{n}$ = 0.142 cm^3 g^{-1}	126
89	1975	?	?	RI	20		546 nm: → $\tilde{n}$ = 0.150 cm^3 g^{-1}	135
90	1976	1	1D4	LS	r.t.		546 nm: → $\overline{M}_w$	136
91	1976	1	1D4	RI	25, 62		546 nm: → $\tilde{n}$ = 0.150, 0.160 cm^3 g^{-1}	136

CH_4O + C_6H_6: Methanol + Benzene

92	1974	1	3D4	ED	r.t.		$\phi(CH_4O)$ = 0.1—0.6: → PS	124
93	1974	1	3D4	RI	25		$\phi(CH_4O)$ = 0.1—0.6; 546 nm: → $\tilde{n}$	124

Table 3 (continued)
IES ON PEO IN NONAQUEOUS AND MIXED SOLVENTS

Entry	Year	Polymer samples No.	Polymer samples Range of $\overline{M}$	ques	Temperature (t/°C)	Polymer concentration	Comments	Ref.
					C_2HCl_5: Pentachloroethane			
94	1968	1	3D5		40—101	4—50%	→ T_1 (^1H)	91
95	1970	1	4D3		34.5	10%	→ T_1 (^1H)	100
					$C_2H_2Br_4$: 1,1,2,2-Tetrabromoethane			
96	1970	1	4D3		34	10%	→ T_1 (^1H)	100
97	1979	1	9D3		23—35	1%	End groups spin labeled: → τ_R	123
98	1979	1	9D3		30	10%	→ T_1 (^{13}C); τ_R	123
					$C_2H_2Cl_4$: 1,1,2,2-Tetrachloroethane			
99	1970	1	4D3		34	10%	→ T_1 (^1H)	100
100	1979	1	9D3		23—35	1%	End groups spin labeled: → τ_R	123
101	1979	1	9D3		30	10%	→ T_1 (^{13}C); τ_R	123
					C_2H_3N: Acetonitrile (Methyl Cyanide)			
102	1966	6	150—4D4		25		546 nm: → $\overline{M}_w$; A_2	76
103	1966	8	62—1D3		25		546 nm: → $\bar{n}$ = 0.135 $cm^3\ g^{-1}$	76
104	1966	9	106—8D3		25		→ $\overline{M}_n$; A_2; aggregation	76
105	1967	10	62—4D4		25		→ $\bar{v}_2^{\circ}$ = 0.851 $cm^3\ g^{-1}$ ($M \to \infty$)	84
106	1967	9	62—4D4		25		→ [η]; K_a = 1M^{-1} (footnote b)	84
107	1975	?	?		20		546 nm: → $\bar{n}$ = 0.135 $cm^3\ g^{-1}$	135
					$C_2H_4Br_2$: 1,2-Dibromoethane			
108	1979	1	9D3		23—35	1%	End groups spin labeled: → τ_R	123
109	1979	1	9D3		30	10%	→ T_1 (^{13}C); τ_R	123

				$C_2H_4Cl_2$: 1,2-Dichloroethane				
110	1964	6	200—2D3	BPE	84		→ $\overline{M}_n$	128
111	1964	5	200—1500	OP	20?		→ $\overline{M}_n$; A_2	137
112	1979	1	9D3	ESR	23—35	1%	End groups spin labeled: τ_R	123
113	1979	1	9D3	NMR	30	10%	→ T_1 (^{13}C); τ_R	123
				C_2H_6O: Ethanol				
114	1966	5	335—4150[a]	HS	27	w_2 = 0.03—0.96	→ Δ H	129
115	1971	5	62—600	HD	30	ϕ_2 = 0.1—0.8	→ Δ H	133
116	1971	5	62—600	HS	30	ϕ_2 = 0.4—0.7	→ Δ H	133
				C_3H_6O: Acetone				
117	1961	1	4D3	OP	25	1—8%	Permeable membranes	70, 71
118	1961	1	4D3	TD	25	0.4—1.6%	→ D^o	70
119	1961	1	4D3	VI	25	1.5—4.5%	→ [η]; (k')	70
120	1964	7	194—2960	VI	25		→ [η]; MHS ($M \leq$ 1532: 0.5, 0.156)	75
121	1966	4	3D3—3D4	SE	25	0.5—2.0%	→ $\overline{M}_w$	24
122	1969	7	1580—1D6	VI	25		→ [η]; MHS (0.67, 3.2d2)	21
123	1972	?	Cross-linked	SV	32—67	ϕ_2 = 0.36	→ Q[c]	138
124	1972	4	400—4D3	VI	25—68		→ [η]; d[η]/dt [c]	138
				$C_3H_6O_2$: Methyl Acetate				
125	1966	1	6D3	RI	25		546 nm: → $\bar{n}$ = 0.111 cm^3 g^{-1}	76
126	1967	9	62—4D4	SV	25		→ $\bar{v}_2^{\circ}$ = 0.866 cm^3 g^{-1} ($M \to \infty$)	84
127	1967	10	62—4D4	VI	25		→ [η]; K_o indeterminate[b]	84
				C_3H_7NO: Dimethylformamide				
128	1959	10	1D3—3D4	VI	25		→ [η]; MHSR (0.73; 2.4d2; 2)	67
129	1961	3	4D3—2D4	OP	25		Permeable membranes	70,71
130	1961	3	4D3—2D4	VI	25	2—4.5%	→ [η]	70
131	1964	7	2D3—3D4	OP	20		→ $\overline{M}_n$	128
132	1964	12	200—3D4	VI	20		→ [η]; MHSM (0.66; 4.65d2; 0.75)	73,128

Table 3 (continued)
STUDIES ON PEO IN NONAQUEOUS AND MIXED SOLVENTS

Entry	Year	Polymer samples: No.	Polymer samples: Range of $\overline{M}$	Techniques used	Temperature (t/°C)	Polymer concentration	Comments	Ref.
133	1964	12	200—3D4	VI	20		$\rightarrow k'$	125
134	1966	1	1D4	RI	25		546 nm: $\rightarrow \bar{n} = 0.044\ cm^3\ g^{-1}$	76
135	1966	6	200—2D3	VPO	25		$\rightarrow \overline{M}_n$, A_2; no aggregation	76
136	1966	?	2D3—6.5D4	VI	25		$\rightarrow [\eta] = [\eta](H_2O)/1.125$	24
137	1967	9	62—4D4	SV	25		$\rightarrow \bar{v}_2^\circ = 0.878\ cm^3\ g^{-1}\ (M \rightarrow \infty)$	84
138	1967	18	62—4D4	VI	25		$\rightarrow [\eta]$; $K_o = 0^h$	84
139	1974	1	4D3	NMR	25—150	30%	100 MHz; ^{13}C-H satellite side-band spectra: $\rightarrow J$	42
C_3H_8O: 2-Propanol (Isopropanol)								
140	1957	1	1D3	LS	20?		546 nm: $\rightarrow \overline{M}_w$	31
141	1957	1	1D3	RI	20?		546 nm: $\rightarrow \bar{n} = 0.091\ cm^3\ g^{-1}$	31
C_4H_8O: Butanone (Methyl Ethyl Ketone)								
142	1957	3	1D3—8D3	LS	20?		546 nm: $\rightarrow \overline{M}_w$	31
143	1957	8	800—1D4	RI	20?		546 nm: $\rightarrow \bar{n} = 0.097\ cm^3\ g^{-1}\ (M \rightarrow \infty)$	31
144	1975	1	1D4	RI	25		546 nm: $\rightarrow \bar{n} = 0.094\ cm^3\ g^{-1}$	135,136
C_4H_8O: Tetrahydrofuran								
145	1975	1	1D4	RI	25		546 nm: $\rightarrow \bar{n} = 0.068\ cm^3\ g^{-1}$	135,136
$C_4H_8O_2$: Dioxane								
146	1939	7	$P = 2$—186	VI	20—60	4.4%	$\rightarrow \eta$	118
147	1957	1	1D4	RI	20?		546 nm: $\rightarrow \bar{n} = 0.045\ cm^3\ g^{-1}$	31
148	1957	13	62—2D4	VI	20		$\rightarrow [\eta]$; MHSR (0.71; 3.5d2; 0.75)	61,62
149	1961	1	4D3	OP	25		Permeable membranes	70,71
150	1961	1	4D3	VI	25		$\rightarrow [\eta]$	70

151	1964	Reevaluation of VI data of Rempp[61] (entry 148): → MHSM (0.57; 5.01d2; 0) (20°)						73
152	1964	7	194—1532	VI	25		→ [η]; MHS ($M \leq 326$: 0.5; 0.138)	75
153	1965	5	335—4150[a]	HS	27	w_2 = 0.02—0.94	→ ΔH	119
154	1966	6	100—3D3	VPO	25		→ $\overline{M}_n$, A_2; aggregation	76
155	1967	9	62—4D4	SV	25		→ $\bar{v}_2^{\circ}$ = 0.886 cm³ g⁻¹ ($M \to \infty$)	84
156	1967	10	62—4D4	VI	25		→ [η]; K_o = 2 M⁻¹ (footnote b)	84
157	1972	4	430—1100	VI	25—55		→ [η]; d[η]/dt [c]	138
158	1974	1	6D5	LS	30, 45		436 nm: → $\overline{M}_w$; no aggregation	126
159	1974	1?	~1D6?	RI	30, 45		436 nm: → $\tilde{n}$ = 0.0492, 0.0607 cm³ g⁻¹	126
				C_4H_9NO: Dimethylacetamide				
160	1978	5	1D4—2D4	GPC	80		→ $\overline{M}_n$; $\overline{M}_w/\overline{M}_n$ (= 1.2 ± 0.2)	115
				$C_4H_{10}O$: Butanol				
161	1971	5	62—600	HD	30	ϕ_2 = 0.2—0.9	→ ΔH	133
162	1971	5	62—600	HS	30	ϕ_2 = 0.4—0.7	→ ΔH	133
				C_5H_5N: Pyridine				
163	1968	6	150—1D3	NMR	35	5%	→ δ(¹H)	38
164	1974	1	4D4	NMR	25—110	30%	100 MHz; ¹³C-H satellite side-band spectra: → J	42
165	1978	1	2D4	ESR	0—30	w_2 = 0.04	End groups spin labeled: → τ_R [d]	116
				C_6H_5Br: Bromobenzene				
166	1977	1	9D3	ESR	23—35	1%	End groups spin labeled: → τ_R	123
167	1977	1	9D3	NMR	30	10%	→ T_1(¹³C); τ_R	123
				C_6H_5Cl: Chlorobenzene				
168	1966	14	300—3D4	VPO	65	1—7%	→ $\overline{M}_n$	24
169	1970	1	4D3	NMR	34.5	10%	→ T_1 (¹H)	100
170	1974	1	4D3	NMR	25—130	30%	100 MHz; ¹³C-H satellite side-band spectra: → J	42
171	1979	1	9D3	NMR	30	10%	→ T_1 (¹³C); τ_R	123

Table 3 (continued)
STUDIES ON PEO IN NONAQUEOUS AND MIXED SOLVENTS

Entry	Year	Polymer samples No.	Range of $\overline{M}$	Techniques used	Temperature (t/°C)	Polymer concentration	Comments	Ref.
						C_6H_6: Benzene		
172	1957		150—1D3	EDM	20		→ μ	140
173	1957	4	P = 2—8	RI	20?		436 nm: → $\bar{n}$ = −0.086 ± 0.059 $cm^3\ g^{-1}$	31
174	1957	4	P = 2—8	TD	20	0.3—0.8%	→ $D°$	61
175	1957	11	100—2D4	VI	20		→ [η]; MHSR (0.68; 4.85d2; 0)	61,62
176	1959	2	1D5, 2D5	VI	25		→ [η]	18
177	1961	1	4D3	OP	25		Permeable membranes	70,71
178	1961	1	4D3	VI	25		→ [η]	70
179	1964	10	200—9D3	BPE	80		→ $\overline{M}_n$	128
180	1964	9	200—3D4	VI	20		→ [η]; MHSM (0.66; 4.65d2; 0)	73
181	1964	10	200—2D4	OP	20?		→ $\overline{M}_n$; A_2	137
182	1964	9	200—3D4	VI	20		→ k'	125
183	1964	9	194—7770	VI	25		→ [η]; MHS ($M \leqslant$ 326: 0.5; 2.129)	75
184	1966	1	3D3	RI	25		546 nm: → $\bar{n}$ = −0.016 $cm^3\ g^{-1}$	76
185	1966	10	200—1D4	VPO	25		→ $\overline{M}_n$; A_2; aggregation	76
186	1966	5	335—4150[a]	HS	27	w_2 = 0.01—0.96	→ Δ H	129
187	1967	9	8D4—5D6	VI	25		→ [η]; MHS (0.686; 3.97d2)	130
188	1967	6	62—4D4	SV	10, 37		→ $\bar{v}_2°$ = 0.858, 0.878 $cm^3\ g^{-1}$ ($M \to \infty$)	84
189	1967	8	200—4D4	VI	10—37		→ [η]; K_o = $4M^{-1}$ (25°)[b]	84
190	1967	8	3D5—2D6	VI	30		→ [η]; MHS (0.64; 6.14d2)	141
191	1968	15	150—2D4	NMR	35	5, 50%	→ δ(^{1}H)	38
192	1969	2	350, 4D3	IR		ϕ_2 = 0.05—1.0	2.5—4.0, 6.5—12.5 μm	96
193	1969	1	400	NMR	35	5—100%	→ δ(^{1}H)	96
194	1971	1	6D3	SV	46, 70	w_2 = 0.1—0.8	→ Volume change of mixing	23
195	1971	2	6D3, 8D5	VP	46—70	w_2 = 0.3—0.97	→ TP	23
196	1974	2	3D4, 7D5	VI	15—50		→ [η]; d[η]/dt [c]	107
197	1974	2	1D5, 6D5	LS	30, 54		436, 546 nm: → $\overline{M}_w$; A_2; aggregation	126
198	1974	?	1D5—1D6?	RI	30, 54		436 nm: → $\bar{n}$ = −0.0181, −0.0128 $cm^3\ g^{-1}$	126

199	1976	2[f]	7D3[f]	NMR	30, 35	w_2 = 0.05—0.3	C_6H_6 and C_6D_6: → T_1 (^1H); T_1 (^{13}C)	142
200	1978	1	2D4	ESR	5—32	w_2 = 0.01—0.22	End groups spin labeled: → τ_R[d]	116
				$C_6H_{10}O$: Cyclohexanone				
201	1957	11	62—1D4	VI	20		→ [η]; MHSR (0.69; 3.55d2; 0.5)	61,62
202	1961	1	4D3	VI	25		→ [η]	70
203	1964	Reevaluation of VI data of Rempp[61] (entry 201): → MHSM (0.68; 4.58d2; 0.45) (20°)						73
				$C_6H_{12}O$: 2-Methylpentan-4-one (Methyl Isobutyl Ketone)				
204	1969	7	1580—1D6	VI	50		→ [η]; MHS (0.52; 0.12)	21
				C_6H_{14}: Hexane — See $CHCl_3$ + C_6H_{14}				
				C_7H_8: Toluene				
205	1960	4	400—4D3	VI	35	0.5—0.9%	→ [η]; *k′*; MHS (0.70; 1.45d2)	68
206	1964	7		VI	30		→ [η]; MHSM (0.66; 4.08d2; 0.40)	73
207	1964	7	400—3D4	VI	30		→ *k′*	125
208	1972	1?	Cross-linked	SV	27—70	$\phi_2 \approx 0.3$	→ *Q* [c]	138
				C_7H_8O: Benzyl Alcohol				
209	1961	1	4D3	OP	25		Permeable membranes	70
210	1961	1	4D3	VI	25		→ [η]; (*k′*)	70
211	1967	9	62—4D4	SV	25		→ $\bar{v}_2^\circ$ = 0.865 $cm^3\ g^{-1}$ ($M \to \infty$)	84
212	1967	11	62—4D4	VI	25		→ [η]; K_o indeterminate[b]	84
				C_7H_{16}: Heptane — See $CHCl_3$ and C_7H_{16}				
				C_8H_{10}: *p*-Xylene				
213	1966	14	300—3D4	VPO	65	0.5—3%	→ $\overline{M}_n$	24
				$C_8H_{18}O_3$: Bis(2-Ethoxyethyl) Ether (Diethylene Glycol Diethyl Ether)				
214	1969	7	1580—1D6	VI	50		→ [η]; MHS (0.51; 0.14)	21

Table 3 (continued)
STUDIES ON PEO IN NONAQUEOUS AND MIXED SOLVENTS

Entry	Year	Polymer samples No.	Polymer samples Range of $\overline{M}$	Techniques used	Temperature (t/°C)	Polymer concentration	Comments	Ref.
						$C_{10}H_7Br$: 1-Bromonaphthalene		
215	1979	1	9D3	ESR	23—35	1%	End groups spin labeled: → τ_R	123
216	1979	1	9D3	NMR	30	10%	→ $T_1(^{13}C)$; τ_R	123
						$C_{10}H_7Cl$: 1-Chloronaphthalene		
217	1968	4	P = 4—7	NMR	35, 130	5%	→ $\delta(^1H)$	38
218	1971	1	4D3	NMR	34.5	10%	→ $T_1(^1H)$	100

[a] See Reference 80 for revision of molecular weight values.
[b] K_o: association constant for the aggregation of polymer molecules in the solvent.[84]
[c] See Reference 139 for further discussion of the results from Reference 138.
[d] See Reference 116 for general discussion of segmental motion of PEO in solution.
[e] See Reference 107 also for general discussion of behavior of PEO in solution.
[f] Polymer samples are PEO, and mixture of 20% PEO and 80% fully deuterated PEO.

1. Aqueous Solution

Table 2 summarizes the results of various studies that have been carried out on the behavior of PEO in aqueous solution. Although in general the consideration of the effects of additives such as salts is delayed until later (see Chapter 2 of Volume II), some studies have also been included in the table in two aqueous salt solutions, 0.45 M K_2SO_4 at 35°C (entries 15, 16, 64, and 82) and 0.39 M $MgSO_4$ at 45°C (entry 15), since these seem to be essentially ideal systems at the temperatures indicated.

For the *partial specific volume*, $\bar{v}_2^\circ$, of PEO in aqueous solution, there is a total of seven values (for the temperature range 20 to 30°C) in Table 2 (entries 11, 17, 25, 43, 51, 55, and 77); that from entry 17 (0.820 $cm^3\ g^{-1}$) is well out of the range of the others, while that from entry 55 is apparently a repeat of that for entry 51. The mean of the remaining five values is 0.834 ± 0.004 $cm^3\ g^{-1}$, for a mean temperature of 25°C.

For the *specific refractive index increment*, $\tilde{n}$, there are eight values (in entries 6 — twice, 30, 39 — twice, 49, and 53 — twice) for a variety of molecular weights (mainly, around D4), temperatures (in many cases not specified but probably all within the range 20 to 30°C) and wavelengths (mainly, 436 and 546 nm, with some unspecified); the value of $\tilde{n}$ seems to be largely independent of these parameters, and a simple pooled average gives a value of 0.136 ± 0.003 cm^3g^{-1}.

Considering the relations between *intrinsic viscosity* and molecular weight, Figure 1 shows that there is an essentially linear correlation between log K and a for different temperatures, which is the behavior frequently observed for the same polymer with different solvents;[127] however, as the figure also shows the extrapolation to a = 0.5 gives a much higher value of K_θ than the range of those accepted for the polymer. The plots in Figure 2 then also show that there are essentially linear correlations between both a and temperature, and log K and temperature. Extrapolation of the first of these two plots to lower temperatures gives t = − 12 ± 3°C for a = 0.50, indicating this (supercooled) state to be an ideal (Θ_-) system, as previously suggested on similar grounds;[109] the continuation of this plot in the opposite direction, i.e., to temperatures above 35°C, is speculative but nevertheless shows the trend expected if the system is to attain the known Θ_+ state at 96(±3)°C (see Section II.G.2).

These changes in the MHS parameters with temperature are more marked than might be expected, although there does not seem to be sufficient comparable data available for other systems to confirm this; nevertheless, they are consistent with the polymer having a quasirandom coil conformation which steadily expands over the range 10 to 35°C. However, there is evidence from Raman spectra,[37] NMR spectra,[96] and calorimetry[105] that PEO in aqueous solution retains to a greater or lesser extent the helical conformation that it has in the crystalline solid state, in which case much higher values of a would be expected. It would seem that any helical sections (right or left handed) are relatively short, allowing the overall conformation to be a quasirandom coil; the marked variation in MHS parameters may represent changes in the average length of these sections, and the "turnover" point at around 40°C may represent the melting point of these helical sections.

From the practical viewpoint, on the basis of the data given in Figures 1 and 2 the following pairs of values of the MHS parameters a and K/cm^3g^{-1} are recommended: 10°C, 0.65, 7·0d2; 15°C, 0.68, 4·4d2; 20°C, 0.71, 2·8d2; 25°C, 0.75, 1·7d2; 30°C, 0.78, 1·1d2; 35°C, 0.82, 6·8d3. Where low molecular weight samples are involved, Rempp's extension of the MHS equation (i.e., the MHSR equation) is applicable with $[\eta]_o$ = 2 cm^3g^{-1} for both 20 and 25°C.

2. Nonaqueous Solution

Table 3 summarizes studies on the behavior of PEO in nonaqueous solution, including mixed solvent media. The entries indeed serve to indicate those common liquids which are solvents for PEO; further details of the solubility behavior are given in Section II.G.3.

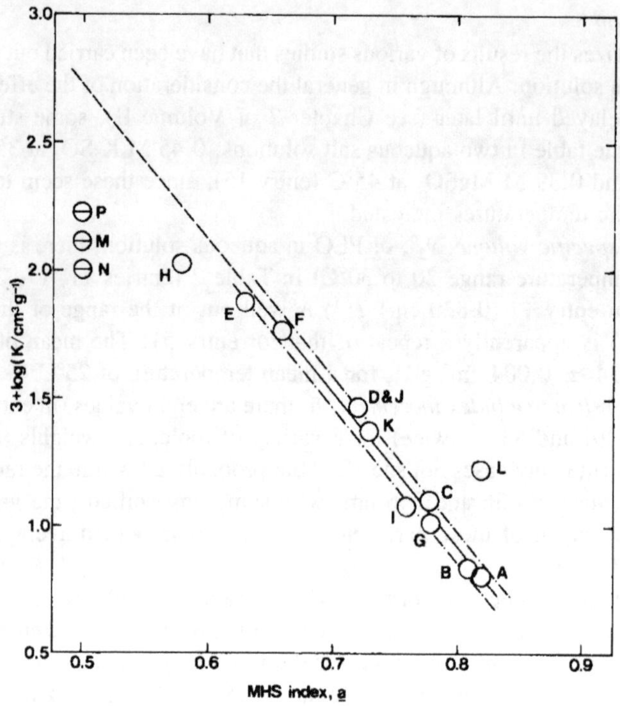

FIGURE 1. Correlation between the MHS parameters a and K (logarithmic scale) for PEO in aqueous solution (plain circles) and in theta solvents (barred circles). (A,B) Bailey and Callard (35, 45°C);[65] (C) Bailey et al. (30°C);[63] (D) Crouzet and Marchal (25°C);[106] (E) Hammes and Roberts (10°C);[89] (F) Hammes and Schimmel (10°C);[85] (G) Merrill et al. (25°C);[230] (H) Müh (20°C);[73] (I) Nabi (25°C);[93] (J) Rempp (20°C);[61,62] (K) Ring et al. (25°C);[24] (L) Thomas and Charlesby (35°C);[68] (M) Bailey and Callard (0.45M K_2SO_4, 35°C);[65] (N) Bailey and Callard (0.39M $MgSO_4$, 45°C);[65] (P) Beech and Booth (Stockmayer-Fixman extrapolation for data in non-aqueous solution).[21] The three data points B, H, and L are deviant in one or more of the plots in this figure and in Figure 2, so that for consistency all three have been omitted in fitting the best straight lines to these three plots.

Common liquids which are *nonsolvents* for PEO include hexane and other aliphatic hydrocarbons, and diethyl ether.

Data for partial specific volume, $\bar{v}_2°$, are available for seven liquids, as follows (values in $cm^3\ g^{-1}$): carbon tetrachloride, 0.868; methanol, 0.846 (from entry 76 — the value of 0.785 from entry 66 is anomalous); acetonitrile, 0.851; methyl acetate, 0.866; dimethylformamide, 0.878; dioxane, 0.886; benzene, 0.858. It is interesting to note that the highest value is that for dioxane, and that it is essentially the same as the specific volume for the bulk polymer in the amorphous state (see Section II.D); this is what would be expected for this liquid which is the cyclic dimer analog of the polymer chain. The values for the other liquids range from 0.878 for dimethylformamide down to the lowest value, 0.846, for methanol, indicating progressively greater extents of compaction of the liquid by the polymer, although still not to the extent seen with water, where $\bar{v}_2° = 0.834$ (see Section F.I.).

Data are also available for *specific refractive index increment*, $\tilde{n}$, in the same seven liquids as well as chloroform, carbon tetrachloride + methanol, and methanol + benzene. For methanol, there are 11 values at temperatures from 20 to 62°C and wavelengths of 436 and 546 nm; the values do not seem to show any consistent variation with wavelength, and little if any with temperature, and the 8 values for the temperature range 20 to 30°C give an

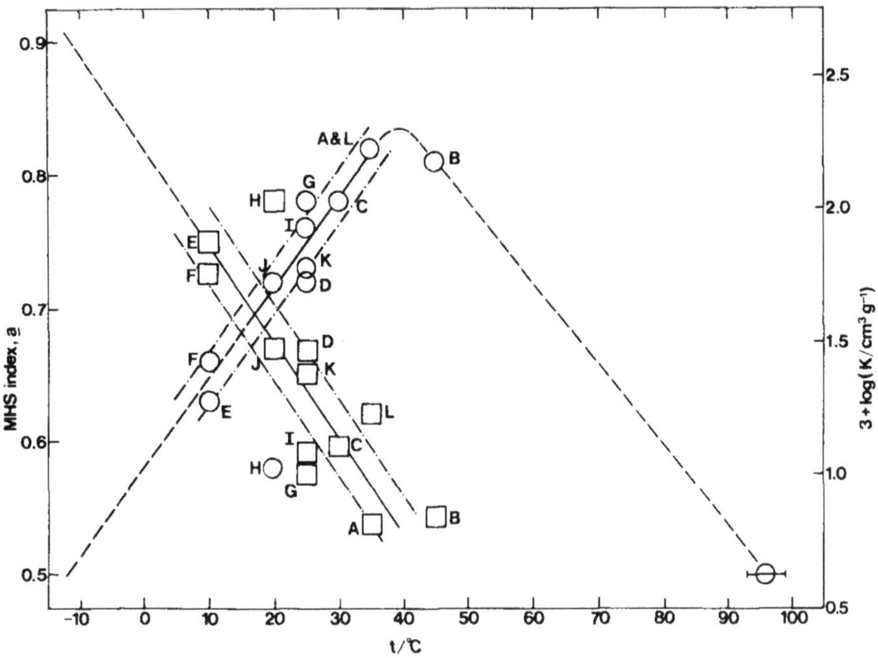

FIGURE 2. Variation of the MHS parameters a (circles and left-hand ordinate scale) and K (logarithmic scale — squares and right-hand ordinate scale) for PEO in aqueous solution, with Celsius temperature t. Letter labels as in Figure 1; the barred circle at t = 96 ± 3°C, a = 0.50 represents the Θ_+ state for this system (see Section II.G.2).

overall average of 0.147 ± 0.003 cm^3 g^{-1}. For the liquids and mixtures in general, the value of $\bar{n}$ decreases as usual with increase in the refractive index of the solvent, n_1; a fuller examination of this correlation and its connection with the refractive index of (amorphous) polymer requires that the partial specific volume values also be taken into account.

For the intrinsic viscosity-molecular weight relations, the present compilation in Table 3 extends the data previously discussed[109] while at the same time confirming the correlation between the two parameters a and K also seen in the case of water (Figure 1). However, the data are not sufficiently self-consistent to give a clear impression as to whether the temperature variation of these MHS parameters is as marked for organic solvents as it is with water; nevertheless, in the cases of carbon tetrachloride, benzene, and dimethylformamide there does seem to be a definite tendency for the parameter a to increase with rise in temperature, which is indeed also the direction of increasing solvent power as judged from the solubility behavior (see Section II.G.3).

From the practical viewpoint it is necessary to have trustworthy MHS parameters for estimating molecular weights from intrinsic viscosity values; the parameters seem to be quite well established in the cases of carbon tetrachloride and benzene. For carbon tetrachloride there are three sets of values in Table 3 (entries 4, 5, and 12 — although the first two of these relate to the same set of experimental data) which give average values: a = 0.62 ± 0.02; K = 6.4 (±0.5)d2 cm^3 g^{-1}; for benzene there are four sets of values (entries 175, 180, 187, and 190) which give the average values: a = 0.67 ± 0.02; K = 5(±1)d2 cm^3 g^{-1}. With each solvent both Rempp's additional parameter $[\eta]_0$, and Müh's μ, are zero, so that the simple form of the MHS equation seems to be usable down to low molecular weights.

G. Phase Equilibria

1. Uptake of Water Vapor

The uptake of water from the vapor state by PEO over a range of humidity values has

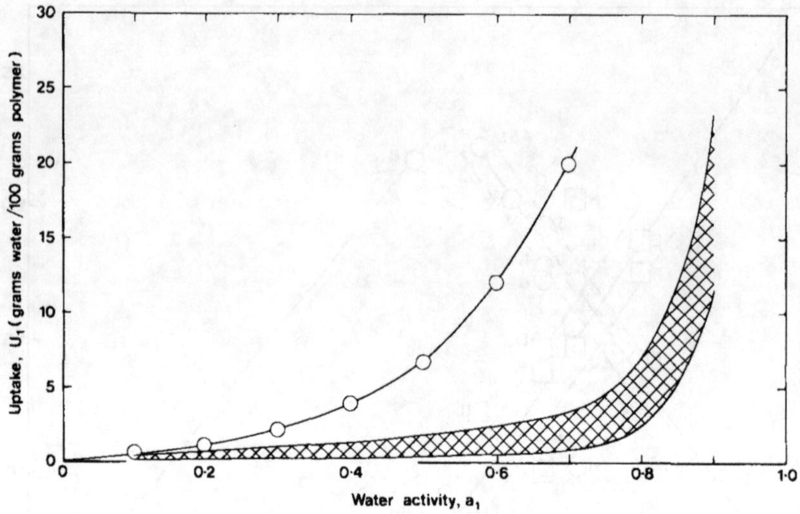

FIGURE 3. Equilibrium uptake of water, U_1, by poly(ethylene oxide) vs. water activity, a_1. ○ Low molecular weight polymer ("Methoxypolyglycol 750" [Me(EO)$_{16}$OMe], $\overline{M}$ = 750, 25°C);[59] shaded band: high molecular weight polymer ("*Polyox*", $\overline{M} \approx$ 1D6, temperature not stated).[2,3,10]

been reported for two widely different samples, as shown in Figure 3: first, for a so-called "Methoxypolyglycol 750"[59] which may be assumed to be a methoxy-ended PEO with this molecular weight (and hence with an average degree of polymerization P = 16), and second, for high molecular weight ($\overline{M} \approx$ D6) and hence highly crystalline PEO, where the data are presented graphically in the form of a broad band.[2,3,10]

The uptakes observed are much less in the second case than in the first, because of the highly crystalline nature of the high molecular weight polymer.

2. *Phase Separation in Aqueous Solution*

Aqueous solutions of PEO show no sign of phase separation when cooled down from normal temperatures to 0°C, although there are indications from the viscosity behavior that there may be a Θ_- point at about −10°C (see Section F.1). However, like aqueous solutions of many ethers, these solutions have an inverse solubility-temperature behavior, since precipitation occurs when they are heated to temperatures around the normal boiling point of 100°C. Phase diagrams have been reported for these systems by Bailey and Callard,[65] Malcolm and Rowlinson,[60] and Nakayama.[101] The ideal temperature Θ_+ value for this phase separation has been determined by Boucher and Hines[115] using plots of $1/t_c$ vs. log ϕ_2 extrapolated to log ϕ_2 = 0 (i.e., pure polymer, with ϕ_2 = 1), giving Θ_+ = 369 ± 3 K (96 ± 3°C). The critical temperatures from the published phase diagrams[60,65,101] are concordant with this value.

3. *Phase Separation and Solubility Behavior in Nonaqueous Solution*

In contrast with the behavior in aqueous solution, most nonaqueous solutions of PEO show the normal dependence of solubility upon temperature, i.e., the polymer dissolves on heating, and precipitates out of solution again on cooling. With the melting point of the polymer at about 66°C, this means that at temperatures below this value it is frequently the crystalline polymer that precipitates out, rather than a liquid phase as is the case with the precipitation from aqueous solutions that takes place close to 100°C. Studies of the dissolution and precipitation temperatures have been reported by the Union Carbide group for 34 liquids, and their results are summarized in Table 4; further details, such as the time required for

Table 4
DISSOLUTION AND PRECIPITATION TEMPERATURES FOR PEO[a]

Entry	Liquid	Dissolution temperature/°C	Precipitation temperature/°C 0.5%	1%	2%
1	CCl_4: Carbon tetrachloride	S	3	NP	NP
2	CH_2Cl_2: Dichloromethane (methylene chloride)	S	20	NP	—
3	CH_4O: Methanol[b]	S	20	20	—
4	C_2HCl_3: Trichloroethylene	S	NP	NP	—
5	C_2H_3N: Acetonitrile (Methyl cyanide)	S	NP	NP	—
6	$C_2H_4Cl_2$: Ethylene dichloride (1,2-dichloroethane)	S	NP	NP	—
7	C_2H_6O: Ethanol	45	31	31	—
8	$C_2H_6O_2$: Ethylene glycol	Insoluble: 0—100°C			
9	$C_2H_8N_2$: Ethylene diamine	40	26	26	—
10	$C_3H_4O_3$: Ethylene carbonate	S	—	20	—
11	C_3H_6O: Acetone	35	22	20	20
12	C_3H_7NO: Dimethylformamide	S	13	12	—
13	C_3H_8O: 2-Propanol (Isopropanol)	50	37	37	—
14	$C_3H_8O + H_2O$: 2-Propanol ($\phi = 0.91$) + water	S	3	2	—
15	$C_3H_8O_3$: Glycerol	Insoluble: 0—100°C			
16	C_4H_8O: Butanone (Methyl ethyl ketone)	S	—	20	—
17	C_4H_8O: Tetrahydrofuran	S	—	18	—
18	$C_4H_8O_2$: Dioxane	40	4	4	—
19	$C_4H_8O_2$: Ethyl acetate	40	—	25	—
20	$C_4H_{10}O$: Butanol	50	—	33	—
21	$C_4H_{10}O_2$: 1,3-Butanediol	Insoluble: 0—100°C			
22	$C_4H_{10}O_2$: Ethylene glycol monoethyl ether	45	28	28	—
23	$C_4H_{10}O_2$: Ethylene glycol dimethyl ether	40	—	27	—
24	$C_4H_{10}O_3$: Diethylene glycol	Insoluble: 0—100°C			
25	C_6H_6: Benzene	S	3	2	—
26	$C_6H_{12}O$: 2-Methylpentan-4-one	50	—	40	—
27	$C_6H_{12}O_2$: Butyl acetate	50	—	34	—
28	$C_6H_{12}O_3$: Ethylene glycol monoethyl ether acetate	35	25	25	25
29	$C_6H_{14}O_2$: Ethylene glycol monobutyl ether	50	—	33	—
30	$C_6H_{14}O_2$: Ethylene glycol diethyl ether	50	—	46	—
31	$C_6H_{14}O_3$: Diethylene glycol monoethyl ether	50	32	32	—
32	C_7H_8: Toluene	30	—	20	—
33	C_7H_8O: Anisole	40	0	0	0
34	C_8H_{10}: Xylene	30	—	20	—

[a] Collected data from References 1 and 7 (0.5 and 2%), and 2 and 3 (1%) for high molecular weight ($M \approx$ D6) PEO. In each case, the systems were heated to 100°C to dissolve the polymer as fully as possible and then cooled back to 25°C. For the *dissolution temperature* the system was then heated in steps of 5° up to 100°C, observing for complete dissolution at each stage; "S" indicates that the polymer was already dissolved at the starting temperature. For the *precipitation temperature* the system was then slowly cooled down to 0°C, noting the initial appearance of precipitate; "NP" indicates that no precipitation was observed even at the lowest temperature; a blank (—) indicates that no observations were made for that system. See the original references for further details such as the time required initially to dissolve the polymer, and the appearance at 100 and at 25°C.

[b] See Reference 134 for the phase diagram for the system PEO/methanol.

the initial dissolution of the polymer, and the appearance of the systems at 100°C and then at 25°C, will be found in the original references.[1-3,7] For these systems, the dissolution temperatures are less significant from the fundamental viewpoint than the precipitation temperatures, since the dissolution is relatively slow and the values were only taken at 5°

intervals. The precipitation temperatures show a number of interesting regularities when correlated with the molecular structures of the liquids; e.g., the addition of an ethoxy ($-CH_2CH_2O-$) group to the molecule of the liquid generally leaves the value of t_p essentially unchanged, which is indeed what might be expected from this group being the monomer unit of the polymer itself and hence being thermodynamically "neutral" towards the polymer chain. These and other correlations which exist indicate that it should be possible to predict the value of t_p from the structure of the liquid.

The solubility behavior of PEO in methanol has been studied in more detail by Napper and Netschey;[134] the phase diagram obtained by them shows that t_p rises monotonically with polymer concentration, which is characteristic of the separation of a crystalline phase.

The solubilities of monodisperse short-chain PEO samples (with $P = 6, 18, 42, 90$, and 186) in methanol + diethyl ether and dioxane + diethyl ether at 0°C have been determined by Lovell and Hibbert.[145]

4. *Liquid/Liquid Partition (Distribution) Behavior*

The only published work on the liquid/liquid partition behavior of PEO seems to be that done many years ago by Schulz,[146] who studied the partition of two different molecular-weight samples between water and benzene + chloroform mixtures over the range ϕ(benzene) = 0.48—0.54. The value of the organic/water partition coefficient K_p was found to be very sensitive to the composition of the mixture, with log K_p a steeply falling linear function of ϕ(benzene); extrapolation in each direction shows that the polymer would be expected to be essentially all in the aqueous phase for water/benzene, but all in the organic phase for water/chloroform. The latter situation is opposite to that shown, e.g., by PVP (see Chapter IV), and should provide a method for separating the two polymers. The partition coefficient is also dependent on the molecular weight of the polymer, a fact which may be used in its fractionation.[24,25]

H. Chemical Behavior

PEO shows a variety of forms of chemical instability which have to be considered and wherever possible counteracted in the use and the study of the polymer.

For the high molecular weight polymer ($M \approx$ D6), Bortel and Lamot[147] have shown that when freshly prepared with strontium carbonate, the standard commercial catalyst, the material is unstable even when stored in the dark and in the absence of air; the molecular weight was found to fall off to an essentially constant value of 6D5 after a period of 10 to 20 days, whether the starting molecular weight was 1.5D6 or 3D6. Similar effects were observed in the presence of daylight, and in the presence of air. The breakdown is accompanied by the formation of free radicals (as detected by ESR) and the production of carbonyl groups (as shown by IR); it is inhibited by phenothiazine.[147] It would appear that these effects result from the presence of "weak links" along the polymer chains at an average interval of about 6D5. Although the degradation can be inhibited by phenothiazine (and possibly other additives), it is probably wiser to allow it to go essentially to completion before using the polymer for any scientific studies.

In aqueous solution and in the presence of oxygen, PEO undergoes the autoxidation reactions characteristic of ethers; the process is catalyzed by traces of metal ions and also speeded up by light.[148,149] In practice, this may largely be prevented by adding antioxidants such as hindered phenols,[21] although the presence of these may affect the conformational size of the polymer in solution (particularly, aqueous solution). These autoxidation reactions are also important in connection with the stability of nonionic surfactants whose molecules contain PEO chains.[150,151]

The autoxidation of PEO and related compounds in aqueous solution initiated by X-rays has been studied in detail in a series of eight papers by Crouzet et al.[152] The effect of X-rays on deoxygenated aqueous solutions of the polymer has been studied by King and Ward,[153,154] who showed the marked effect of even traces of oxygen on the effects observed. Electron-pulse irradiation has been used by Borgwardt and co-workers,[155,155a] and by Matheson and co-workers[155b] to investigate the kinetics of attack of hydroxyl radicals on PEO in aqueous solution, and of the combination of the polymer radicals so produced.

The degradation of PEO in solution by high-speed stirring has been investigated by Minoura and co-workers;[141] the solvents used included water and a variety of nonaqueous liquids, and the effects of polymer molecular weight and concentration, nature of the solvent, and speed of stirring on the kinetics of the process were studied.

The presence of the hydroxyl end groups on PEO chains enables various forms of deliberate chemical modification to be carried out on the polymer. For example, by reaction with phenylazobenzoic anhydride the hydroxyl groups may be converted into colorimetrically active end groups for estimation of $\overline{M}_n$ of the polymer.[26] They also provide the means for the attachment of spin-labeled groups for ESR studies of the segmental mobility of the polymer.[116,123] Where their presence is undesirable, as may be the case in the measurements of physical properties of the oligomers, they may be converted into methoxy groups.[29]

I. Biological Behavior

In connection with the microbial degradation of synthetic polymers in the environment, Hasoya and co-workers[156] have examined the ability of bacteria to assimilate PEO samples with M = 400 and 6000 as well as the shortest chain oligomers (P = 1—4). The biological aspects of PEO have also been considered in relation to their presence as components of the chains of nonionic detergents.[150]

J. Analysis

There seem to be no standard methods for determining the concentration of PEO in solution, either for concentrated or for dilute solutions. For relatively concentrated solutions presumably determination of the dry content by evaporation suffices, as long as this involves heating above the melting point of the polymer (~66°C) to ensure complete removal of the solvent, although prolonged heating at high temperatures should be avoided because of the likelihood of autoxidation as well as some loss (with oligomers) of the shortest chain components; the main drawback of this method is that it is relatively nonspecific. Refractometry has also been used to assay aqueous solutions of PEO;[88] again this has the drawback of being largely nonspecific, although it has the advantage of speed and convenience.

For dilute aqueous solutions of PEO, Guttman and Higuchi[157] have suggested turbidimetric titrations with phenol (added as its concentrated aqueous solution) to identify and assay the polymer. This should indeed be a sensitive method, which could be extended to estimating the average molecular weight and molecular weight distribution of the polymer; however, the method would suffer interference from other water-soluble polymers such as PPO or PVP which are also precipitated by phenols.[157]

Other analytical aspects of PEO have been discussed in relation to their presence in the chains of the nonionic detergents.[150]

III. POLY(PROPYLENE OXIDE) — PPO

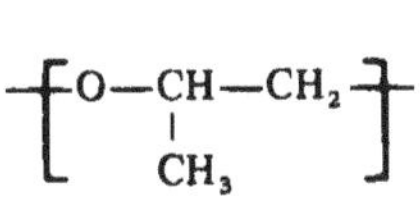

$M_0(C_3H_6O)$ = 58.08

Other names and abbreviations are polyoxypropylene glycol; polypropylene glycol; poly(propylene glycol); PPG; *Polyglycol P* (Dow). For other trade names see Reference 6.

A. General Features

Poly(propylene oxide) may be viewed as derived from PEO by the substitution of a methyl group on each chain unit; comparison of the properties and behavior of the two polymers thus serves to show the effects of such a substitution. This change is indeed sufficient to render all high molecular weight samples of the polymer insoluble in water (although still soluble in a range of organic solvents), and only oligomers of PPO ($M < 2000$) are appreciably water soluble. These oligomers are therefore important in the present context, but as with PEO their behavior is complicated both by the relative shortness of their chains (which means that many concepts and relationships usable for high polymers become inapplicable), and also by the greater prominence of the end groups, which are mainly hydroxyl groups (although effects from these can be obviated by methylation).

The presence of the substituent methyl group on each monomer unit also converts the carbon atom to which it is attached into a chiral center, and hence leads to the possibility of stereoregular forms of the polymer. This is discussed further in connection with the bulk polymer in Section III.D below.

For simplicity, the term "poly(propylene oxide)" is used here to cover all forms of the polymer including its oligomers, with the distinction then being made where necessary between "low molecular weight" ($M < 2000$, i.e., appreciably water soluble) and "high molecular weight" ($M > 2000$, i.e., completely insoluble in water) materials.

The chemical aspects of PPO have been reviewed by St. Pierre,[5] while Kastens[6] has discussed the properties of commercial grades and their technological applications.

B. Production and Purification

The low molecular weight PPO grades which are soluble in water (i.e., $M < 2000$) are readily obtainable commercially, either directly from the original producers such as Dow or Union Carbide, or indirectly through general chemical suppliers. Higher molecular weight materials may be prepared by "extending" the chains of these oligomers using agents for linking the hydroxyl end groups, such as toluene diisocyanate.[158] Alernatively, these longer chain materials may be synthesized directly from the monomer; potassium hydroxide as heterogeneous catalyst gives medium molecular weights (4D3—2D4),[159] while extremely high molecular weights (~D6) can be obtained either by using zinc diethyl plus water in dioxane,[27,160-164] or by heterogeneous catalysis with strontium carbonate[27,159] or ferric chloride.[159]

Branched PPO may be prepared by starting the polymerization of the monomer with a trihydroxy compound such as glycerol or trimethylolpentane,[165] while cross-linked material (for use, say, in studies on swelling behavior) may be obtained from low molecular weight polymer by linking with trifunctional agents.[166]

Where the presence of the hydroxyl end groups is undesirable, as may be particularly the case for physical studies with the oligomers, they may be converted into methoxy groups by conventional methylation methods.[29,167,168]

There seem to be few good nonsolvents for the purification of PPO by precipitation from solution, to free it of low molecular weight contaminants. For fairly high molecular weight PPO one possible precipitant is water (with, say, methanol as the solvent), but this is clearly inappropriate for the (water-soluble) oligomers. It is likely that these can be purified using the inverse solubility-temperature behavior of aqueous solutions of these oligomers (see Section III.F), i.e., heating a solution of suitable concentration to the temperature required to precipitate out all the polymer, allowing to stand and then decanting off the supernatent containing the bulk of the contaminants.

C. Fractionation and Characterization

Most of the fractionation procedures for PPO that have been published relate to high molecular weight polymer, and many of these are concerned with its separation into a crystallizable (isotactic or isotactic-rich) and a noncrystallizable (essentially atactic) portion.[27,158-161,163,164]

However, Moacanin[159] has used preparative GLC to separate oligomers ($P < 11$), and molecular distillation for fractionation of somewhat higher molecular weights ($M < 2000$). Once again, as with the purification procedures discussed above, it should be possible to fractionate the oligomers using the inverse solubility-temperature behavior of their aqueous solutions.

For high molecular weight PPO, conventional methods of determination of $\overline{M}$ (LS, OP, etc.) are applicable as usual; however, with the oligomeric water-soluble materials these methods become less readily applicable, and the more classical colligative methods such as BPE, FPD, and VPO (giving $\overline{M}_n$ values) have to be used (see Tables 5 and 6). End-group methods have also been applied, although as with PEO their application needs to take into account the likely presence of other groups, especially terminal unsaturation, replacing some of the hydroxyl end groups. This unsaturation has been determined by iodine number estimation[169] and by mercuric acetate methods;[159,165] hydroxyl end groups have been determined chemically using acetic anhydride[159] or phthalic anhydride,[165] or alternatively using IR absorption.[159,169]

Conerning the molecular structure of the PPO chain, the crystalline high molecular weight material is stated to be isotactic in its stereoregularity.[46] Nothing definite seems to be known about the tacticity of the short-chain polymers; however, it has been stated that head-to-head links are common in these.[163]

D. Bulk Polymer

Information on the density, ρ, and its reciprocal specific volume, ν, of PPO of various ranges of molecular weights and temperatures is available from several sources.[11,27,29,164,165,168-170]

For the amorphous polymer, the value of ν has been shown to be a linear function of the reciprocal of the average molecular weight:

$$\nu = \nu_\infty - k/\overline{M} \qquad (1)$$

where $\nu_\infty = 1.0013$ cm^3 g^{-1} and $k = 2.5$ cm^3 g^{-1} at 20°C using $\overline{M}_n$ values,[170] with the corresponding values 1.00 cm^3 g^{-1} and 4 cm^3 g^{-1} at 25°C using $\overline{M}_w$.[165] The limiting value $\nu_\infty = 1.00$ cm^3 g^{-1} (i.e., for very high $\overline{M}$) is attained at relatively low values of $\overline{M}$, but the variation is still noticeable in the range of very low molecular weights. The thermal expansion coefficient, α, for high molecular weight amorphous polymer is 7.8d4 K^{-1}.[29,168]

The relation between bulk viscosity at 25°C and molecular weight for PPO has been obtained by Moacanin.[159] Unfortunately this is not likely to be a suitable method for estimating the molecular weights of oligomers, despite its apparent simplicity, because the viscosity is so sensitive to the nature of the end groups — e.g., methylation of the hydroxyl groups markedly reduces the viscosity.[167]

The low-resolution IR spectrum has been reported by Nyquist,[32] while the region 1200 to 1500 cm^{-1} has been studied in greater detail both for crystalline and for molten PPO by Allen and co-workers.[161]

Calorimetric studies on amorphous and crystalline samples of high molecular weight PPO give a value for the glass transition temperature of $T_g = 198$ K ($t_g = -75$°C).[27] Thus amorphous samples are in the rubbery state at normal temperatures.

The NMR behavior of PPO has been studied by Connor et al.[167] who measured the proton spin-lattice relaxation times for amorphous and crystalline samples of an extremely wide range of molecular weights (250—1.3D6). The dynamic mechanical properties of the polymer have been measured and discussed.[29,40]

Studies on the crystal structure of PPO have been reviewed by Tadokoro.[46] The X-ray diffraction measurements show the stereoregularity to be isotactic in type, with a density for the wholly crystalline state of 1.154 $cm^3\ g^{-1}$.[46] Thus the volume percent crystallinity of a sample may be calculated from the density assuming a linear relation between the two.[27,167]

The refractive index value $n_D^{20} = 1.456$ has been quoted for an amorphous, high molecular weight sample of the polymer.[167] The dielectric relaxation properties of a similar sample have been studied over a wide range of pressures, temperatures, and frequencies by Williams.[171]

E. Solution Behavior

Again, as with PEO, it is useful to consider studies in aqueous solution separately from those in nonaqueous media, particularly because of the limitation of aqueous solubility to oligomers of the polymer.

1. Aqueous Solution

The results of various studies on the behavior of PPO in aqueous solution are summarized in Table 5. The work carried out on these systems has been necessarily restricted by the limitation to oligomers, but the studies summarized provide a fair characterization of the behavior of these systems. The *partial specific volume* (entries 1 and 9) is close to 0.914 $cm^3\ g^{-1}$, a reduction of about 8% below the specific volume of the bulk polymer, and indicative as with PEO of the compacting effect of the polymer on the solvent. The *specific refractive index increment* (one value only, entry 7) is 0.132 $cm^3\ g^{-1}$. The *intrinsic viscosity* is (for this limited range of molecular weights) linearly related to molecular weight, i.e., according to the MHSR equation with $a = 1.0$ (entries 6 and 10).

2. Nonaqueous Solution

The results of various studies on PPO in nonaqueous and mixed solvent systems are summarized in Table 6, which also gives an indication of the types of liquids which are solvents for this polymer. For greater completeness, studies on samples with $M > 2000$ (i.e., water insoluble) have been included, as well as those on low molecular weights which are of particular interest in this context because of their water solubility.

F. Phase Equilibria

The main topic of interest in this area in the present context is the equilibrium aqueous solubility of the oligomers of PPO. This solubility falls off markedly with increase in molecular weight, in that whereas for $M = 400$ the polymer is indefinitely soluble in water, for $M = 1200$ it is soluble to the extent of only 2%, while for $M \geqslant 2000$ the solubility is less than 0.1%.[116]

As with ethers in general, PPO oligomers show inverse solubility-temperature behavior in aqueous solution. Some rather fragmentary quantitative information is available concerning this behavior. Malcolm and Rowlinson[60] have published the phase diagram for $M = 400$, which shows the critical temperature as $t_c = 50°C$ (at $w_2 = 0.3$). Phase diagrams published by Kastens[6] give $t_c = 44°C$ ($w_2 = 0.25$) for $M = 400$, and $t_c = -7°C$ ($w_2 = 0.6$) for $M = 1200$. Saito and Otsuka[182] have obtained precipitation temperature values for $c_2 = 0.05\%$: $t_p = 15°C$ for $M = 3000$ and $t_p = 35°C$ for $M = 1200$. These data are supported by more qualitative observations from other sources; thus Sandell and Goring[173] noted that polymers with $M = 400$ and 425 at concentrations between 0.2 and 1% started to precipitate between 50 and 70°C, while for $M = 1025$ the precipitation began between 20 and 25°C. Similarly, Lakhanpal and co-workers[80] in their heat of mixing studies found that homogeneous mixtures could not be obtained at 27°C for $M = 425$ above $w_2 = 0.775$, or for $M = 2025$ above $w_2 = 0.66$; however, the last composition value seems a surprisingly high one for such a molecular weight and it is most likely that the quoted M value of 2025 is a mistake for 1025.

Table 5
STUDIES ON PPO IN AQUEOUS SOLUTION

Entry	Year	Polymer samples: No.	Polymer samples: Range of $\overline{M}$	Techniques used	Temperature (t/°C)	Polymer concentration	Comments	Ref.
1	1957	1	400	SV	50	$w_2 = 0$—1.0	→ ΔV (mixing): $\bar{v}_2$-$\bar{v}_2^{\circ}$ = 0.085 cm^3 g^{-1}	60
2	1957	1	400	VP	30, 50	$w_2 = 0.6$—0.99	→ TP	
3	1960	3	Cross-linked	SV	22	$\phi_2 = 0.96$	→ $\underline{Q}$	166
4	1968	1	2D3	FPD	0		→ $\overline{M}_n$	168
5	1968	3	150—2025[a]	HS	27	$w_2 = 0.01$—0.79	→ ΔH	80
6	1970	6	130—940	VI	25	1—10%	→ [η]; k'; MHSR (1.0; 4.2d3; 3.3)	170,172
7	1970	1	1025	RI	24?	1, 2%?	λ = ?: → $\tilde{n}$ = 0.132 cm^3 g^{-1}	173
8	1970	1	1025	SE	24	1, 2%	→ $\overline{M}_w$; no aggregation	173
9	1970	1	1025	SV	24?	1, 2%?	→ $\bar{v}^{\circ}_2$ = 0.913 cm^3 g^{-1}	173
10	1970	5	130—940	VI	10—59	1—10%	→ [η]; (k'); d [η]/dt ; MHSR (1.0; 2.4d3—3.1d3; 2.3—3.9)	173

[a] The molecular weight value of 2025 quoted for their third polymer sample by Lakhanpal and co-workers[80] would appear to be an error for 1025, in view of the relatively high miscibility with water of the sample used.

Table 6
STUDIES ON PPO IN NONAQUEOUS AND MIXED SOLVENTS

Entry	Year	Polymer samples No.	Polymer samples Range of $\overline{M}$	Techniques used	Temperature (t /°C)	Polymer concentration	Comments	Ref.
					CCl_4:Carbon Tetrachloride			
1	1964	3	1.1D4—1.4D4	NMR	35, 80		60 MHz: → δ(^{1}H); tacticity not determinable	161
2	1967	6	425—4D3	BPE	76		→ $\overline{M}_n$	174
3	1968	1	2050	HS	5.5	w_2 = 0.1—0.9	→ Δ H	168
4	1968	1	2050	SV	5.5	w_2 = 0.1—0.8	→ ΔV(mixing); thermal pressure coefficients	168
5	1973		Theoretical interpretation of HS data from Reference 168 (entry 3)					175
					CS_2: Carbon Disulfide			
6	1964	3	1.1D4—1.4D4	NMR	35, 80		60 MHZ: → δ(^{1}H); tacticity not determinable	161
7	1965	4	250—2D3	NMR	−150 to +50	ϕ_2 = 0.165—1.0	→ T_1 (^{1}H)	167
					$CHCl_3$:Chloroform			
8	1968	1	2050	HS	5.5	w_2 = 0.1—0.9	→ ΔH	168
9	1968	1	2050	SV	5.5	w_2 = 0.01—0.8	→ ΔV(mixing); thermal pressure coefficients	168
					CH_4O: Methanol			
10	1959	11	8D3—7D4	LS	~22	0.3—2%	436, 546 nm: → $\overline{M}_w$; A_2	158
11	1959	4[a]	8D3—4D4[a]	RI	~22		436, 546 nm: → $\tilde{n}$ = 0.152, 0.150 cm^3 g^{-1}	158
12	1959	9[b]	3D4—7D4[a]	RI	~22		436, 546 nm: → $\tilde{n}$ = 0.148, 0.145 cm^3 g^{-1}	158
13	1959	15	1D3—7D4	VI	25	0.3—2%	→ [η]; k'; MHS (0.55; 7.69d2)	158
14	1960	3	Cross-linked	SV	−82 to +47	ϕ_2 = 0.05—0.15	→ Q ; dQ/dt ; thermodynamic parameters	166
15	1960	3	Cross-linked	RF	25—100	ϕ_2 = 0.1—0.3	→Thermodynamic parameters	176
16	1960	4	150—3350	HS	27	w_2 = 0.1—0.9	→ Δ H	169,177
17	1960	4	150—3350	VP	−25 to +25	w_2 = 0.12—0.54	→ Thermodynamic parameters	169,177
18	1964	1[c]	1D6[c]	RI	25		436, 546 nm: → $\tilde{n}$ = 0.118 cm^3 g^{-1}	160

19	1964	1[c]	1D6[c]	RI	25		436, 546 nm: → $\bar{n}$ = 0.115 cm^3 g^{-1}	160
20	1966	3	132—2D3	SV	25?	2—8%	→ $\bar{v}_2$	159
21	1966	38	132—1D5	VI	25		→ [η]; MHS (M>600: 0.55; 7.69d2); MHSR (132<M<2D3: 1.0; 1.75d3; 1.8)	159
22	1967	6	74—3375	TD	20		→ $D°$	174
23	1967	19	3D2—5D3	SE	20		→ $\overline{M}_w$; A_2	165
24	1967	19	5D2—3D3	SV	20	1—4%	→ $\bar{v}_2°$ = 1.000 cm^3 g^{-1} ($M \to \infty$)	165
25	1967	6[d]	5D2—3D3[d]	VI	20		→ [η]; k'; MHS (0.64; 4.06d2)	165
26	1967	13[e]	3D2—5D3[e]	VI	20		→ [η]; k'; MHSR (0.56; 5.55d2; 1.0)	165
27	1967	13[f]	3D2—5D3[f]	VPO	37	0.5—4%	→ $\overline{M}_n$; no aggregation	165
28	1970	2	130, 1025	SV	25	5%	→ $\bar{v}_2$ = 0.948, 0.961 cm^3 g^{-1}	172
29	1970	2	130, 1025	VI	25	2—8%	→ [η]	172

$C_2Cl_3F_3$: 1,1,2-Trichloro-1,2,2-Trifluoroethane ("Isceon 113")

30	1964	?	'D6	RI	25		436, 546 nm: → $\bar{n}$ = 0.118, 0.115 cm^3 g^{-1}	160

C_2H_6OS: Dimethyl Sulfoxide

31	1970	2	130,1025	VI	1025		→ [η]	172

C_3H_6O: Acetone

32	1967	8	74—4D3	LS	25	0.2—20%	436 nm: → $\overline{M}_w$; A_2	178
33	1967	4	74—4D3	RI	25	1—40%	436 nm → $\bar{n}$ = 0.085—0.101 cm^3 g^{-1}	178
34	1967	8	74—4D3	VI	25	2—24%	→ [η]; k'	178
35	1971	6	134—4D3	SAXS	20		1.54 Å?: → $\hat{R}$	179
36	1971	8	76—4D3	VI	24		→ [η]; MHS (132≤M≤2D3: 1.0; 1.58d3; 2.01)	179
37	1971	8	76—4D3	LS	25		436 nm: → $\overline{M}_w$; A_2	180
38	1971	5	76—3025	RI	20		436 nm: → $\bar{n}$ = 0.085—0.1005 cm^3 g^{-1}	180
39	1971	7	76—4D3	SAXS	20		1.54 Å: → A_2; R	180
40	1971	5	76—3025	SV	20		→ $\bar{v}_2°$ = 0.940—0.980 cm^3 g^{-1}	180

C_3H_7NO: Dimethylformamide

41	1966	9	300—2D3	VI	25		→ [η]; MHSR (M>600: 0.56; 7.55dz)	159

C_3H_8O: 2-Propanol (Isopropanol)

42	1970	6	130—940	VPO	37		→ $\overline{M}_n$	170

Table 6
STUDIES ON PPO IN NONAQUEOUS AND MIXED SOLVENTS

Entry	Year	Polymer samples No.	Polymer samples Range of $\overline{M}$	Techniques used	Temperature (t /°C)	Polymer concentration	Comments	Ref.
					C_4H_8O: Butanone (Methyl Ethyl Ketone)			
43	1959	9	3D4—7D4	RI	22		436 nm: → $\bar{n}$ = 0.094 cm^3 g^{-1}	158
					C_4H_8O: Tetrahydrofuran			
44	1967	6[d]	5D2—3D3[d]	VI	20		→ [η]; k'; MHS (0.62; 5.50d2)	165
45	1967	5[f]	3D2—5D3[f]	VI	20		→ [η]; k'; MHSR (0.64; 3.05d2; 1.60)	165
46	1967	5[g]	4D2—5D3[g]	VI	20		→ [η]; k'; MHSR (0.68; 2.12d2; 1.60)	165
47	1971	3	4D2—1.2D3	GPC			→ MWD ($\overline{M}_n$; $\overline{M}_w$; etc.)	180
					$C_4H_8O_2$: Dioxane			
48	1960	3	Cross-linked	SV	12—45	ϕ_2 = 0.04—0.15	→ Q	166
49	1960	3	Cross-linked	RF	25—100	ϕ_2 = 0.05—0.2	→ Thermodynamic parameters	176
					$C_4H_{10}O_2$: 1,2-Dimethyoxyethane (Ethylene Glycol Dimethyl Ether)			
50	1966	7	132—2D3	VI	25		→ [η]; MHSR (300 ≤M ≤2D3:1.0; 1.98d3; 1.33)	159
					$C_5H_{10}O$: Tetrahydropyran			
51	1970	2	130, 1025	SV	25	5%	→ $\tilde{v}_2$ = 0.987, 0.996 cm^3 g^{-1}	172
52	1970	2	130, 1025	VI	25		→ [η]	172
					C_6H_5Cl: Chlorobenzene			
53	1964	?	~D6	RI	25		436, 546 nm: → $\tilde{n}$ = − 0.066, − 0.064 cm^3 g^{-1}	160

C_6H_6: Benzene

54	1959	9	3D4—7D4	RI	~22		436 nm: → $\tilde{n}$ = − 0.031 cm³ g⁻¹ (footnote h)	158
55	1959	19	1D3—7D4	VI	25	0.3—2%	→ [η]; k′; MHS (0.64; 4.13d2)	158
56	1959	6	8D3—7D4	VI	39.5	0.3—2%	→ [η]; k′; MHS (0.66; 4.27d2)	158
57	1960	3	Cross-linked	SV	9—48	ϕ_2 = 0.05—0.15	→ Q; dQ/dt; thermodynamic parameters	166
58	1960	3	Cross-linked	RF	25—100	ϕ_2 = 0.05—0.2	→ Thermodynamic parameters	176
59	1964	?	~D6	RI	25		436, 546 nm: → $\tilde{n}$ = − 0.053, − 0.045 cm³ g⁻¹	160
60	1964	3	1D4—6D5	VI	25		→ [η]; MHS (0.77; 1.12d2)	160
61	1966	3	132—2D3	SV	25?	2—8%	→ $\tilde{v}_2$	159
62	1966	16	132—1D5	VI	25		→ [η]; MHS (0.64; 4.13d2); MHSR (132 ≤M ≤2D3:1.0; 2.72d3; 0.62)	159
63	1967	8	4D5—2.5D6	VI	25		→ [η]; $\bar{M}_v$	163
64	1967	5	134—3D3	TD	20		→ D°	174
65	1967	6	425—3D3	VPO	20	0.5—3%	→ $\bar{M}$	174
66	1967	6	134—4D3	VI	25	2—20%	→ [η]; k′	178
67	1967	5[d]	8D2—3D3[d]	VI	20		→ [η]; k′ MHS (0.79; 1.11d2)	165
68	1970	2	130, 1025	SV	25	5%	→ $\tilde{v}_2$ = 0.994, 0.992 cm³ g⁻¹	172
69	1970	2	130, 1025	VI	25	2—10%	→ [η]	172
70	1971	2[i]	1.5D6[i]	VP	40—75	w_2 = 0.5—0.9	→ Thermodynamic parameters	164
71	1971	2[i]	1.5D6[i]	VP	38—75	w_2 = 0.3—0.9	→ Heat of fusion of polymer, etc.	181
72	1971	6	134—4D3	VI	25		→ [η]; MHS (M >400: 0.65; 4.15d2)	179

C_6H_{14}: Hexane

73	1964	16	3D4—4D6	LS	46		436, 546 nm: → $\bar{M}_w$; A_2; $\hat{R}$	160
74	1964	2	2D5, 1D6	RI	25—57		436, 546 nm: → $\tilde{n}$ = 0.0775—0.104 cm³ g⁻¹	160
75	1964	16	3D4—4D6	VI	46		→ [η]; MHS (0.67; 1.97d2)	160
76	1964	6	1.0D6—1.4D6	LS	46	0.1—0.4%	436 nm: → $\bar{M}_w$; $\bar{M}_z$; A_2; $\hat{R}$	161
77	1964	6	1.0D6—1.4D6	VI	46		→ [η] (correlation with crystallizability)	161
78	1966	5	8D5—9D5	LS	46		546 nm: → $\bar{M}_w$; A_2; $\hat{R}$	162
79	1966	2	8D5, 9D5	VI	40—60		→ [η]; d[η]/dt (= 0)	162
80	1967	5[d]	6D2-D3[d]	VI	20		→[η]; MHS (0.79; 1.11dz)	165

Table 6
STUDIES ON PPO IN NONAQUEOUS AND MIXED SOLVENTS

Entry	Year	Polymer samples No.	Polymer samples Range of $\overline{M}$	Techniques used	Temperature (t /°C)	Polymer concentration	Comments	Ref.
					C_7H_8: Toluene (See Also C_7H_8 + C_8H_{18})			
81	1964	3	1D4—6D5	VI	25		→ [η]; MHS (0.75; 1.29d2)	160
82	1964	4	3D5—6D5	OP	25		→ $\overline{M}_n$; A_2	161
83	1966	22	132—1D5	VI	25		→[η]; MHS?; MHSR (300 ≤M ≤2D3:1.0; 2.72d3; 0.62)	159
84	1967	5[d]	8D2—3D3[d]	VI	20		→ [η]; k'; MHS (0.72; 2.08d2)	165
85	1967	6[j]	3D2—5D3[j]	VI	20		→ [η]; k'; MHS (0.72; 2.08d2)	165
86	1967	5[g]	4D2—5D3[g]	VI	20		→ [η]; k'; MHS (0.70; 2.02d2)	165
87	1967	13[f]	3D2—5D3[f]	VPO	20	0.5—4%	→ $\overline{M}_n$; aggregation	165
					C_7H_8 + C_8H_{18}: Toluene + 2,2,4-Trimethylpentane (Iso-Octane)			
88	1959	8	2D3—7D4	VI	39.5		$\phi(C_7H_8)$ = 0.417 (Θ system): → [η]; k'; MHS (0.50; 0.1075)	158
					C_7H_{16}: Heptane			
89	1964	?	~D6	RI	40		436, 546 nm: → $\bar{n}$ = 0.046 cm^3 g^{-1}	160
					C_8H_{18}: 2,2,4-Trimethylpentane (Iso-Octane) (See Also C_7H_8 + C_8H_{18})			
90	1964	?	~D6	RI	35		436, 546 nm: → $\bar{n}$ = 0.0655 cm^3 g^{-1}	160
91	1966	5	8D5—9D5	LS	46—89		546 nm: → $\overline{M}_w$;A_2;$\hat{R}$; Θ (=50 ± 1°C)	162
92	1966	4	8D5—3D6	VI	40—80		→ [η]; d[η]/dt (> 0)	162

93	1967	3	8D5—9D5	LS	50	Θ system; 436, 546 nm?: → $\overline{M}_w$; (A_2 = 0?)	163
94	1967	7	4D5—2.5D6	VI	50	Θ system: → [η]; MHS (0.50; 0.14)	163

[a] Low molecular weight fractions.
[b] High molecular weight fractions.
[c] Polymer samples not otherwise differentiated.
[d] Linear samples only.
[e] Samples branched with glycerol or 1,1,1-trimethylolpropane.
[f] Samples linear or branched with glycerol.
[g] Samples branched with 1,1,1-trimethylolpropane.
[h] Negative sign on $\bar{n}$ value omitted in original reference.
[i] Two samples differ in crystallizability, i.e., stereoregularity.
[j] Samples branched with glycerol.

It is useful to derive from these data at least an estimate of the value of the ideal temperature, Θ_+, for this system. To obtain a third t_c value, i.e., in addition to those already quoted for $M = 400$ and 1200, we use the fact that Saito and Otsuka[182] observed a 20° difference between the t_p at 0.05% values for $M = 1200$ and 3000; this suggests that there should be a similar difference between the t_c values, on which basis this gives $t_c = 27°C$ for $M = 3000$. A plot of $1/T_c$ vs. $1/M^{1/2}$ for these three data points is essentially linear, and extrapolation to $1/M^{1/2} = 0$ (i.e., $M \rightarrow \infty$) gives $\Theta_+ = 220$ K (= −53°C). More work needs, of course, to be done in this area to define this value more closely, but the estimate gives an orientation point for discussing the precipitation behavior of PPO in aqueous solution, and for fitting this into the context of the water-soluble polymers in general; it indicates, e.g., by comparison with PEO, that the substitution of the methyl group leads to the Θ_+ value being reduced by about 150°.

IV. OTHER POLYOXIDES

As already noted, PEO is apparently anomalous in its behavior in that it is the only member of the series of unsubstituted polyoxides:

$$-\!\!\left[\text{O}-(\text{CH}_2)_m\right]\!\!-$$

which is water soluble. It is therefore relevant to discuss briefly the behavior of the other members, particularly the neighboring ones, i.e., $m = 1$ and $m = 3$, of which the former might also be expected to be readily water soluble, and the latter might be expected to show at least marginal water solubility in the same fashion as PPO (with which it is isomeric). General reviews are available of the solution behavior,[45,183] melting points,[183] and crystal structures[46] of the members of this series. Dale[184] has also discussed, from the viewpoints of both small-molecule and of polymeric compounds, the conformational consequences of replacing a methylene group by an ether oxygen.

A. Polyoxymethylene — POM

$$-\!\!\left[\text{O}-\text{CH}_2\right]\!\!-$$

With this polymer only the very shortest oligomers, i.e., poly(formaldehyde), are at all water soluble;[185] the true high polymer is completely insoluble not only in water but also in most common organic solvents, being soluble only in rather exotic solvents such as hot (90°C) aqueous phenol solution (composition not stated)[186] and hexafluoro-acetone sesquihydrate.[187] The polymer is highly crystalline, with a high melting point (180°C), while X-ray diffraction shows that the chains form tight helices in the crystal lattice involving strong intramolecular forces between oppositely oriented C-O dipoles along the chain.[46] All these features point to strong cohesive forces in the solid due to its highly polar character, and indicate that its water insolubility is not anomalous but simply one aspect of the general behavior of the polymer. In contact with water there is a limited uptake, presumably into the small amorphous regions in the material.[188]

The methyl derivative of POM, poly(acetaldehyde):

$$-\!\!\left[\text{O}-\text{CH}(\text{CH}_3)\right]\!\!-$$

is also insoluble in water although it is soluble in such organic media as acetone, chloroform, and ethyl acetate,[18] indicating that the cohesive forces in the solid are less than those in POM itself. The crystal structure (of the isotactic form) has been determined.[46] Since this

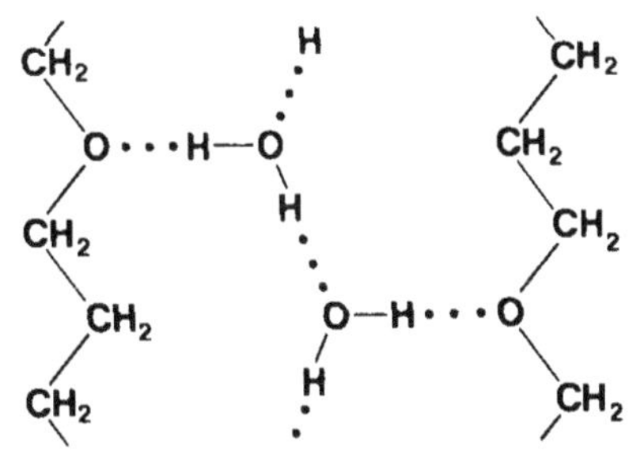

FIGURE 4. Crystal structure of poly(trimethylene oxide) hydrate ("modification I" of the polymer), $[O(H_2O)CH_2CH_2CH_2]$, from the studies of Tadokoro and co-workers.[193,194] The crystal lattice is pseudorhombic ($\beta = 91°$). The polymer molecules are arranged in flat sheets, parallel to the *ac* lattice plane, spaced at $b/2 = 3.63$ Å. The sheets are formed of ribbon-like parallel pairs of flat zigzag polymer chains held together by hydrogen-bonded chains of water molecules, all the atoms involved (except for the methylene hydrogens) being within the plane of the sheet as shown. The repeat distance along the ribbon is $c = 4.80$ Å, and the parallel ribbons are centered within the plane at spacing $a = 12.3$ Å.

polymer is isomeric with PEO, its water insolubility is perhaps more remarkable than in the case of POM.

B. Poly(Trimethylene Oxide) — P3MO

$$-\!\!\left[O-(CH_2)_3\right]\!\!-$$

This polymer, also called polyoxacyclobutane (POCB), is the next higher member of the polyoxide series from PEO. It exists in three crystalline forms, one of which (so-called modification I) is in fact a hydrate (see below); the crystal structures[46] and infrared spectra[190] of these forms have been determined, and calorimetric measurements have been carried out on the most stable form (modification III).[191]

The polymer is insoluble in water, but soluble in a number of organic media, such as carbon tetrachloride, methanol (and methanol + water), acetone, butanone, and benzene.[183,192] Comparing this behavior with that of its isomeric analog, PPO, in the case of this latter polymer it is only the amorphous, oligomeric members that are at all water soluble; with P3MO, the behavior of the oligomers does not seem to have been specifically studied, while because of the simplicity of the chain even the solid oligomers are likely to be highly crystalline (in contrast to PPO where their amorphous character suggests they are atactic and hence not crystallizable).

The most remarkable feature of this polymer is that it has a crystalline form produced in contact with water, the so-called modification I, which is in fact a monohydrate (i.e., with one molecule of water per C_3H_6O chain unit) in which the water molecules lie in chains between pairs of polymer chains in a hydrogen-bonded structure, as shown in Figure 4.[193,194] The infrared spectrum of this monohydrate, and also that of the monodeuterate have been investigated;[190] the dissociation isotherms and pressures have also been determined over the temperature range −25 to +20°C, giving an enthalpy change for the dissociation of 13.6 kcal mol^{-1} (57 kJ mol^{-1}), presumably representing the breaking of the two hydrogen bonds per water molecule released.[195] This is the first, and apparently so far the only well-defined crystal hydrate of a synthetic polymer to have been discovered; the study of this and any other hydrates that may exist should be useful in suggesting types of hydration likely to occur in aqueous solutions of polyoxides and polyethers.

C. Higher Polyoxides

The greater the ratio of methylene groups to oxygen atoms in the chain, the lesser the general likelihood of any appreciable water solubility for the polyoxides; nevertheless, studies on the higher polyoxides are still useful in the present context in providing comparisons with the behavior of the lower polyoxides such as PEO (especially, for their behavior in nonaqueous solution), as well as in enabling general theories of polyoxide behavior to be tested. The next higher member, with $m = 4$, is poly(tetramethylene oxide) (P4MO), also called polytetrahydrofuran (PTHF); the crystal structure of the solid has been determined,[46] and the solution behavior also studied.[196] Likewise, for the higher members still, with m = 6, 8, 10, and 12, crystal structures have been determined,[197] and the behavior in solution studies for the members with $m = 6$[198] and $m = 10$.[199]

V. OXIDE COPOLYMERS

The production of copolymers between the oxide type of monomers widens the range of possibilities for such polymer properties as water solubility. The synthetic routes to copolymers of EO with a variety of other monomers have been discussed by Bailey and Koleske.[7] The two most important simple types of copolymer are the random and the block types. With random copolymers, Bailey and Callard[65] have presented data (also reproduced in References 2 and 7) on the precipitation points of PEO-PO (presumably, essentially random copolymers) at 1 and 0.1% total concentration; the t_p value was about 98°C for pure PEO at both concentrations, but then fell essentially linearly with increase in mole percent PO content, reaching 60°C for 34 mol % PO at 1% concentration, and 48°C for the same PO content at 0.1% concentration. Such behavior is entirely in line with that of pure PPO itself (see Section III.F) and with the molecular structures of the individual component monomers; more detailed studies of this type may enable the t_p values of samples of pure PPO which are insoluble above 0°C to be obtained by extrapolation from the data for the corresponding but more soluble PEO-PO copolymers.

Block copolymers of EO and PO are produced commercially in various grades as surface-active agents; the most commonly encountered are those marketed under the trade name *Pluronic* (Wyandotte),[6,150] and with the pharmaceutical name Poloxamer.[13] These materials have the sandwich structure $(EO)_A(PO)_B(EO)_A$. The *Pluronic* grades are labeled with a letter and two (or three) digit code, in which the initial letter indicates the physical state (L = liquid, P = paste, F = flake, i.e. solid), the first digit (or first two digits) gives the (approximate) molecular weight of the PPO section divided by 300, and the last digit is one tenth the weight percentage of EO in the material; the Poloxamer code lacks the initial letter, and the first two digits give the molecular weight of the PPO section divided by 100, with the last digit as for the *Pluronic* code. Thus, *Pluronic* F68 (Poloxamer 188) is a solid material with a PPO core molecular weight of about 1800 (i.e., $B \approx 30$) and 80% w/w EO, giving a combined molecular weight for the two outer PEO sections of 6600 (i.e., $A \approx 75$) and a total molecular weight of 8400. As this example indicates, these materials are in many cases only oligomers rather than true polymers, but they are nevertheless included here and discussed later in Volume II (especially Chapters 2 and 4) on the same basis as the PPO oligomers. They show the same inverse solubility-temperature behavior as the other polyoxides; however, the behavior in aqueous solution of certain grades is complicated by reversible aggregation into micelles, the tendency for which is likely to be very sensitive to the sizes of the individual PEO and PPO sections of the molecule.[200]

VI. POLY(VINYL METHYL ETHER) — PVME

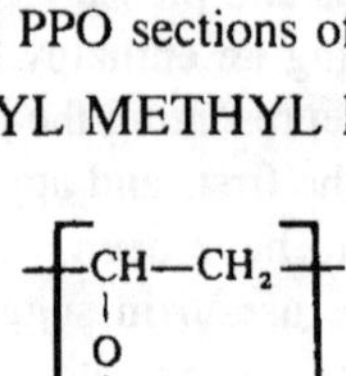

$M_0(C_3H_6O) = 58.08$

Other names, abbreviations, and trade names are poly(methoxyethylene); poly(methyl vinyl ether); PMVE; *Gantrez* M (e.g., grades M-093, M-154, M-555, M-556, M-574) (GAF Corporation).[201]

A. General Features

Poly(vinyl methyl ether) and PEO share the distinction of being the only two members of this group of polymers whose high molecular weight forms are water soluble at normal temperatures. Despite this distinction, the properties of PVME, and particularly its behavior in aqueous solution, have apparently been little studied; the published information on its behavior is rather fragmentary and in many cases only available from the technical literature.[201] In relation to other polymers, it is isomeric both with PPO and with P3MO, while it is the methyl ether derivative of another water-soluble polymer, poly(vinyl alcohol).

In addition to its water solubility, commercially available PVME is soluble in a wide variety of organic media, including alcohols, amines, esters, ethers, ketones, alicyclic and aromatic hydrocarbons, and chlorinated hydrocarbons; however, it is insoluble in noncyclic aliphatic hydrocarbons, glycols, and glycerol.[201]

B. Production and Purification

The monomer is polymerized by cationic catalysts. In commercial practice this is carried out using boron trifluoride-water in dioxane.[202] Friedel-Crafts type of catalysts have also been used.[203] Details of the commercially important grades are given in the technical literature from the main manufacturer, GAF Corporation;[201] the polymer is also generally available from chemical suppliers.

Purification may be carried out along the same lines as the fractionation procedures, i.e., precipitation from benzene solution with aliphatic hydrocarbons, or heating an aqueous solution of the polymer.[203] The polymer may be obtained in film form by dissolving in methanol and evaporating to dryness at room temperature.

C. Fractionation and Characterization

The polymer has been fractionated successfully by precipitation from benzene solution with hexane, heptane, or decane, and also by gradually raising the temperature of an aqueous solution.[203] It has been characterized by light scattering and viscometry (see Section VI.E); from the literature values for the MHS parameters in benzene solution[203] and from the data quoted in the technical literature for the specific viscosities at 1% in the same solvent, the commercially available grades have average molecular weights ranging from 2D4 to 1D5.[201]

D. Bulk Polymer

Information is available in the technical literature on many of the physical properties of the polymer in the bulk state, including the following[201]

1. Density, ρ (20°C): 1.05 g cm^{-3}
2. Refractive index, n_D^{25}: 1.47
3. Glass transition temperature, t_g: -34°C

Crystalline PVME has an isotactic configuration; the structure of the crystal and other properties of the polymer have been discussed by Tadokoro.[46] However, the commercially available polymer is stated to be amorphous[201] and hence presumably essentially atactic. From the t_g value quoted above, it is in the rubbery state at normal temperatures, so that small-molecule substances (e.g., solvents) will diffuse fairly freely into and out of the polymer.

E. Solution Behavior

The polymer has been characterized in butanone and benzene solutions by Manson and

Arquette.[203] Light-scattering studies on 13 whole polymers and fractions with $\overline{M}$ = 1D4 − 4.5D6 in butanone gave the values of $\overline{M}_w$, A_2, and $\hat{L}$; the specific refractive index increment in this solvent (25°C, 436 nm) was 0.0944 cm³ g⁻¹. *Viscometry* in butanone and in benzene at 30°C gave the values of [η] and k', and led to the following sets of MHS parameters for this range of $\overline{M}$:

1. Butanone, 30°C: a = 0.57 ± 0.02; K = 9.7(±1.3)d2 cm³ g⁻¹ (these values have been recalculated from the original data)
2. Benzene, 30°C: a = 0.60; K = 7.6 d2 cm³ g⁻¹

Andersson[72] has obtained the value 0.983 cm³ g⁻¹ for the *partial specific volume*, $\tilde{v}_2°$, of the polymer in *n*-amyl alcohol (pentanol) at 23°C, and has also measured the effects of pressure up to 4.5 katm.

Information for aqueous solution behavior of PVME seems to be more fragmentary. Dobry[204] quoted a value [η] = 0.4 (presumably, dl g⁻¹, i.e., 40 cm³ g⁻¹, and presumably in water at normal temperatures) for a sample of PVME with $\overline{M}_n$ = 1.66D5. Horne and co-workers,[205] in their studies on cloud points in aqueous solutions, quote inherent viscosity, i.e., $[\ln(\eta/\eta_1)]/c_2$ values for 1% solutions as 0.35 for *Gantrez* M-154 and 0.57 for *Gantrez* M-155 (this is clearly an error for grade M-555), which however seem to relate to *benzene* solutions since they correspond to the specific viscosity values quoted in the technical literature for the two grades M-154 and M-555.[201] On the other hand, the precipitation point behavior (see Section VI.F), which gives alternative t_{p+} values of 32 and 34.5°C for relatively high molecular weight samples of PVME, suggests that the corresponding Θ_+ value is not too greatly below this, and we may therefore suggest a value of about 30°C for this quantity. This indicates that water at normal temperatures is close to being an ideal system for PVME.

F. Phase Equilibria

Considering absorption of water vapor by the bulk polymer, the commercial grade of amorphous PVME is a relatively nonhygroscopic material, with an equilibrium uptake of only about 5% at 90% humidity.[201]

Like other ethers, PVME in water shows inverse solubility-temperature behavior.[201,203,205] The technical literature[201] gives t_{p+} = 32°C for a 10% solution of an unspecified commercial grade, while Horne and co-workers[205] found t_{p+} = 34.5°C for 2.5% solutions of two different commercial grades from the same range. The technical literature[201] also gives the phase diagram from 10 up to 80% w/w concentration; at concentrations immediately above 10% the value of t_{p+} is essentially constant, but subsequently it falls off with rise in concentration up to about 70% where there is a fairly sharp minimum in t_{p+} followed by an upturn. For precipitated systems the dissolution of the polymer on cooling is slow and frequently does not begin to occur until well after the cloud point.[205] As already noted above, this precipitation behavior from aqueous solution has been used to fractionate the polymer.

Because of the relatively high molecular weights of the polymer samples used in these cloud-point studies, the actual Θ_+ value (i.e., in the limit of $M \rightarrow \infty$) should not be greatly lower, and we may therefore suggest a value of about 30°C for this quantity. It is difficult to compare this rationally with the values already quoted for PEO (+96°C) and for PPO (~−53°C) because of the differences between the structures of their monomer units and that of PVME; however, simply on the basis of the monomer *molecular* formula it is evident that for the change PEO → PVME the addition of a CH_2 group leads to the Θ_+ value being reduced by about 70°, whereas although the monomer units of PPO and PVME are isomeric, their Θ_+ values differ by about 85°. This suggests that, broadly speaking, a polyether is likely to be more water soluble (higher Θ_+ value) than the corresponding polyoxide.

The technical literature[201] also gives the corresponding phase diagrams for the precipitation behavior in water (ϕ = 0.75 and 0.90) + 2-propanol, and also the values for the temperature of phase separation (if any) in equal volume mixtures of water with various organic media. The freezing-point diagram for aqueous solutions of PVME[201] parallels that for the precipitation points, with a sharp minimum at about 70°C, although there is no evident separation of any polymer during the freezing process.

Data for the partition of PVME between water and organic solvents have been given in the technical literature,[201] which yield the following values of the organic solvent/water partition coefficients: nitro-ethane, 125 ± 8; ethyl acetate, 167 ± 14; butanone, 77 ± 3; benzene, 167 ± 14; cyclohexanone, 56 ± 2; toluene, 200 ± 20. These high values indicate the relatively hydrophobic character of the polymer; they also serve to show how the polymer may be separated by extraction from other, more hydrophilic, polymers (such as PVAL or PVP) in aqueous solution. The data for butanone and benzene are in line with the values of 0.57 and 0.60 obtained for the MHS a parameter in these solvents (see Section VI.E).

G. Chemical Behavior

The polymer is stable in salt and alkali solutions (except for showing readily reversible precipitation effects), but it is unstable in mineral acids.[201] Presumably, on analogy with the behavior of other ethers, this results from the breaking of the ether link, eventually giving poly(vinyl alcohol) via intermediate extents of conversion; if so, this would provide a useful link between these two different water-soluble polymers, while the intermediate copolymers (PVAL-VME) would also be interesting materials in their own right.

H. Analysis

Where analysis of PVME in its dilute aqueous solutions is desired, this could be carried out by warming the solution up to beyond the cloud point and measuring the turbidity produced; progressive measurements through the cloud-point region would indeed give an estimate of the average molecular weight and molecular weight distribution of the polymer. In each case it would be necessary to calibrate with known concentrations of characterized polymer fractions. The limitation of such procedures is that other polymers such as PPO (i.e., its oligomers) would interfere since they also show this cloud-point behavior.

Alternatively, turbidimetric methods (at a fixed temperature) with phenolic precipitants may be applicable, as suggested by Guttman and Higuchi[157] for other water-soluble polymers, although here the interference from other water-soluble polymers would again be severe.

VII. HIGHER POLYETHERS

Although poly(vinyl ethyl ether) and its higher homologs are water insoluble, they are still significant in the present context in two respects. First, their behavior in nonaqueous solution serves as a basis of comparison with that of PVME in similar solvents, while the study of the way in which the behavior along the series changes with alkyl chain size also gives by extrapolation at least estimates of the behavior of the ether of zero alkyl chain length, i.e., poly(vinyl alcohol) in the hypothetical state where its hydroxyl hydrogen atom is unable to act as a hydrogen-bond donor. Second, there is much current interest in the behavior in aqueous solution of the 1:1 alternating copolymers formed between these ethers and maleic acid, where variation in the length of the alkyl chain gives a variable hydrophobic character leading to different strengths of conformational supercoiling in the nonionized state; these materials are considered along with the other polyelectrolytes in Chapter 1 of Volume II.

The crystal structures of the stereoregular forms of these higher polyethers have been reviewed by Tadokoro,[46] while the light scattering, refractive index, and viscosity behavior in butanone for the ethyl, isopropyl, and butyl compounds were studied by Manson and

Arquette in parallel with their corresponding (but more extensive) measurements on the methyl compound.[203]

VIII. POLY(ETHYLENE IMINE) — PEI

$$\left[NH{-}CH_2{-}CH_2 \right]$$

Linear polymer: $M_0(C_2H_5N) = 43.07$

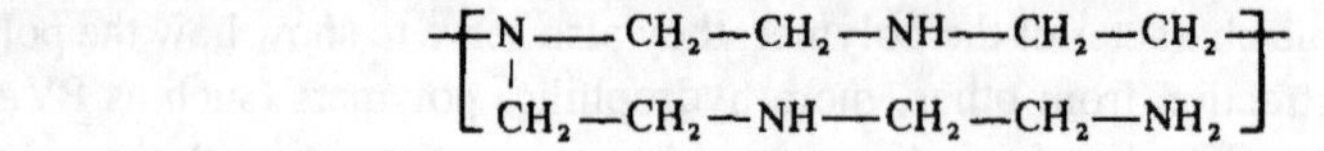

Other chemical names and trade names are polyaziridine; polyethylene imine; polyethyleneimine; polyethylenimine; poly(ethylenimine); poly(iminoethylene) (IUPAC); *Montrek* (Dow); *Polyamin* (BASF); *Polymin* (Fluka).

A. General Features

The general properties, behavior, and applications of this polymer have been reviewed.[206] In the present context, one particular feature of interest is the close structural relation of PEI to PEO; any differences in the behavior of these two polymers should be mainly the effect of replacing the oxygen atoms (with their purely accepting ability towards hydrogen bonds) by the NH groups (with both donating and accepting properties). Further, it is enlightening to compare the behavior of PEI in the nonionized state (which is the aspect considered in this section) with its behavior as a polyelectrolyte, i.e., when it is protonated (see Chapter 1 of Volume II). Unfortunately, these apparently simple comparisons are complicated by the fact that commercial grades of PEI are not linear but highly branched; for these materials the ratio of primary:secondary:tertiary groups in the molecule is essentially 1:2:1.[206] (Note that in general there must be an equal number of primary and tertiary groups if the molecular formula (C_2H_5N) is to be preserved, but the ratio of either of these to the secondary groups is in principle variable.) The structure given at the beginning of this section is a purely schematic method of representing this ratio of the three types of groups; it is believed that the true structure is very highly branched, i.e., into a bush-like form.

B. Production and Purification

Linear PEI can be synthesized by special means, via the isomerization polymerization of 2-oxazolines.[202] However, the commercial production of PEI by the simple (and more economic) acid-catalyzed polymerization of ethylene imine itself, because of chain transfer reactions to the NH groups during the polymerization, leads to the highly branched structures already mentioned.[202,206] The commercially available grades range in $\overline{M}$ from 600 (i.e., $\overline{P} = 10$) up to 5D4 or 1D5; these grades are generally given a number representing $\overline{M}/100$, e.g., PEI 6 represents the grade with $\overline{M} = 600$.

C. Fractionation and Characterization

Swanson and co-workers[207,208] have successfully fractionated the polymer using GPC with sodium acetate solution[207] or dilute alkali (pH = 9.6)[208] as eluants. Van den Berg and co-workers[209,210] did not find this method to be satisfactory except as a method of removing low molecular weight components, for which they considered exhaustive dialysis to be more convenient; they preferred to use extractive fractionation of the hydrochloride salt with aqueous hydrochloric acid/ethanol as the solvent/nonsolvent system.

D. Bulk Polymer

True linear PEI is highly crystalline, with $t_g = -23.5°C$ and $t_m = 58.5°C$.[202] The branched commercial grades of PEI are amorphous.[206] The densities of the short-chain, liquid materials range from 1.031 g cm^{-3} for PEI 6 up to 1.039 g cm^{-3} for PEI 18.

E. Solution Behavior

The commercial grades of PEI are all soluble in water and in methanol, ethanol, and 2-propanol; PEI grades 6 to 18 are soluble in butanol, but grades 600 and above are not. All grades are insoluble in tetrahydrofuran and toluene,[206] and also in benzene.[116]

In aqueous solution, the pH of the solution must be made sufficiently high if the ionization is to be suppressed. At pH 7 the ionization is 10%, while even at pH 10.5 it is still 2.5%. To some extent with the more concentrated solutions the ionization is self suppressed (the pH of 5% solution is 10 to 11[116]), but for dilute solution studies it is necessary to use a suitably alkaline medium as solvent. The results of various studies on the behavior of nonionized PEI in aqueous solution are summarized in Table 7. The low values of intrinsic viscosity, and the low value (0.39 — entry 8) for the MHS exponent *a*, together with other data on conformational size, confirm the compactness of the polymer molecule in solution.

F. Chemical Behavior

Thermogravimetric analyses on PEI grades 12 and 1000 in air and in nitrogen atmospheres up to 500°C show that the polymer possesses good thermal stability, with little weight loss even up to 200°C in air.[206]

Since the polymer contains amine and imine groups it will absorb carbon dioxide, and hence needs to be protected from such contamination from the atmosphere both in storage and in use. The presence of primary, secondary, and tertiary amine groups on the molecule of PEI makes it readily susceptible to a diversity of forms of chemical modification.[206] Klotz and co-workers have prepared a number of acyl and alkyl derivatives and studied their catalytic action as synthetic enzyme-like materials ("synzymes").[212-217]

G. Analysis

Swanson and co-workers[207,208] have analyzed PEI solutions by a colorimetric method using a complex formed with cupric ions, and also by nitrogen determination.

Dick and Ham[218] found that titration with hydrochloric acid in aqueous solution gave a very poorly defined end point; however, with *p*-toluenesulfonic acid in glacial acetic acid the end point was well defined, but corresponded to only about 80% neutralization, which was ascribed to the reduced basicity of the residual amino groups as their adjacent groups became protonated.

Van den Berg and co-workers[209,210] also found that potentiometric and conductimetric titrations in aqueous solution were not successful; they therefore used refractive index measurements as the routine method of analysis, supported by an absolute gravimetric method involving freeze-drying solutions acidified with hydrochloric acid to give the solid hydrochloride (use of unacidified solutions gave irreproducible results because of the hygroscopic nature of the free base).

POSTSCRIPT

II.A. Reviews of the properties and applications of low molecular weight PEO,[220] and high molecular weight PEO[221]

II.B. Fractionation with benzene/heptane at 40°C[222]

II.D. Raman spectra (800—1,500 cm^{-1}), on solid and molten polymer (25 to 72°C)[223]

II.E. Crystalline complexes with: KI:[224] $MgCl_2$;[225] $NaBPh_4$, NH_4SCN, KSCN, $X(SCN)_2$ (X = Mg-Ba), BaI_2[226-228]

II.F.1. Discussion of the behavior of PEO in aqueous solution (including phase separa-

Table 7
STUDIES ON NONIONIZED PEI IN AQUEOUS SOLUTION

Entry	Year	Polymer samples No.	Polymer samples Range of $\overline{M}$	Techniques used	Temperature (t/°C)	Polymer concentration	Comments	Ref.
1	1971	5	8D3—2D4	SE			+ NaOH (0.08, 0.8 mM) (pH9.6,10.9) + NaCl (0.1M): → $\overline{M}_w$	207
2	1971	5	8D3—2D4	TD			+ NaOH (0.08,0.8mM) (pH9.6,10.9) + NaCl (0—5mM): → $D°$	207
3	1973	4	600—1D5	USA	25	2—9%	"i";[a] 1—190 MHz	211
4	1973	16	6D3—3D6	LS	24 ± 1	0.05—0.8%	436 nm: → $\overline{M}_w$; $(\overline{M}_w/\overline{M}_n)$[b,c]	209,210
5	1973	4	5D4—8D5	OP	25	0.05—0.8%	→ $\overline{M}_n$; $(\overline{M}_w/\overline{M}_n)$[b]	209,210
6	1973	?	?	RI	25		589 nm: → $\bar{n}$ = 0.216 $cm^3\ g^{-1}$ (footnote b)	209,210
7	1973	4	2D4—3D6	SS		0.2%	→ MWD; no aggregation[b]	209,210
8	1973	7	3D4—3D6	VI	25	0.05—0.8%	→ [η]; MHS (0.39; 0.15)[b,c]	209,210
9	1974	6	2D3—2D4	ED			Dialysis membrane pore size 48 Å: → molecular size	208
10	1974	1	2D4	LS		0.16%	λ = ?: → Rayleigh scattering ratio[d]	208
11	1974	3	5D3—2D4	OP			→ $\overline{M}_n$; $(\overline{M}_w/\overline{M}_n)$[d]	208
12	1974	8	2D3—2D4	SE			→ $\overline{M}_w$; $(\overline{M}_w/\overline{M}_n)$[d]	208
13	1974	6	2D3—2D4	TD			→ $D°$ (footnote d)	208
14	1974	6	2D3—2D4	VI			→ [η][d]	208

[a] No alkali added to suppress ionization; e.g., if pH = 7, true $i \approx 0.1$[206]; however, ionization may be largely self suppressed at the relatively high concentration used (~5%).

[b] Solvent: either 0.01M KOH (i.e., pH ≈ 12)[209] or 0.01M NaOH + 0.01M NaCl (i.e., pH ≈ 12).[210]

[c] Identical results stated to have been obtained using 0.1M NaOH + 0.1M NaCl as solvent.[210]

[d] Solvent: 42.5 μM NaOH + 0.109M NaCl (i.e., pH ≈ 9.6).

tion);[229] viscosity, light scattering, and drag reduction in turbulent flow;[230] drag reduction in turbulent flow;[231] viscosity, drag reduction, and degradation in turbulent flow;[232] Raman spectra;[223] viscosity, and neutron scattering[233]

II.F.2. Chloroform (density and heats of mixing);[234] chloroform (Raman spectra);[223] dimethylformamide (GPC for molecular weight distribution);[232] benzene (thermodynamics and viscosity);[235] benzene (thermodynamics)[236]

II.G. Uptake of organic vapors studied by GLC;[237,238] phase diagrams in water and in *t*-butyl acetate[222]

II.H. Effects of radiation;[239,240] thermal degradation, and shear degradation in turbulent flow;[232] wet oxidation[241]

II.J. Analysis of dilute solutions by nephelometry with tannic acid[242]

III.A. Wet oxidation[241]

III.E.2. Benzene (viscosity and MHS parameters)[235]

V. Block copolymers *(Pluronics)* — review,[243] viscosity,[244] light-scattering,[245] sedimentation[246]

VI. Uptake of organic vapors studied by GLC[247]

VIII. Methods of chemical modification[248]

REFERENCES

1. **Powell, G. M. and Bailey, F. E.,** Poly(ethylene oxide), in *Kirk-Othmer Encyclopedia of Chemical Technology*, 2nd Suppl. Vol., Standen, A., Ed., Interscience, New York, 1960, 597.
2. **Stone, F. W. and Stratta, J. J.,** Ethylene oxide polymers, in *Encyclopedia of Polymer Science and Technology*, Vol. 6, Mark, H. F., Gaylord, N. G., and Bikales, N. M., Eds., John Wiley & Sons, New York, 1967, 103.
3. **Harding, R. H. and Rose, J. K.,** Ethylene oxide polymers, in *Water-Soluble Resins*, 2nd ed., Davidson, R. L. and Sittig, M., Eds., Van Nostrand Reinhold, New York, 1968, chap. 10.
4. **Gaylord, N. G., Ed.,** *Polyethers. Part I. Polyalkylene Oxides and Other Polyethers*, Wiley-Interscience, New York, 1963.
5. **St. Pierre, L. E.,** Polymers from 1,2-epoxides. I. Chemistry, in *Polyethers. Part I. Polyalkylene Oxides and Other Polyethers*, Gaylord, N. G., Ed., Wiley-Interscience, New York, 1963, chap. 3.
6. **Kastens, A. S.,** Polymers from 1-2-epoxides. II. Applications and technology, in *Polyethers. Part I. Polyalkylene Oxides and Other Polyethers*, Gaylord, N. G., Ed., Wiley-Interscience, New York, 1963, chap 4.
7. **Bailey, F. E. and Koleske, J. V.,** *Poly(Ethylene Oxide)*, Academic Press, New York, 1976.
8. **Hill, F. N., Bailey, F. E., and Fitzpatrick, J. T.,** High molecular weight polymers of ethylene oxide. I. Polymerization with alkaline earth carbonate catalysts, *Ind. Eng. Chem.*, 50, 5, 1958.
9. **Bailey, F. E., Powell, G. M., and Smith, K. L.,** High molecular weight polymers of ethylene oxide. II. Solution properties, *Ind. Eng. Chem.*, 50, 8, 1958.
10. **Smith, K. L. and Van Cleve, R.,** High molecular weight polymers of ethylene oxide. III. Plastic properties, *Ind. Eng. Chem.*, 50, 12, 1958.
11. **Anon.,** *Polyethylene Glycols*, Dow Chemical Co., Midland, Mich., 1962.

11a. **Anon.,** *Dow Products*, Dow Chemical Co., Midland, Mich., 1978.

12. **Anon.,** *Polyethylene Glycols*, Shell Chemical Co., London (undated).
13. **Reynolds, J. E. F., and Prasad, A. B., Eds.,** *Martindale: The Extra Pharmacopoeia*, 28th ed., The Pharmaceutical Press, London, 1982.
14. **Tadokoro, H., Chatani, Y., Yoshihara, T., Tahara, S., and Murahashi, S.,** Structural studies on polyethers, $[-(CH_2)_m-O-]_n$. II. Molecular structure of polyethylene oxide, *Makromol. Chem.*, 73, 109, 1964.
15. **Fordyce, R., Lovell, E., and Hibbert, H.,** Studies on reactions relating to carbohydrates and polysaccharides. LVI. The synthesis of the higher polyoxyethylene glycols, *J. Am. Chem. Soc.*, 61, 1905, 1939.
16. **Fordyce, R. and Hibbert, H.,** Studies on reactions relating to carbohydrates and polysaccharides. LVII. The synthesis of 90-membered oxyethylene and 186-membered oxyethylene glycols, *J. Am. Chem. Soc.*, 61, 1910, 1939.
17. **Staudinger, H. and Lohmann, H.,** Über hochpolymere Verbindungen. 81. Mitteilung. Über eukolloides Polyäthylenoxyd, *Justus Liebigs Ann. Chem.*, 505, 41, 1933; *Chem. Abstr.*, 27, 4790, 1933.

18. **Miller, R. A. and Price, C. C.,** Polyethers. VII. Aluminium catalysts for polymerization of ethylene oxide, *J. Polym. Sci.*, 34, 161, 1959.
19. **Afifi-Effat, A. M. and Hay, J. N.,** Enthalpy and entropy of fusion and the equilibrium melting point of polyethylene oxide, *J. Chem. Soc. Faraday Trans.* 2, 68, 656, 1972.
20. **Elias, H.-G.,** Konstitution und Lösungseigenschaften von Makromolekülen. I. Ermittlung von Θ-Lösungen, *Makromol. Chem.*, 50, 1, 1961.
21. **Beech, D. R. and Booth, C.,** Unperturbed dimensions of poly(ethylene oxide), *J. Polym. Sci. Part A-2*, 7, 575, 1969.
22. **Booth, C. and Price, C.,** The fractional crystallization of poly(ethylene oxide), *Polymer (London)*, 7, 85, 1966.
23. **Booth, C. and Devoy, C. J.,** Thermodynamics of mixtures of poly(ethylene oxide) and benzene, *Polymer (London)*, 12, 309, 1971.
24. **Ring, W., Cantow, H.-J., and Holtrup, W.,** Molekulargewichte und Molekulargewichtsverteilungen von Polyäthylenoxiden, *Eur. Polym. J.*, 2, 151, 1966.
25. **Schulz, G. V. and Nordt, E.,** Fraktionierung polymolekularer Stoffe durch Verteilung zwischen zwei flüssigen Phasen, *J. Prakt. Chem.*, 155, 115, 1940.
26. **Kämmerer, H. and Grover, P. N.,** UV-Photometrisch bestimmte relative molare Massen von anionisch erzeugten Polyäthylenoxiden mit unterschiedlichen Endgruppen, *Makromol. Chem.*, 177, 965, 1976.
27. **Beaumont, R. H., Clegg, B., Gee, G., Herbert, J. B. M., Marks, D. J., Roberts, R. C., and Sims, D.,** Heat capacities of propylene oxide and of some polymers of ethylene and propylene oxides, *Polymer (London)*, 7, 401, 1966.
28. **Braun, W., Hellwege, K.-H., and Knappe, W.,** Enthalpie von Polyoxyäthylenen im Temperaturbereich von 15—100°C, *Kolloid-Z. Z. Polym.*, 215, 10, 1967.
29. **Allen, G., Gee, G., Mangaraj, D., Sims, D., and Wilson, G. J.,** Intermolecular forces and chain flexibilities in polymers. II. Internal pressures of polymers, *Polymer* 1, 467, 1960.
30. **Allen, G. and Sims, D.,** Intermolecular forces and chain flexibilities. IV. Internal pressures of polyethylene glycol in the region of its melting point, *Polymer*, 4, 105, 1963.
31. **Rempp, P.,** Contribution à l'étude des solutions de molécules en chaine à squelette oxygéné. I. Préparation de substance-modèle; détermination des mass moléculaires et étude de la polydispersité des polyoxyéthylèneglycols, *J. Chim. Phys.*, 54, 421 and 466, 1957.
32. **Nyquist, R. A.,** *Infrared Spectra of Plastics (Polymers) and Resins*, 2nd ed., The Dow Chemical Co., Midland, Mich., 1961.
33. **Kuroda, Y. and Kubo, M.,** CH_2 rocking vibrations and the structure of polyethylene glycols, *J. Polym. Sci.*, 36, 453, 1959.
34. **Miyazawa, T., Fukushima, K., and Ideguchi, Y.,** Molecular vibrations and structure of high polymers. III. Polarized infrared spectra, normal vibrations, and helical conformation of polyethylene glycol, *J. Chem. Phys.*, 37, 2764, 1962.
35. **Yoshihara, T., Tadokoro, H., and Murahashi, S.,** Normal vibrations of the polymer molecules of helical conformation. IV. Polyethylene oxide and polyethylene-d_4 oxide, *J. Chem. Phys.*, 41, 2902, 1964.
36. **Angood, A. C. and Koenig, J. L.,** Infrared studies of chain folding in polyethylene oxide, *J. Appl. Phys.*, 39, 4985, 1968.
37. **Koenig, J. L. and Angood, A. C.,** Raman spectra of poly(ethylene glycols) in solution, *J. Polym. Sci. Part A-2*, 8, 1787, 1970.
38. **Liu, K.-J.,** Nuclear magnetic resonance studies of polymer solutions. IV. Polyethylene glycols, *Macromolecules*, 1, 213, 1968.
39. **Connor, T. M.,** Proton spin-lattice relaxation in polyethylene oxide. Spin-diffusion and methyl group rotation, *Polymer*, 7, 426, 1966.
40. **Allen, G., Booth, C., and Price, C.,** The effect of the side group on the properties of the poly(epoxides). VI. The physical properties of poly(epoxides), *Polymer*, 8, 414, 1967.
41. **Connor, T. M. and McLauchlan, K. A.,** High resolution nuclear resonance studies of the chain conformation of polyethylene oxide, *J. Phys. Chem.*, 69, 1888, 1965.
42. **Matsuzaki, K. and Ito, H.,** Conformational analysis of poly(ethylene oxide) and model compounds by nuclear magnetic resonance spectroscopy, *J. Polym. Sci. Polym. Phys. Ed.*, 12, 2507, 1974.
43. **Takahashi, Y. and Tadokoro, H.,** Structural studies of polyethers, $[-(CH_2)_m-O-]_n$. X. Crystal structure of poly(ethylene oxide), *Macromolecules*, 6, 672, 1973.
44. **Mark, J. E. and Flory, P. J.,** The configuration of the polyoxyethylene chain, *J. Am. Chem. Soc.*, 87, 1415, 1965.
45. **Mark, J. E.,** Configurational partition functions and entropies for polyoxide chains having the repeat unit $[(CH_2)_y-O-]$, *J. Chem. Phys.*, 67, 3300, 1977.
46. **Tadokoro, H.,** Structure of crystalline polyethers, *Macromol. Rev.*, 1, 119, 1966.
47. **Anon.,** Polyox Water-Soluble Resins: Forming Association Compounds, Tech. Bull. F-43272, Union Carbide Corp., New York, 1971.

48. **Blumberg, A. A., Pollack, S. S., and Hoeve, C. A. J.,** A poly(ethylene oxide)-mercuric chloride complex, *J. Polym. Sci. Part A,* 2, 2499, 1964.
49. **Doscher, T. M., Myers, G. E., and Atkins, D. C.,** The behavior of nonionic surface active agents in salt solutions, *J. Colloid Sci.,* 6, 223, 1951.
50. **Kono, K., Takeda, Y., Nogami, H., and Nagai, T.,** Complexes of polyethylene oxide with guanidine hydrochloride and phenobarbital. Preliminary structural studies by differential scanning calorimetry, polarized infrared spectroscopy and X-ray fiber photography, *Chem. Pharm. Bull.,* 22, 165, 1974.
51. **Iwamoto, R., Saito, Y., Ishihara, H., and Tadokoro, H.,** Structure of poly(ethylene oxide) complexes. II. Poly(ethylene oxide)-mercuric chloride complex, *J. Polym. Sci. Part A-2,* 6, 1509, 1968.
52. **Bailey, F. E. and France, H. G.,** Molecular association complexes of polymers. Urea and thiourea complexes of high molecular weight poly(ethylene oxide), *J. Polym. Sci.,* 49, 397, 1961.
53. **Barker, G. E. and Ranauto, H. J.,** Solid nonionic surface-active agents, *J. Am. Oil Chem. Soc.,* 32, 249, 1955.
54. **Parrod, J. and Kohler, A.,** Combinations d'insertion du polyoxyéthylène-glycol dans l'urée, *C. R. Acad. Sci.,* 246, 1046, 1958.
55. **Tadokoro, H., Yoshihara, T., Chatani, Y., and Murahashi, S.,** A preliminary report of structural studies on polyethylene oxide-urea complex, *J. Polym. Sci. Part B (Polym. Lett.),* 2, 363, 1964.
56. **Hild, G., Kohler, A., and Parrod, J.,** Sur la structure des combinaisons d'addition de l'urée et de macromolécules du type polyoxyéthylène, *Bull. Soc. Chim. France,* 2284, 1966.
57. **Schulz, G. V.,** Osmotische Molekulargewichtsbestimmungen in polymerhomologen Reihen hochmolekularer Stoffe. 141. Mitteilung über hochpolymere Verbindungen, *Z. Phys. Chem. Abt. A,* 176, 317, 1936.
58. **Huggins, M. L.,** Theory of solutions of high polymers, *J. Am. Chem. Soc.,* 64, 1712, 1942.
59. **Katchman, B. and McLaren, A. D.,** Sorption of water vapour by proteins and polymers. IV, *J. Am. Chem. Soc.,* 73, 2124, 1951.
60. **Malcolm, G. N. and Rowlinson, J. S.,** The thermodynamic properties of aqueous solutions of polyethylene glycol, polypropylene glycol and dioxane, *Trans. Faraday Soc.,* 53, 921, 1957.
61. **Rempp, P.,** Contribution à l'étude des solutions de molécules en chaine à squelette oxygéné. II. Viscosités intrinsèques et coefficients de diffusion de translation des polyoxyéthylèneglycols, *J. Chim. Phys.,* 54, 432 and 466, 1957.
62. **Sadron, C. and Rempp, P.,** Viscosités intrinsèques des solutions de chaînes courtes, *J. Polym. Sci.,* 29, 127, 1958.
63. **Bailey, F. E., Kucera, J. L., and Imhof, L. G.,** Molecular weight relations of poly(ethylene oxide), *J. Polym. Sci.,* 32, 517, 1958.
64. **Alexandrowicz, Z.,** Light-scattering measurement of the osmotic activity of a polyethylene glycol solution, *J. Polym. Sci.,* 40, 107, 1959.
65. **Bailey, F. E. and Callard, R. W.,** Some properties of poly(ethylene oxide) in aqueous solution, *J. Appl. Polym. Sci.,* 1, 56, 1959.
66. **Bailey, F. E. and Callard, R. W.,** Thermodynamic parameters of poly(ethylene oxide) in aqueous solutions, *J. Appl. Polym. Sci.,* 1, 373, 1959.
67. **Ritscher, T. A. and Elias, H.-G.,** Das Problem der Membran bei osmotischen Messungen an Hochpolymeren. III, *Makromol. Chem.,* 30, 48, 1959.
68. **Thomas, D. K. and Charlesby, A.,** Viscosity relationship in solutions of polyethylene glycols, *J. Polym. Sci.,* 42, 195, 1960.
69. **Elias, H.-G.,** Bestimmung des Molekulargewichts in der Ultrazentrifuge nach dem Archibald-Verfahren, *Angew. Chem.,* 73, 209, 1961.
70. **Elias, H.-G.,** Osmose an permeablen Membranen. I. Der Lösungsmittel-Einfluss auf den Stavermann-Effekt, *Z. Phys. Chem. (Frankfurt am Main),* 28, 303, 1961.
71. **Elias, H.-G.,** Osmose an permeablen Membranen. II. Anomale Osmosen an ionenfreien Membranen, *Z. Phys. Chem. (Frankfurt am Main),* 28, 322, 1961.
72. **Andersson, G. R.,** A study of the pressure dependence of the partial specific volume of macromolecules in solution by compression measurements in the range 1—8000 atm, *Arkiv Kemi,* 20, 513, 1963.
73. **Müh, G.,** Über den Zusammenhang zwischen dem Molekulargewicht und der Viskositätszahl von Lösungen kurzkettiger Polyäthylenglykole. 2. Mitteilung über Polyäthylenglykole, *Kolloid-Z. Z. Polym.,* 196, 140, 1964.
74. **Rodriguez, F. and Goettler, L. A.,** The flow of moderately concentrated polymer solutions in water, *Trans. Soc. Rheol.,* 8, 3, 1964.
75. **Rossi, C. and Cuniberti, C.,** Intrinsic viscosities of low poly(ethylene) oxides POEG in different solvents, *J. Polym. Sci. Part B (Polym. Lett.),* 2, 681, 1964.
76. **Elias, H.-G. and Lys, H.,** Virialkoeffizienten und Assoziation von Polyäthylenglykolen, *Makromol. Chem.,* 92, 1, 1966.
77. **Hammes, G. G. and Knoche, W.,** Ultrasonic-absorption measurements in mixed solvents: water-dioxane, *J. Chem. Phys.,* 45, 4041, 1966.

78. **Hammes, G. G. and Lewis, T. B.,** Ultrasonic absorption in aqueous polyethylene glycol solutions, *J. Phys. Chem.*, 70, 1610, 1966.
79. **Lakhanpal, M. L., Kapoor, V., Sharma, R. K., and Sharma, S. C.,** Studies on polyoxyethylene glycols: III. Heats of mixing of polyoxyethylene glycols-water systems, *Indian J. Chem.*, 4, 59, 1966.
80. **Lakhanpal, M. L., Singh, H. G., Singh, H., and Sharma, S. C.,** A comparative study of heats of mixing of polyoxyethylene glycol-water & polyoxypropylene glycol-water systems, *Indian J. Chem.*, 6, 95, 1968.
81. **Mark, J. E. and Flory, P. J.,** Dipole moments of chain molecules. I. Oligomers and polymers of oxyethylene, *J. Am. Chem. Soc.*, 88, 3702, 1966.
82. **Teramoto, A., Kusamizu, S., Tanaka, H., Murakami, Y., and Fujita, H.,** Viscosities of poly(ethylene oxide)-water systems over the complete range of composition, *Makromol. Chem.*, 90, 78, 1966.
83. **Bailey, F. E. and Koleske, J. V.,** Configuration and hydrodynamic properties of the polyoxyethylene chain in solution, in *Nonionic Surfactants (Surfactant Science Series, Vol. 1)*, Schick, M. J., Ed., Marcel Dekker, New York, 1967, chap. 23.
84. **Elias, H.-G.,** Multimerisation: Assoziation und Aggregation. III. Einfluss der offenen Assoziation auf die ungestörten Dimensionen von Polyäthylenglykolen, *Makromol. Chem.*, 103, 214, 1967.
85. **Hammes, G. G. and Schimmel, P. R.,** An investigation of water-urea and water-urea-polyethylene glycol interactions, *J. Am. Chem. Soc.*, 89, 442, 1967.
86. **Kagemoto, A., Murakami, S., and Fujishiro, R.,** The heats of dilution of poly(ethylene oxide)/water solutions, *Makromol. Chem.*, 105, 154, 1967.
87. **Nichol, L. W., Ogston, A. G., and Preston, B. N.,** The equilibrium sedimentation of hyaluronic acid and two synthetic polymers, *Biochem. J.*, 102, 407, 1967.
88. **Edmond, E. and Ogston, A. G.,** An approach to the study of phase separation in ternary aqueous systems, *Biochem. J.*, 109, 569, 1968.
89. **Hammes, G. G. and Roberts, P. B.,** Cooperativity of solvent-macromolecule interactions in aqueous solutions of polyethylene glycol and polyethylene glycol-urea, *J. Am. Chem. Soc.*, 90, 7119, 1968.
90. **Lakhanpal, M. L., Chhina, K. S., and Sharma, S. C.,** Thermodynamic properties of aqueous solutions of polyoxyethylene glycols, *Indian J. Chem.*, 6, 505, 1968.
91. **Liu, K.-J. and Ullman, R.,** Proton magnetic relaxation in polyethylene oxide solutions, *J. Chem. Phys.*, 48, 1158, 1968.
92. **Nabi, G.,** Rheological and solution properties of poly(ethylene oxide), *Pak. J. Sci. Res.*, 20, 19, 1968.
93. **Nabi, G.,** Light-scattering studies of aqueous solutions of poly(ethylene oxide), *Pak. J. Sci. Res.*, 20, 136, 1968.
94. **Couper, A. and Stepto, R. F. T.,** Diffusion of low-molecular weight poly(ethylene oxide) in water, *Trans. Faraday Soc.*, 65, 2486, 1969.
95. **Hawley, S. A. and Dunn, F.,** Ultrasonic absorption in aqueous solutions of dextran, *J. Chem. Phys.*, 50, 3523, 1969.
96. **Liu, K.-J. and Parsons, J. L.,** Solvent effects on the preferred conformation of poly(ethylene glycols), *Macromolecules*, 2, 529, 1969.
97. **Hammes, G. G. and Roberts, P. B.,** Ultrasonic absorption in aqueous polyethylene glycol solutions, *J. Chem. Phys.*, 52, 5496, 1970.
98. **Hawley, S. A. and Dunn, F.,** Ultrasonic relaxation spectra in aqueous solutions of dextran and polyethylene glycol, *J. Chem. Phys.*, 52, 5497, 1970.
99. **Kessler, L. W., O'Brien, W. D., and Dunn, F.,** Ultrasonic absorption in aqueous solutions of polyethylene glycol, *J. Phys. Chem.*, 74, 4096, 1970.
100. **Liu, K.-J. and Anderson, J. E.,** Solvent effects on the proton magnetic relaxation of poly(ethylene oxide), *Macromolecules*, 3, 163, 1970.
101. **Nakayama, H.,** Temperature dependence of the heats of solution of poly(ethylene glycol) and of related compounds, *Bull. Chem. Soc. Jpn.*, 43, 1683, 1970.
102. **Napper, D. H.,** Flocculation studies of sterically stabilized dispersions, *J. Colloid Interface Sci.*, 32, 106, 1970.
103. **Napper, D. H.,** Steric stabilization and the Hofmeister series, *J. Colloid Interface Sci.*, 33, 384, 1970.
104. **Vink, H.,** Precision measurements of osmotic pressure in concentrated polymer solutions, *Eur. Polym. J.*, 7, 1411, 1971.
105. **Maron, S. H. and Filisko, F. E.,** Heats of solution and dilution for polyethylene oxide in several solvents, *J. Macromol. Sci. -Phys.*, B6, 79, 1972.
106. **Crouzet, C. and Marchal, J.,** Charactérisation de reactions primaires de degradation oxydante au cours de l'autoxydation des polyoxyéthylènes a 25°C: l'étude en solution aqueuse avec amorçage par radiolyse du solvent. I. Étude viscometrique: stabilité des hydroperoxydes du polyoxyéthylène et cinétique de production des coupures, *Makromol. Chem.*, 166, 69, 1973.
107. **Bluestone, S., Mark, J. E., and Flory, P. J.,** The interpretation of viscosity-temperature coefficients for poly(oxyethylene) chains in a thermodynamically good solvent, *Macromolecules*, 7, 325, 1974.

108. **Jones, A. A. and Stockmayer, W. H.**, Effect of urea on magnetic relaxation in aqueous solutions of poly(ethylene oxide), *J. Phys. Chem.*, 78, 1528, 1974.
109. **Molyneux, P.**, Synthetic polymers, in *Water: A Comprehensive Treatise*, Vol. 4, Franks, F., Ed., Plenum Press, New York, 1975, chap. 7.
110. **Nomura, H., Kato, S., and Miyahara, Y.**, Ultrasonic absorption in polymer solutions, *Mem. Fac. Eng., Nagoya Univ.*, 27, 72, 1975.
111. **Chew, B. and Couper, A.**, Diffusion, viscosity and sedimentation of poly(ethylene oxide) in water, *J. Chem. Soc. Faraday Trans. 1*, 72, 382, 1976.
112. **McComb, W. D. and Ayyash, S.**, Effect of drag-reducing polymers on pulsation noise of small air bubbles, *Nature (London)*, 262, 47, 1976.
113. **Virk, P. S.**, Conformational effects in drag reduction by polymers, *Nature (London)*, 262, 46, 1976.
114. **Rogers, J. A. and Tam, T.**, Solution behavior of polyethylene glycols in water using vapor pressure osmometry, *Can. J. Pharm. Sci.*, 12, 65, 1977.
115. **Boucher, E. A. and Hines, P. M.**, Properties of aqueous salt solutions of poly(ethylene oxide): thermodynamic quantities based on viscosity and other measurements, *J. Polym. Sci. Polym. Phys. Ed.*, 16, 501, 1978.
116. **Braun, D. and Törmälä, P.**, Segmental motions of spin-labelled poly(ethylene oxide) in dilute solution, *Makromol. Chem.*, 179, 1025, 1978.
117. **Douhéret, G. and Morénas, M.**, Thermodynamic and physical behaviour of some water + polyethyleneglycol mixtures. II. Dielectric properties, *Can. J. Chem.*, 57, 608, 1979.
118. **Fordyce, R. and Hibbert, H.**, Studies on reactions relating to carbohydrates and polysaccharides. LVIII. The relation between chain length and viscosity of the polyoxyethylene glycols, *J. Am. Chem. Soc.*, 61, 1912, 1939.
119. **Lakhanpal, M. L., Lal, M., and Sharma, R. K.**, Studies on polyoxyethylene glycols. I. Heats of mixing of polyoxyethylene glycols with dioxane and carbon tetrachloride, *Indian J. Chem.*, 3, 547, 1965.
120. **Moldovan, L. and Strazielle, C.**, Étude par diffusion de la lumière et viscosité du système ternaire: polyoxyéthylène/tétrachlorure de carbone/méthanol, *Makromol. Chem.*, 140, 201, 1970.
121. **Patterson, G. D. and Flory, P. J.**, Optical anisotropies of polyoxyethylene oligomers, *J. Chem. Soc. Faraday Trans. 2*, 68, 1111, 1972.
122. **Bothorel, P. and Fourche, G.**, Molecular optical anisotropy of n-alkanes and polyoxyethylene oligomers by depolarized Rayleigh scattering, *J. Chem. Soc. Faraday Trans. 2*, 69, 441, 1973.
123. **Lang, M.-C., Lauprêtre, F., Noël, C., and Monnerie, L.**, Molecular motion of polyethylene oxide in dilute solutions studied by electron spin resonance and nuclear magnetic relaxation, *J. Chem. Soc. Faraday Trans. 2*, 75, 349, 1979.
124. **Hert, M. and Strazielle, C.**, Étude des systèmes ternaires polymère/mélange de solvants en équilibre thermodynamique avec le solvant binaire correspondant, *Makromol. Chem.*, 175, 2149, 1974.
125. **Müh, G.**, Über die Huggins-Konstanten von Polyäthylenglykollösungen. (Vierte Mitteilung über Polyäthylenglykole), *Kolloid-Z. Z. Polym.*, 196, 64, 1964.
126. **Carpenter, D. K., Santiago, G., and Hunt, A. H.**, Aggregation of polyoxyethylene in dilute solutions, *J. Polym. Sci. Polym. Symp.*, 44, 75, 1974.
127. **Elias, H.-G.**, Konstitution und Lösungseigenschaften von Makromolekülen. II. Die Beziehungen zwischen den Konstanten *K* and *a* der Staudingerschen Viskositätsgleichung, *Kunstst.-Plast. (Solothurn Switz.)*, 8, 441, 1961.
128. **Müh, G.**, Die Molekulargewichtsbestimmung handelsüblicher Polyäthylenglykole und die Berechnung ihrer Molekulargewichtsverteilung aus der Reaktionskinetik. I. Mitteilung über Polyäthylenglykole, *Makromol. Chem.*, 77, 64, 1964.
129. **Lakhanpal, M. L., Taneja, H. L., and Sharma, R. K.**, Studies on polyoxyethylene glycols. II. Heats of mixing of polyoxyethylene glycols with benzene, methanol and ethanol, *Indian J. Chem.*, 4, 12, 1966.
130. **Allen, G., Booth, C., Hurst, S. J., Jones, M. N., and Price, C.**, The effect of the side group on the properties of the poly(epoxides). II. Intrinsic viscosity and light scattering measurements on poly(ethylene oxide), poly(styrene oxide), and poly(*t*-butyl ethylene oxide), *Polymer*, 8, 391, 1967.
131. **Liu, K.-J.**, Nuclear magnetic resonance studies of polymer solutions. V. Cooperative effects in the ion-dipole interaction between potassium iodide and poly(ethylene oxide), *Macromolecules*, 1, 308, 1968.
132. **Liu, K.-J. and Anderson, J. E.**, Nuclear magnetic relaxation in poly(ethylene oxide)-salt solutions, *Macromolecules*, 2, 235, 1969.
133. **Kagemoto, A., Itoi, Y., Baba, Y., and Fujishiro, R.**, The heats of dilution of the oligomeric ethylene oxide/alcohol solutions, *Makromol. Chem.*, 150, 255, 1971.
134. **Napper, D. H. and Netschey, A.**, Studies of the steric stabilization of colloidal particles, *J. Colloid Interface Sci.*, 37, 528, 1971.
135. **Seow, P. K., Gallot, Y., and Skoulios, A.**, Synthèse et charactérisation de copolymères séquencés poly(méthacrylate d'alkyle)/poly(oxyéthylène), *Makromol. Chem.*, 176, 3153, 1975.

136. **Thierry, A. and Skoulios, A.,** Copolymères greffés: poly(méthacrylate d'alkyl-*g*-oxyde d'éthylène). I. Synthèse et charactérisation, *Makromol. Chem.*, 177, 319, 1976.
137. **Müh, G.,** Über die Abhängigkeit. 2. Virialkoeffizienten des osmotischen Drucks vom Molekulargewicht. (3. Mitteilung über Polyäthylenglykole). *Ber. Bunsenges. Phys. Chem.*, 68, 445, 1964.
138. **De Candia, F., Vittoria, V., Bianchi, U., and Patrone, E.,** Temperature coefficient of unperturbed polymer chain dimensions from thermoelastic and solution properties. I. Poly(ethylene oxide), *Macromolecules*, 5, 493, 1972.
139. **Mark, J. E. and Flory, P. J.,** The temperature coefficient of the unperturbed dimensions of polyoxyethylene, *Macromolecules*, 6, 300, 1973.
140. **Marchal, J. and Benoit, H.,** Comparaison entre les résultats fournis par la détermination des dimensions et du moment électrique moyen des chaînes macromoléculaires, *J. Polym. Sci.*, 23, 223, 1957.
141. **Minoura, Y., Kasuya, T., Kawamura, S., and Nakano, A.,** Degradation of poly(ethylene oxide) by high-speed stirring, *J. Polym. Sci. Part A-2*, 5, 125, 1967.
142. **Heatley, F. and Walton, I.,** Contributions to the proton relaxation of poly(ethylene oxide) in benzene solutions, *Polymer*, 17, 1019, 1976.
143. **Evans, R. and Napper, D. H.,** Steric stabilization. I. Comparison of theories with experiment, *Kolloid-Z. Z. Polym.*, 251, 409, 1973.
144. **Napper, D. H.,** The steric stabilization of hydrosols by nonionic macromolecules, *J. Colloid Interface Sci.*, 29, 168, 1969.
145. **Lovell, E. L. and Hibbert, H.,** Studies on reactions relating to carbohydrates and polysaccharides. LIX. The precipitability of pure hemicolloidal polyoxyethylene glycols, *J. Am. Chem. Soc.*, 61, 1916, 1939.
146. **Schulz, G. V.,** Über die Löslichkeit und Fällbarkeit hochmolekularer Stoffe, *Z. Phys. Chem. Abt. A.*, 179, 321, 1937.
147. **Bortel, E. and Lamot, R.,** Untersuchung des Abbaus hochmolekulare Polyäthylenoxide in Festzustand, *Makromol. Chem.*, 178, 2617, 1977.
148. **McGary, C. W.,** Degradation of poly(ethylene oxide), *J. Polym. Sci.*, 46, 51, 1960.
149. **Vink, H.,** Degradation of some polymers in aqueous solution, *Makromol. Chem.*, 67, 105, 1963.
150. **Schick, M. J., Ed.,** *Nonionic Surfactants (Surfactant Science Series, Vol. 1)*, Marcel Dekker, New York, 1967.
151. **Hamburger, R., Azaz, E., and Donbrow, M.,** Autoxidation of polyoxyethylenic non-ionic surfactants and of polyethylene glycols, *Pharm. Acta Helv.*, 50, 10, 1975.
152. **Crouzet, C., Decker, C., and Marchal, J.,** Charactérisation de réactions primaires de dégradation oxydante au cours de l'autoxydation des poly(oxyéthylène)s à 25°C: étude en solution aqueuse avec amorçage par radiolyse du solvant. VIII. Étude cinétique en fonction du pH compris entre 1 et 13, *Makromol. Chem.*, 177, 145, 1976.
153. **King, P. A. and Ward, J. A.,** Radiation chemistry of aqueous poly(ethylene oxide) solutions. I, *J. Polym. Sci. Part A-1*, 8, 253, 1970.
154. **Ward, J. A.,** Radiation chemistry of aqueous poly(ethylene oxide) solutions. II. Highly purified samples, *J. Polym. Sci. Part A-1*, 9, 3555, 1971.
155. **Borgwardt, U., Schnabel, W., and Henglein, A.,** Pulsradiolytische Messung der Geschwindigkeitskonstanten der Kombination von Polyäthylenoxid-Radikalen in wässriger Lösung, *Makromol. Chem.*, 127, 176, 1969.

155a. **Behzadi, A., Borgwardt, U., Henglein, A., Schamberg, E., and Schnabel, W.,** Pulsradiolytische Untersuchung der kinetik diffusionkontrollierter Reaktionen des OH-Radikals mit Polymeren und Oligomeren in wässriger Lösung, *Ber. Bunsenges. Phys. Chem.*, 74, 649, 1970.

155b. **Matheson, M. S., Mamou, A., Silverman, J., and Rabani, J.,** Reactions of hydroxyl radicals with polyethylene oxide in aqueous solution, *J. Phys. Chem.*, 77, 2420, 1973.

156. **Hosoya, H., Miyazaki, N., Sugisaki, Y., Takanashi, E., Tsurufuji, M., Yamasaki, M., and Tamura, G.,** Bacterial degradation of synthetic polymers and oligomers with the special reference to the case of polyethylene glycol, *Agric. Biol. Chem.*, 42, 1545, 1978.
157. **Guttman, D. and Higuchi, T.,** Possible complex formation between macromolecules and certain pharmaceuticals. X. The interaction of some phenolic compounds with polyethylene glycols, polypropylene glycols, and polyvinylpyrrolidone, *J. Am. Pharm. Assoc. Sci. Ed.*, 45, 659, 1956.
158. **Moacanin, J.,** Diisocyanate-linked polymers. I. Dilute-solution properties of toluene diisocyanate-extended polypropylene glycol, *J. Appl. Polym. Sci.*, 1, 272, 1959.
159. **Moacanin, J.,** The effect of thermodynamic interactions on solution viscosities of low molecular weight poly(propylene oxides), *Am. Chem. Soc. Div. Polym. Chem. Preprints*, 7, 1061, 1966.
160. **Allen, G., Booth, C., and Jones, M. N.,** Polypropylene oxide. I. An intrinsic viscosity/molecular weight relationship, *Polymer*, 5, 195, 1964.
161. **Allen, G., Booth, C., and Jones, M. N.,** Polypropylene oxide. II. Dilute solution properties and tacticity, *Polymer*, 5, 257, 1964.

162. **Allen, G., Booth, C., and Price, C.**, The dilute solution properties of poly(propylene oxide), *Polymer*, 7, 167, 1966.
163. **Allen, G., Booth, C., and Price, C.**, The effect of the side group on the properties of the poly(epoxides). III. Unperturbed dimension of poly(propylene oxide), *Polymer*, 8, 397, 1967.
164. **Booth, C. and Devoy, C. J.**, Thermodynamics of mixtures of poly(propylene oxide) and benzene, *Polymer*, 12, 320, 1971.
165. **Scholtan, W. and Lie, S. Y.**, Molekulargewichtsabhängigkeit der Vikositätszahl, des spezifischen Volumens und des zweiten Virialkoeffizienten bei niedermolekularen Polypropylenglykolen, *Makromol. Chem.*, 108, 104, 1967.
166. **Conway, B. E. and Tong, S. C.**, Studies on polyoxypropylene glycols. III. Equilibrium swelling of cross-linked polyurethane elastomers, *J. Polym. Sci.*, 46, 113, 1960.
167. **Connor, T. M., Blears, D. J., and Allen, G.**, Proton spin-lattice relaxation in polypropylene oxides, *Trans. Faraday Soc.*, 61, 1097, 1965.
168. **Kershaw, R. W. and Malcolm, G. N.**, Thermodynamics of solutions of polypropylene oxide in chloroform and in carbon tetrachloride, *Trans. Faraday Soc.*, 64, 323, 1968.
169. **Lakhanpal, M. L. and Conway, B. E.**, Studies on polyoxypropylene glycols. I. Vapor pressures and heats of mixing in the systems: polyglycols-methanol, *J. Polym. Sci.*, 46, 75, 1960.
170. **Sandell, L. S. and Goring, D. A. I.**, A comparison of the intrinsic viscosities of oligomeric propylene glycols with the behavior predicted for models in aqueous solution at 25°, *Macromolecules*, 3, 50, 1970.
171. **Williams, G.**, Dipole relaxation of polypropylene oxide as a function of frequency, temperature and pressure, *Trans. Faraday Soc.*, 61, 1564, 1965.
172. **Sandell, L. S. and Goring, D. A. I.**, Solvent-induced conformational expansion of oligomeric propylene glycols, *Macromolecules*, 3, 54, 1970.
173. **Sandell, L. S. and Goring, D. A. I.**, The effect of temperature on the conformation of oligomeric propylene glycols in aqueous solution, *Makromol. Chem.*, 138, 77, 1970.
174. **Meyerhoff, G.**, Polypropyleneglykol in Lösung, Zahlenmittelmolekulargewicht und Reibungsverhalten, *Makromol. Chem.*, 107, 124, 1967.
175. **Huggins, M. L.**, Thermodynamic properties of liquids, including solutions. IX. Thermodynamic properties of polymer solutions, *Polym. J.*, 4, 502, 1973.
176. **Conway, B. E.**, Studies on polyoxypropylene glycols. IV. Elastic properties of derived elastomers, *J. Polym. Sci.*, 46, 129, 1960.
177. **Conway, B. E. and Lakhanpal, M. L.**, Studies on polyoxypropylene glycols. II. Derived thermodynamic functions for the systems: polyglycols-methanol, *J. Polym. Sci.*, 46, 93, 1960.
178. **Meyerhoff, G. and Moritz, U.**, Lichtstreuungsmessungen an Oligomeren; Lösungsverhalten von Polypropylenglykol, *Makromol. Chem.*, 109, 143, 1967.
179. **Meyerhoff, G.**, Reibungsverhalten und Trägheitsradien von oligomeren Polypropylenglykolen, *Makromol. Chem.*, 145, 189, 1971.
180. **Meyerhoff, G., Moritz, U., Kirste, R. G., and Heitz, W.**, Lösungsverhalten von Polypropylenglykol. Untersuchungen mittels Röntgenkleinwinkelstreuung und Gelchromatographie, *Eur. Polym. J.*, 7, 933, 1971.
181. **Booth, C., Devoy, C. J., and Gee, G.**, Heat of fusion of isotactic poly(propylene oxide). Appendix: heat of fusion of poly(ethylene oxide), *Polymer*, 12, 327, 1971.
182. **Saito, S. and Otsuka, T.**, Dissolution of some polymers in aqueous solutions of urea, of its related compounds, and of tetraalklammonium salts, *J. Colloid Interface Sci.*, 25, 531, 1967.
183. **Takahashi, Y. and Mark, J. E.**, Random-coil dimensions of poly(trimethylene oxide), an unusually compact chain molecule, *J. Am. Chem. Soc.*, 98, 3756, 1976.
184. **Dale, J.**, The conformational consequences of replacing methylene groups by ether oxygen, *Tetrahedron*, 30, 1683, 1974.
185. **Bevington, J. C.**, Polymers from aldehydes and other carbonyl compounds, in *Polyethers. Part I. Polyalkylene Oxides and Other Polyethers*, Gaylord, N. G., Ed., Wiley-Interscience, New York, 1963, chap. 2.
186. **Kokle, V. and Billmeyer, F. W.**, Unperturbed dimensions of polyoxymethylene, *J. Polym. Sci. Part B (Polym. Lett.)*, 3, 47, 1965.
187. **Stockmayer, W. H. and Chan, L.-L.**, Solution properties of polyoxymethylene, *J. Polym. Sci. Part A-2*, 4, 437, 1966.
188. **Williams, J. L., Hopfenberg, H. B., and Stannett, V.**, Water transport and clustering in poly(vinyl chloride), poly(oxymethylene), and other polymers, *J. Macromol. Sci.-Phys.*, B3, 711, 1969.
189. **Delzenne, G. and Smets, G.**, Stabilité et dégradation du polyacétaldéhyde. I. Dégradation homolytique du polyacétaldéhyde, *Makromol. Chem.*, 18/19, 82, 1956.
190. **Makino, D., Kobayashi, M., and Tadokoro, H.**, Structural studies of polyethers, $[-(CH_2)_m-O-]_n$. VII. Infrared spectra of polyoxacyclobutane, *J. Chem. Phys.*, 51, 3901, 1969.

191. **Yoshida, S., Sakiyama, M., and Seki, S.,** Calorimetric studies of solid polyoxacyclobutane (POCB), *Rep. Prog. Polym. Phys. Jpn.*, 12, 247, 1969.
192. **Yamamoto, K., Teramoto, A., and Fujita, H.,** Limiting viscosity number versus molecular weight relations for polyoxacyclobutane, *Polymer,* 7, 267, 1966.
193. **Tadokoro, H., Takahashi, Y., Chatani, Y., and Kakida, H.,** Structural studies of polyethers, $[-(CH_2)_m-O-]_n$. V. Polyoxacyclobutane, *Makromol. Chem.*, 109, 96, 1967.
194. **Kakida, H., Makino, D., Chatani, Y., Kobayashi, M., and Tadokoro, H.,** Structural studies of polyethers. $[-(CH_2)_m-O-]_n$. VIII. Polyoxacyclobutane hydrate (modification I), *Macromolecules,* 3, 569, 1970.
195. **Yoshida, S., Sakiyama, M., and Seki, S.,** Isotherms and dissociation pressures of crystalline hydrate of polyoxacyclobutane (POCB), *Rep. Prog. Polym. Phys. Jpn.*, 12, 245, 1969.
196. **Evans, J. M. and Huglin, M. B.,** Unperturbed dimensions of polytetrahydrofuran, *Makromol. Chem.*, 127, 141, 1969.
197. **Kobayashi, S., Tadokoro, H., and Chatani, Y.,** Structural studies on polyethers, $[-(CH_2)_m-O-]_n$. VI. The higher members with m = 6 − 10, 12, *Makromol. Chem.*, 112, 225, 1968.
198. **Yamamoto, K. and Fujita, H.,** Limiting viscosity number versus molecular weight relations for polyhexamethylene oxide, *Polymer,* 7, 557, 1966.
199. **Yamamoto, K. and Fujita, H.,** Limiting viscosity number versus molecular weight relations for polydecamethylene oxide, *Polymer,* 8, 517, 1967.
200. **Prasad, K. N., Luong, T. T., Florence, A. T., Paris, J., Vaution, C., Seiller, M., and Puisieux, F.,** Surface activity and association of ABA polyoxyethylene-polyoxypropylene block copolymers in aqueous solution, *J. Colloid Interface Sci.*, 69, 225, 1979.
201. **Anon.,** Gantrez M Polymer: Poly(Vinyl Methyl Ether), Tech. Bull., GAF Corp., New York, 1970.
202. **Elias, H.-G.,** *Macromolecules,* Vol. 2, John Wiley & Sons, Plenum Press, New York, 1977.
203. **Manson, J. A. and Arquette, G. J.,** Molecular weight relations and chain extensions of some vinyl alkyl ethers, *Makromol. Chem.*, 37, 187, 1960.
204. **Dobry, A.,** Sur l'incompatabilité des macromolécules en solution aqueuse, *Bull. Soc. Chim. Belg.*, 57, 280, 1948.
205. **Horne, R. A., Almeida, J. P., Day, A. F., and Yu, N.-T.,** Macromolecule hydration and the effect of solutes on the cloud point of aqueous solutions of polyvinyl methyl ether: a possible model for protein denaturation and temperature control in homeothermic animals, *J. Colloid Interface Sci.*, 35, 77, 1971.
206. **Davis, L. E.,** Polyethylenimine, in *Water-Soluble Resins,* 2nd ed., Davidson, R. L. and Sittig, M., Eds., Van Nostrand Reinhold, New York, 1968, chap. 11.
207. **Kindler, W. A. and Swanson, J. W.,** Adsorption kinetics in the polyethylenimine-cellulose fiber system, *J. Polym. Sci. Part A-2,* 9, 853, 1971.
208. **Hostetler, R. E. and Swanson, J. W.,** Diffusion into and adsorption of polyethylenimine on porous silica gel, *J. Polym. Sci. Polym. Chem. Ed.*, 12, 29, 1974.
209. **Van den Berg, J. W. A.,** De Interactie tussen Polyethyleenimine en Natriumdodecylsulfaat in Water, Doctoral thesis, University of Leiden, Netherlands, 1971.
210. **Van den Berg, J. W. A., Bloys van Treslong, C. J., and Polderman, A.,** Polyethyleneimine. I. Fractionation; Mark-Houwink relation, *Recl. Trav. Chim. Pays-Bas,* 92, 3, 1973.
211. **Baumgartner, E., Atkinson, G., and Emara, M.,** Ultrasonic absorption in aqueous polyelectrolyte solutions. II. Polyethylenimine, *J. Am. Chem. Soc.*, 95, 5881, 1973.
212. **Klotz, I. M., Royer, G. P., and Sloniewsky, A. R.,** Macromolecule-small molecule interactions. Strong binding and cooperativity in a model synthetic polymer, *Biochemistry,* 8, 4752, 1969.
213. **Royer, G. P. and Klotz, I. M.,** Enhanced rates due to apolar interactions between polymer and substrate, *J. Am. Chem. Soc.*, 91, 5885, 1969.
214. **Klotz, I. M., Royer, G. P., and Scarpa, I. S.,** Synthetic derivatives of polyethylenimine with enzyme-like catalytic activity (synzymes), *Proc. Natl. Acad. Sci. U.S.A.*, 68, 263, 1971.
215. **Kiefer, H. C., Congdon, W. I., Scarpa, I. S., and Klotz, I. M.,** Catalytic accelerations of 10^{12}-fold by an enzyme-like synthetic polymer, *Proc. Natl. Acad. Sci. U.S.A.*, 69, 2155, 1972.
216. **Hierl, M. A., Gamson, E. P., and Klotz, I. M.,** Nucleophilic catalysis by polyethylenimines with covalently attached 4-dialkylaminopyridine, *J. Am. Chem. Soc.*, 101, 6020, 1979.
217. **Johnson, R. S. and Klotz, I. M.,** Accelerated deacylation of acyl salicylates and neighboring group effects in derivatives of poly(ethylenimine), *Biopolymers,* 18, 313, 1979.
218. **Dick, C. R. and Ham, G. E.,** Characterization of polyethylenimine, *J. Macromol. Sci.-Chem.*, A4, 1301, 1970.
219. **Davidson, R. L., Ed.,** *Handbook of Water-Soluble Gums and Resins,* McGraw-Hill, New York, 1980.
220. **Powell, G. M.,** Polyethylene glycol, in *Handbook of Water-Soluble Gums and Resins,* Davidson, R. L., McGraw-Hill, New York, 1980, chap. 18.
221. **Braun, D. B.,** Poly(ethylene oxide), in *Handbook of Water-Soluble Gums and Resins,* Davidson, R. L., McGraw-Hill, New York, 1980, chap. 19.

222. **Saeki, S., Kuwahara, N., Nakata, M., and Kaneko, M.**, Upper and lower critical solution temperatures in poly(ethylene glycol) solutions, *Polymer (London)*, 17, 685, 1976.
223. **Maxfield, J. and Shepherd, I. W.**, Conformation of poly(ethylene oxide) in the solid state, melt and solution measured by Raman scattering, *Polymer (London)*, 16, 505, 1975.
224. **Lundberg, R. D., Bailey, F. E., and Callard, R. W.**, Interaction of inorganic salts with poly(ethylene oxide), *J. Polym Sci. Part A-1*, 4, 1563, 1966.
225. **Yokoyama, M., Ishihara, H., Iwamoto, R., and Tadokoro, H.**, Structure of poly(ethylene oxide) complexes. III. Poly(ethylene oxide)-mercuric chloride complex. Type II, *Macromolecules*, 2, 184, 1969.
226. **Yanagida, S., Takahashi, K., and Okahara, M.**, Metal-ion complexation of noncyclic poly(oxyethylene) derivatives. I. Solvent extraction of alkali and alkaline earth metal thiocyanates and iodides, *Bull. Chem. Soc. Jpn.*, 50, 1386, 1977.
227. **Yanagida, S., Takahashi, K., and Okahara, M.**, Metal-ion complexation of noncyclic poly(oxyethylene) derivatives. II. PMR studies of the complexation with alkali and alkaline-earth metal cations, *Bull. Chem. Soc. Jpn.*, 51, 1294, 1978.
228. **Yanagida, S., Takahashi, K., and Okahara, M.**, Metal-ion complexation of noncyclic poly(oxyethylene) derivatives. III. Complexation in aprotic solvent and isolation of their solid complexes, *Bull. Chem. Soc. Jpn.*, 51, 3111, 1978.
229. **Kjellander, R. and Florin, E.**, Water structure and changes in thermal stability of the system poly(ethylene oxide)-water, *J. Chem. Soc. Faraday Trans.*, 1, 77, 2053, 1981.
230. **Merrill, E. W., Smith, K. A., Shin, H., and Mickley, H. S.**, Study of turbulent flows of dilute polymer solutions in a Couette viscometer, *Trans. Soc. Rheol.*, 10, 335, 1966.
231. **Virk, P. S., Merrill, E. W., Mickley, H. S., Smith, K. A., and Mollo-Christensen, E. L.**, The Toms phenomenon: turbulent pipe flow of dilute polymer solutions, *J. Fluid Mech.*, 30, 305, 1967.
232. **Paterson, R. W. and Abernathy, F. H.**, Turbulent flow drag reduction and degradation with dilute polymer solutions, *J. Fluid Mech.*, 43, 689, 1970.
233. **Maconnachie, A., Vasudevan, P., and Allen, G.**, Molecular dynamics of poly(ethylene oxide) in concentrated solution, *Polymer (London)*, 19, 33, 1978.
234. **Malcolm, G. N., Baird, C. E., Bruce, G. R., Cheyne, K. G., Kershaw, R. W., and Pratt, M. C.**, Thermodynamics of polyether solutions, *J. Polym. Sci. Part A-2*, 7, 1495, 1969.
235. **Hoa, Se. Î. and Aelenei, N.**, Thermodynamic and hydrodynamic behaviour of benzenic solutions of poly(ethylene oxide) and poly(propylene oxide), *Bul. Inst. Politehn. Iasi Sect. 2*, 23, 41, 1977.
236. **Sugamiya, K.**, Polar effect in free volume theory for polymer solution thermodynamics, *Makromol. Chem.*, 178, 565, 1977.
237. **Cheng, Y. L. and Bonner, D. C.**, Sorption of solutes by poly(ethylene oxide) using gas-liquid chromatography, *Macromolecules*, 7, 687, 1974.
238. **Klein, J. and Jeberien, H.-E.**, Chain length dependence of thermodynamic properties of poly(ethylene glycol), *Makromol. Chem.*, 181, 1237, 1980.
239. **Charlesby, A., Garratt, P. G., and Kopp, P. M.**, The use of sulphur as a protecting agent against ionizing radiations, *Int. J. Rad. Biol.*, 5, 439, 1962.
240. **Charlesby, A. and Kopp, P. M.**, Radiation protection in aqueous polymer solutions, *Int. J. Rad. Biol.*, 5, 521, 1962.
241. **Imamura, S.-I., Tonomura, Y., Kawabata, N., and Kitao, T.**, Wet oxidation of water-soluble polymers, *Bull. Chem. Soc. Jpn.*, 54, 1548, 1981.
242. **Attia, Y. A. and Rubio, J.**, Determination of very low concentrations of polyacrylamide and polyethyleneoxide flocculants by nephelometry, *Br. Polym. J.*, 7, 135, 1975.
243. **Schmolka, I. R.**, Polyalkylene oxide block copolymers, in *Nonionic Surfactants (Surfactant Science Series, Vol. 1)*, Schick, M. J., Ed., Marcel Dekker, New York, 1967, chap. 10.
244. **Schmolka, I. R. and Bacon, L. R.**, Viscosity characteristics of aqueous solutions of block copolymers of propylene and ethylene oxides, *J. Am. Oil Chem. Soc.*, 44, 559, 1967.
245. **Mankowich, A. M.**, Micellar molecular weights of selected surface active agents, *J. Phys. Chem.*, 58, 1027, 1954.
246. **Dwiggins, C. W., Bolen, R. J., and Dunning, H. N.**, Ultracentrifugal determination of the micellar character of non-ionic detergent solutions, *J. Phys. Chem.*, 64, 1175, 1960.
247. **Su, C. S. and Patterson, D.**, Determination by gas-liquid chromatography of the polystyrene-poly(vinyl methyl ether) interaction, *Macromolecules*, 10, 708, 1977.
248. **Druzin, M. I., Vakova, I. N., Schenniloa, N. I., Koroleva, L. I., Karapetyan, L. P., Valkova, A. K., Zaitseva, I. V., and Pavlova, I. D.**, Modfications of polyethylene imine, *J. Polym. Sci. Polym. Symp.*, 47, 369, 1974.

Chapter 3

NONIONIC POLYMERS — THE ACRYLIC GROUP

"Though I sang in my chains like the sea"
Dylan Thomas: *Fern Hill*

I. INTRODUCTION

The water-soluble polymers of the acrylic group that are dealt with in this chapter comprise poly(acrylic acid), the parent member of the group (Section II); its homolog, poly(methacrylic acid) (Section IV); their amides, PAAm (Section III) and PMAAm (Section V); certain N-substituted derivatives of these latter (Section VI); and various interpolymers, and copolymers with monomers outside the group (Section VII).

Since the polymers within this group have chain units simple in structure, they provide ideal models for the behavior of more complex materials such as the biopolymers. Thus, PAA and PMAA being acids, their behavior in partly and fully ionized form can be used as the prototype for polyelectrolyte behavior, as discussed in Chapter I of Volume II; in the present chapter, however, attention will be confined to the nonionized forms.

These two polyacids along with their amides form a closed group which for the present purposes may be conveniently termed the "acrylic quartet"; comparative studies on the four polymers then reveal in two distinct ways the effects of backbone methylation (PAA → PMAA; PAAm → PMAAm) and of side-group amidation (PAA → PAAm; PMAA → PMAAm) on polymer properties. This is best exemplified by the light scattering and viscosity studies carried out by Silberberg and co-workers,[1] and by the partial specific volume and adiabatic compressibility measurements of Roy-Chowdhury and Kale,[2-4] on these four polymers in aqueous solution. As previously pointed out in a discussion of the first of these,[5] concurrent or simultaneous studies of the whole quartet by one group of workers using the same instruments and techniques have the greatest likelihood of showing up most clearly any structural correlations.

II. POLY(ACRYLIC ACID) — PAA

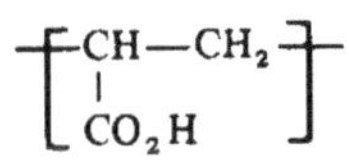

$M_0(C_3H_4O_2) = 72.06$

Other chemical names, abbreviations, and trade names are* polyacrylic acid; poly(1-carboxyethylene) (IUPAC); HPA (i.e., "hydrogen polyacrylate"); *Acrysol* (Rohm & Haas); Carbomer (B.P., U.S.N.F.); *Carbopol* (Goodrich); *Primal* (Rohm & Haas); *Sopronyl.*

A. General Features

Poly(acrylic acid) plays a key role in any discussion of the properties and behavior of water-soluble synthetic polymers because it is the simplest common synthetic polycarboxylic acid. Studies on PAA in its nonionized, partly ionized, and fully ionized forms in solution therefore provide direct information on the interactions amongst carboxylic acid groups and carboxylate ions in the solvent, and between these and molecules of the solvent, counterions

* For a more detailed list see Reference 7.

and any other solutes present. As well as serving as a model for polyelectrolytes in general, studies on PAA help more specifically in the interpretation of the behavior of polymers containing carboxylic acid groups, particularly the acidic polysaccharides and the proteins. From this last aspect, e.g., PAA was proposed many years ago by Kern[8,9] as a simple model for albumin.

In the present chapter, attention is focused upon the behavior of PAA in its nonionized form, while that in the partly and fully ionized states is considered in Chapter 1 of Volume II.

Poly(acrylic acid) is also commercially important, being produced in a variety of grades and forms, as well as its monomer being incorporated into copolymers.

The general features and applications of PAA have been reviewed.[7,10,11] Data on the commercial grades are available from the manufacturer's technical literature (e.g., References 12 and 13).

Although nonionized PAA is commonly considered to be readily water soluble, this really only applies to the atactic form, since the stereoregular forms are much less readily soluble (see Section II.F.2). The atactic polymer is also soluble in methanol, ethanol, dioxane, ethylene glycol, 2-methoxyethanol, and acetic acid, but it is insoluble in chloroform, diethyl ether, alkanes, aromatic hydrocarbons, and other nonpolar solvents.

B. Production and Purification

Poly(acrylic acid) may be prepared in the atactic form by polymerization of the monomer with conventional free radical initiators using the monomer either undiluted[14] or in solution in water,[1,14-16] butanone,[15,17] or dioxane.[4,15] With the undiluted monomer or in butanone solution the polymer precipitates out during the reaction; aqueous solutions can lead at certain compositions to the formation of "popcorn" (proliferous cross-linked) polymer.[14]

Polymerization of the pure monomer in the liquid or the crystalline state initiated by irradiation with gamma rays leads in general to mixtures of the atactic and syndiotactic forms together with some cross-linked material.[18,19] The rigorously dried monomer vapor polymerizes spontaneously at liquid nitrogen temperatures on glass surfaces giving the syndiotactic form.[18,19]

Photochemical polymerization of the monomer in ethanol solution at −78°C using UV light and sensitized by benzoin leads to the syndiotactic form.[20]

Eastmond and co-workers have attempted unsuccessfully to prepare stereoregular forms of PAA by irradiating the crystalline monomer with polarized UV light.[21]

The main alternative route to the production of PAA is the hydrolysis of poly(alkyl acrylates). This method was pioneered by Katchalsky and Eisenberg[22] using acid hydrolysis of poly(methyl acrylate) (PMA); the absence of degradation in the hydrolysis reaction was shown by the fact that the PMA obtained after remethylation of the PAA had the same intrinsic viscosity and infrared spectrum as the original material. Similar techniques have proved particularly useful for obtaining stereoregular forms of PAA; thus Miller and co-workers[23,24] have prepared isotactic PAA by acid hydrolysis of the corresponding poly(*t*-butyl acrylate), while Monjol[25] has prepared the syndiotactic form by alkaline hydrolysis of the corresponding isopropyl ester. The progress of the alkaline hydrolysis of PMA has been studied by Bevington and co-workers[26] using radioactive tracers; generally, however, use of alkali rather than acid for this purpose requires a longer time and can lead to oxidative degradation of the polymer.[22]

The alkaline hydrolysis of polyacryonitrile also seems to lead ultimately to poly(acrylic acid).[26] The alkaline hydrolysis of polyacrylamide does not lead to complete conversion, although that with acid apparently does (see Section III.G).

Poly(acrylic acid) is also directly available commercially under a variety of trade names from the manufacturers, as listed above, or from general chemical suppliers. These materials are presumably linear and essentially atactic. A branched form, soluble in alkali, is produced by B. F. Goodrich under the trade name *Carbopol*;[13] it is used pharmaceutically under the name Carbomer.[27-29]

The atactic polymer may be purified by reprecipitation from aqueous solution by sulfuric acid,[16] or by using a variety of solvent/nonsolvent systems such as methanol/chloroform,[22] methanol/diethyl ether,[4,22] acetic acid/chloroform,[22] or dioxane/benzene.[15,30] The stereoregular polymers may be purified by exhaustive dialysis and isolated by slow precipitation (crystallization) from concentrated aqueous hydrochloric acid at 0°C.[20,25] To obtain the polymer in a conveniently handleable form, it may be freeze dried from aqueous[17,24] or dioxane solution,[15] followed by vacuum drying at 50 to 60°C.[15,30] Even after such treatment the product may contain up to 10% residual solvent.[1,15,17,25] In the case of isotactic PAA the presence of this is beneficial since it makes it easier for the polymer to be dissolved subsequently;[24] the final traces of moisture may be removed if need be by drying at 150°C.[24]

C. Fractionation and Characterization

The atactic and stereoregular (isotactic or syndiotactic) forms of PAA may be separated from one another by extraction with dioxane, in which the former are soluble but the latter are not.[19,24] Separation of the stereoregular form in the residue from cross-linked polymer or other contaminants may be effected by extraction with dioxane containing 20% water, in which both the stereoregular forms are soluble.[19,24]

The atactic form has been successfully fractionated according to molecular weight by gradual precipitation using a variety of solvent/nonsolvent systems, including water/sulfuric acid;[16] water/sodium chloride;[31] aqueous sodium hydroxide/methanol;[32] methanol/diethyl ether;[33] and dioxane/heptane.[15] (The apparent contradiction between the roles played by methanol in the third and fourth of these systems is resolved by the sodium salt actually being the form precipitated out in the case of the third system). Alternatively, the polymer may be fractionated by raising the temperature of its solution in anhydrous dioxane (in which it shows inverse solubility-temperature behavior);[34] this method, however, is not successful when dioxane containing 2% water is used.[34] Gel permeation chromatography has also been used successfully to fractionate 10-g amounts of atactic PAA.[17]

The polymer may be characterized in terms of its stereoregularity, i.e., broadly speaking, whether atactic (*at*), isotactic (*it*), or syndiotactic (*st*) by a variety of means — solubility in anhydrous dioxane (*at* soluble; *it* and *st* insoluble),[19,24] crystallizability (*at* noncrystallizable; *it* and *st* crystallizable), IR spectra and X-ray diffraction patterns (see Section II.D); NMR spectra of the derived methyl esters enable quantitative estimates of the *degree* of stereoregularity to be obtained, which show that (as is commonly the case with polymers) the so-called "isotactic" and "syndiotactic" forms are in fact only relatively *rich* in the named stereoform.[35]

D. Bulk Polymer

The glass transition temperature, t_g, of PAA has been reported variously as 75°C (from stiffness measurements);[36] 106°C (from shear modulus measurements);[37] and 126°C (from DTA measurements).[38] Purely on the basis of the values obtained for PAAm (165 or 188°C — see Section III.D) and for PMAA (185°C — see Section IV.D), it is likely that the highest of these three values for PAA is correct, with the others possibly being low due to retained water. From the practical viewpoint, this would imply that in drying PAA it is necessary to heat the polymer above this temperature to put it into the rubbery state where the absorbed water molecules would be sufficiently mobile.

The infrared absorption spectrum of the bulk polymer has been given for the region 800—1500 cm^{-1} by Bardet and co-workers,[39] for 1000—1800 cm^{-1} by Katchalsky and Eisenberg,[22] and for 1500—4000 cm^{-1} by Boyer-Kawenoki.[40] Miller and co-workers[24] stated that there is no difference between the IR spectra of films of crystallized and uncrystallized isotactic PAA, or between those of the amorphous isotactic and atactic polymers; they also used IR studies to show the absence of anhydride groups even after heating at 150°C.

Monjol[35] has stated that the two stereoregular forms of PAA may be differentiated by means of their IR spectra, in that the isotactic form has a band at 935 cm^{-1} and two bands of equal intensity at 1215 and 1270 $cm,^{-1}$ whereas the syndiotactic form lacks the first band and has only one band at 1240 cm^{-1} in the higher frequency region.

On the basis of this, Bardet and co-workers[39] have identified as syndiotactic the three samples, furnished by industry (but with no further indication of provenance), which they used for their IR studies.

The crystallization process and crystalline forms of stereoregular PAA have been studied by Miller and co-workers,[23,24] and Chapiro and Sommerlatte.[19] The X-ray diffraction patterns have been published for both the isotactic form[24,25] and for the syndiotactic form;[19,25] these patterns differ sufficiently markedly for the two forms to be readily distinguishable. The structure of crystalline isotactic PAA has been reviewed by Chapiro and Sommerlatte.[18]

E. Solution Behavior

It is convenient to discuss the behavior of (nonionized) PAA in aqueous solution separately from that in nonaqueous solution; a full understanding and characterization of its solution behavior naturally requires the correlation of these two areas. Clarification of this behavior may also be obtained by using small-molecule model compounds; for this purpose with PAA, Miyamoto and Cantow[41] have used NMR to study the conformational behavior of nonionized meso and racemic forms of 2,4-dimethylglutaric acid:

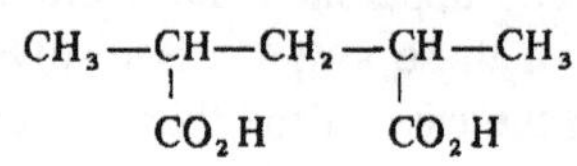

in D_2O, perdeuteroethanol, their mixtures, and perdeuterodimethylsulfoxide.

1. Aqueous Solution

Table 1 summarizes the results obtained in various studies on nonionized PAA in aqueous solution. If the properties of the pure polyacid are to be studied, then its ionization must be suppressed by reducing the pH, which can be done most simply by adding HCl or other suitable strong acid; in a surprising number of cases the necessity for this precaution has been ignored, so that the results obtained are correspondingly vitiated. From the results of the viscosity studies of Braud and co-workers[51] it seems that for quite dilute solutions even 1 mM HCl (i.e., pH = 3) is adequate for this purpose, since they showed that the value of η_c for a 0.01% solution of PAA was reduced by a factor of three on adding this amount of the acid, but that there was little further reduction when the acid concentration was raised to 5 mM. With more concentrated solutions of PAA there is some self suppression of the ionization, although even for a 3% solution the polyacid is still 16% ionized, and even at 10% concentration it is 6% ionized.[53]

The *partial specific volume* of PAA in aqueous solution has been studied, as Table 1 shows, by five groups of workers: Ise and Okubo (entry 12);[46] Friedman and co-workers (entry 14);[48] Roy-Chowdhury and Kale (entry 15);[4] Tondre and Zana (entry 18);[50] and Inzelt and Gróf (entry 25).[53] Unfortunately, in none of these studies was any mineral acid added to suppress the ionization of the polyacid (nor any mention made of the necessity for this), although in many cases the concentrations used were sufficiently high to largely suppress the ionization; the very use of such high concentrations makes the extrapolation to obtain $\bar{v}_2^{\circ}$ uncertain. These data therefore need to be reexamined and reevaluated before any value for $\bar{v}_2^{\circ}$ can be recommended.

The *specific index refractive increment* for PAA in aqueous solution has been determined by three groups of workers, leading to the values for $\tilde{n}$ (in $cm^3\ g^{-1}$) of 0.146,[1] 0.140,[16] and 0.152[49] for wavelengths of 546, 589, and 436 nm, respectively. In these studies only the first[1] involved the addition of acid (0.2 M HCl) to suppress the ionization of the polyacid, so that it is evidently coincidental that the value of $\tilde{n}$ obtained, which is thus the preferred one, also represents the mean of the three values; it was also shown in these same studies that the value of $\tilde{n}$ was independent of temperature over the range 20 to 62°C.[1]

Little work seems to have been done on the *intrinsic viscosity-molecular weight* relationship for nonionized PAA in aqueous solution. Silberberg and co-workers[1] in their extensive studies on the acrylic quartet obtained the value $[\eta] = 111\ cm^3\ g^{-1}$ for a sample with $\overline{M}_v = 4.3D5$ and $\overline{M}_w = 1.1D6$ in the Θ system 0.2 M HCl at 14°C, giving for this system the MHS parameters: $a = 0.50$; $K = 0.169\ cm^3\ g^{-1}$ (from $\overline{M}_v$) or $0.106\ cm^3\ g^{-1}$ (from $\overline{M}_w$).

In practice, where molecular weights are required for samples of PAA in aqueous solution form it is more customary to use the intrinsic viscosity of the ionized form in a system (e.g., salt solution) for which the MHS parameters are available, as discussed in Chapter 1 of Volume II. Alternatively, anhydrous dioxane is often used as the solvent for these determinations (see Section II.E.2).

2. *Nonaqueous and Mixed Solvents*

The results of various studies on PAA in nonaqueous and mixed solvent systems are summarized in Table 2.

The solution behavior of PAA in dioxane was studied by Newmann and co-workers,[15] who showed by phase separation temperature measurements (see Section II.F.3) that it has a Θ_+ value of 30°C; the MHS parameters for this temperature from their work are $a = 0.50$; $K = 8.4\,(\pm 0.6)d2\ cm^3\ g^{-1}$ (this last value being recalculated from the original data).[15] This solvent has been widely used subsequently to obtain molecular weights of samples of PAA. The specific refractive index increment $\tilde{n}$ for PAA in this system is $0.089\ cm^3\ g^{-1}$.[1,15]

The use of dioxane as a solvent for the characterization of PAA has the disadvantage that the stereoregular forms are not soluble in it,[19,24] while even for atactic PAA the dissolution may be difficult because the solvent is such a poor one. For these purposes, dioxane + water mixtures have several advantages; in particular, isotactic PAA is soluble in such mixtures if there is more than about 4% of water present.[24] Miller and co-workers[24] have studied the viscosity behavior of atactic and isotactic samples of PAA in the mixture with 20% (v/v) water, obtaining for atactic samples the MHS parameters (for 30°C): $a = 0.60$; $K = 6.6(\pm 0.2)d2\ cm^3\ g^{-1}$; the higher value of a compared with that in dioxane confirms the better solvent power of this mixture. These parameters have also been used to obtain "apparent molecular weights" (i.e., estimated values) for stereoregular samples.[19,24]

F. Phase Equilibria

1. *Water Vapor Uptake*

The equilibrium uptake of water vapor by poly(acrylic acid) has been measured over a range of values of the water activity a_1 (equivalent to the fractional relative humidity) by Thompson Hughes and Fordyce[37] and Peterson;[56] their combined results are plotted in Figure 1, where the uptake U_1 is expressed on the customary basis of grams water/100 grams (dry) polymer, and plotted on a logarithmic scale for a better representation of the wide span of values. The uptake data show a "toe" (corresponding to about 5% uptake) at low activities, which may represent water which is tenaciously held by the polymer (see Section II.H). Thompson Hughes and Fordyce[37] have also studied the kinetics of the absorption and desorption processes.

2. *Phase Separation in Aqueous Solution*

Silberberg and co-workers[1] found that a sample of PAA (presumably, essentially atactic)

Table 1
STUDIES ON NONIONIZED PAA IN AQUEOUS SOLUTION

Entry	Year	Polymer samples: No.	Polymer samples: Range of $\overline{M}$	Techniques used	Temperature (t/°C)	Polymer concentration	Comments	Ref.
1	1955	?	?	VI		10%	No negative thixotropy	42
2	1957	1	1D6	LS	20—62	0.2—0.8%	+HCl (0.2M); 546 nm: $\rightarrow \overline{M}_w$; A_2; TP	1
3	1957	1	1D6	RI	20—62		+HCl (0.2M); 546 nm: $\rightarrow \bar{n} = 0.146$ cm^3 g^{-1}	1
4	1957	1	1D6	VI	20—80		+HCl (0.2M): $\rightarrow [\eta]$; $[\eta]_\Theta$; K_Θ	1
5	1959	2	5D4, 3D5	LS			+HCl (0.05—0.1M); 436 nm: $\rightarrow M_w$	16
6	1959	2	5D4,3D5	OP			+HCl (0.05—0.1M); concentration osmometer (vs. PEO): $\rightarrow M_n$	16
7	1959	2?	5D4, 3D5?	RI			+HCl?; 589 nm: $\rightarrow n = 0.140$ cm^3 g^{-1}	16
8	1965	1	~1.5D5	UV		0.04—0.9%	pH 3.6; 180—240 nm: $\rightarrow A$, ϵ	43
9	1966	1	~1.5D5	VI	27—45	0.4%	i = ? (No additive): $\rightarrow \eta_c$	43a
10	1966		Review of VI behavior in relation to entanglement					44
11	1967	1	1.4D5	UV	20	0.01—0.06%	+$HClO_4$ (20 mM); 200—220 nm: $\rightarrow A$; ϵ	45
12	1968	1	5D4	SV	15—35	0.4—4%	i = ? (No additive)[a]	46
13	1969	?	?	USA	5—50	0.01—36%	+HCl (0, 0.1M); 1—12 MHz	47
14	1970	3	3D5—5D5	SV	30	0.05—22%	i = ? (No additive)[a]	48
15	1970	1	2D4	SV	25	0.1—1.7%	i = ? (No additive)[a]	4
16	1970	1	2D4	USV	25	0.1—1.7%	i = ? (No additive)	4
17	1971	6?	2D4—2D6?	RI			i = ? (No additive); 436 nm: $\rightarrow \bar{n} = 0.152$ cm^3 g^{-1}	49
18	1972	1	1D5	SV	25	0.4—1%	i = ? (No additive)[a]	50
19	1974	1	3D4	VI	25	0.01%	+HCl (0—7 mM): $\rightarrow \eta_c$	51
20	1974	4	4D4—2D5	VI	30	0.15%	Dynamic viscocity (2—500 kHz)	30
21	1975	1[b]	~2D6[b]	LS		0.02—0.2%	"i = 0" (pH 3.8); 546 nm; studies on gelation	52
22	1975	1[b]	~2D6[b]	VI	20	0.1%	"i = 0" (pH 3.7); rheological studies	52
23	1975	3[c]	3D4—4D5[c]	IR		10%	i = ? (no additive); 800—1500 cm^{-1} (H_2O) and 1500—1800 cm^{-1} (D_2O)	39
24	1975	3[c]	3D4—4D5[c]	RA		10%	i = ? (no additive); λ = 633 nm (excitation), $\Delta\nu$ = 700—1800 cm^{-1}	39
25	1977	1	1.5D5	SV	25—35	3—10%	i = ?(No additive)[a,d]	53
26	1977	1	1.5D5	TD	25, 35	3—10%	i = ? (No additive)[d]: $\rightarrow D$ (H_2O)	53

27	1977	1	1.5D5	VI	25—35	3—10%	i = ? (No additive)[d]: → η	53
28	1978	1	1D5	TD	15, 20	0—5%	i = ? (No additive): → D (H_2O); D (glucose)	54
29	1979	2[e]	~1D6[c]	VI	10—60	1—5%	pH 2.4—3.4; continuous shear rheometry	28
30	1979	2[e]	~1D6[c]	VI	25	3—5%	pH 2.4—3.4; oscillating shear rheometry (8 mHz-25 Hz)	29

[a] See text for discussion of SV values.
[b] *Carbopol* 941.
[c] Syndiotactic polymer.
[d] $i = 0.016$ (c_2 = 3%)—0.006 (c_2 = 10%)
[e] *Carbopol* 940 and 941.

Table 2
STUDIES ON PAA IN NONAQUEOUS AND MIXED SOLVENTS

Entry	Year	Polymersamples No.	Polymersamples Range of $\overline{M}$	Techniques used	Temperature (t/°C)	Polymer concentration	Comments	Ref.
					$H_2O + C_4H_8O_2$: Water + Dioxane			
1	1962	4[a]	4D3—3D4[a]	VI	30	0.05—0.25%	$\phi(H_2O) = 0.2$: → $[\eta]$; k'; MHS (0.60; 6.6d2)	24
2	1962	3[b]	~2D3—3D4[b]	VI	30	0.05—0.25%	$\phi(H_2O) = 0.2$: → $[\eta]$; k'; $\overline{M}$ ("apparent")	24
3	1969	?[c]	1D6—9D6[c]	VI	30		$\phi(H_2O) = 0.2$: → $[\eta]$; $\overline{M}_v$	19
4	1969	?[d]	~5D5—8D6[d]	VI	30		$\phi(H_2O) = 0.2$: → $[\eta]$; $\overline{M}$ ("apparent")	19
					$C_4H_8O_2$: Dioxane			
5	1954	1	8D5	VI	30	0.2—0.5%	Θ system: → $[\eta]$; k'	55
6	1954	5	1D5—1D6	LS		0.04—0.7%	436 nm: → $\overline{M}_w$; A_2; $\hat{L}$	15
7	1954	5?	1D5—1D6?	RI	25		436 nm?: → $\tilde{n}$ = 0.089 $cm^3\ g^{-1}$	15
8	1954	5	1D5—1D6	VI	30		Θ system: → $[\eta]$; k'; MHS (0.50; 8.5d2)[e]	15
9	1962	4	4D4—3D5	VI	30		Θ system: → $[\eta]$; k'	24

Atactic samples.
Isotactic samples.
Atactic samples.
Syndiotactic samples.
K value as originally quoted: recalculation gives K = 8.4 (±0.7)d2 $cm^3\ g^{-1}$.

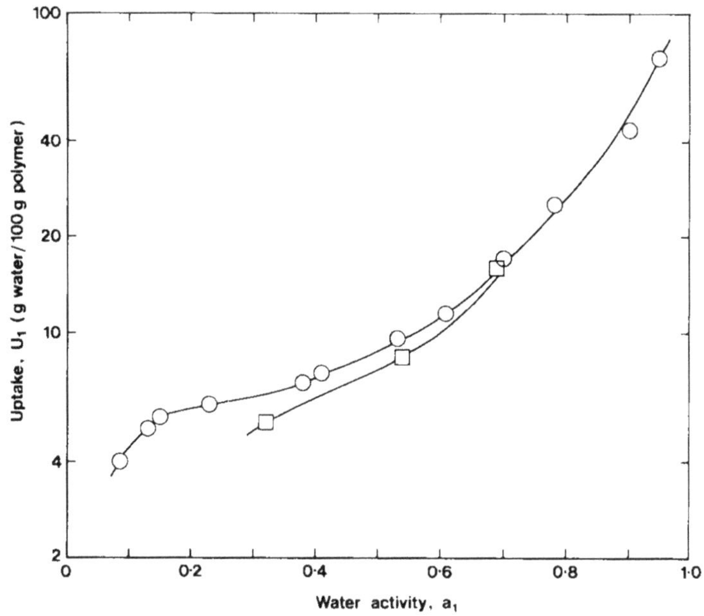

FIGURE 1. Equilibrium uptake of water, U_1 (logarithmic scale) by poly(acrylic acid), vs. water activity, a_1. ○: Peterson (26°C);[56] □: Thompson Hughes and Fordyce (30°C).[37]

with $\overline{M}_w$ = 1.1D6 precipitated out of aqueous 0.2 M HCl on cooling to around 5°C. This behavior is in accord with their light-scattering studies on the same system which gave the ideal temperature as Θ_- = 14°C.[1]

Miller and co-workers[24] found that isotactic PAA precipitates slowly at room temperature from aqueous solutions as dilute as 0.3%; the precipitate was initially amorphous (as shown by X-ray diffraction), but on standing it became crystalline.

3. Phase Separation in Nonaqueous and Mixed Solvents

Flory and co-workers[15,55] have studied the precipitation of PAA from solution in anhydrous dioxane that occurs on warming; this same behavior has been used by Lleras and co-workers[34] to fractionate the polymer (see Section II.C). (It is interesting that this is a system where one component, PAA, has both hydrogen-bond donating and accepting properties and the other, dioxane, has only accepting properties, and that it should show inverse solubility-temperature behavior in the same fashion as occurs in the case of polymers such as PEO, PPO, or PVME in water, where the converse pattern of hydrogen bonding applies.) The precipitation points from the two groups show some inconsistencies, which may be due to differences in the trace water content of the dioxane or in the tacticity of the polymer samples; the data of Flory and co-workers[15,55] give Θ_+ = 30 ± 1°C, which is consistent with their light scattering and viscosity results.[15]

It is evident that the samples used in the above studies must have been essentially atactic, since iso- and syndiotactic forms of PAA are insoluble in anhydrous dioxane.[19,24] The isotactic form dissolves upon the addition of water, e.g., a sample with M = 1D5 dissolved at room temperature with 4% (v/v) water added.[24] Miller and co-workers[24] have studied the relative solubilities of *at*-PAA and *it*-PAA in dioxane containing small amounts of water. For *at*-PAA, over the range 0 to 3% (v/v) water the t_p value rose linearly with water content by about 13° for each 1% water; for a sample of *at*-PAA in dioxane with 4.2% water the t_p value is about 80°C, which is about 20° lower than the (extrapolated) value for *at*-PAA in this mixture.

G. Chemical Behavior

Heating PAA in vacuum, in an effort for example to remove residual solvent (see Section II.B), may lead to cyclic anhydride formation by loss of the elements of water; the kinetics of this reaction have been discussed by Boucher[57] in the general context of intramolecular reactions in polymers. Miller and co-workers, however, found no sign of such anhydride formation after heating at 150°C.[24]

Poly(acrylic acid) may be converted to its methyl derivative, PMA, by treatment with diazomethane; this method of conversion was pioneered by Silberberg and co-workers[1] and has since been used routinely for this purpose, e.g., by Monjol.[35] The conversion PAA → PMA is an important reaction in a number of respects. First, it represents the reversal of the hydrolysis reaction and enables the complete cycle of hydrolysis and remethylation to be carried out, whereby the comparison of the properties of the starting and the final materials discloses any degradation in the reactions involved. Second, the alkyl ester may be more amenable to some types of investigation than PAA itself, e.g., its NMR spectrum is used to determine the degree of tacticity of the backbone of the chain.[35] Third, the esterification reaction when carried out to partial extents enables copolymers to be obtained (i.e., PAA-MA) which may have a different distribution of component monomer units from the copolymers obtained by the partial hydrolysis of PMA, or by copolymerization of AA and MA.

H. Analysis

The simplest method of analysis of relatively concentrated aqueous PAA solutions is the gravimetric estimation of the solids content after evaporation and vacuum or freeze-drying; in the case of PAA, however, as with many other hydrophilic polymers, this leads almost invariably to the retention of an appreciable amount (~10%) of solvent.[1,15,17,56,58]

Use of high temperatures in the vacuum drying could possibly lead to some decomposition of the polymer, for example by decarboxylation[57,58] or anhydride formation,[57] although Miller and co-workers found no sign of the latter (from IR studies) even after heating at 150°C.[24]

It is therefore more customary to assay PAA solutions via their carboxyl group content by titration with alkali (e.g., carbonate-free 0.05 M NaOH);[1,17,30,56] sodium chloride (0.1 M) may be added to sharpen the end point,[15,17,56] which can be detected either with indicators such as phenolphthalein[17,56] or thymolphthalein,[1] or by potentiometry.[17,30,56]

Crisp and co-workers have shown the advantages of conductimetric over potentiometric methods for the detection of this end point, particularly with copolymers of acrylic acid.[58]

Methods which have been used for the assay of dilute PAA solutions included a turbidimetric method (applicable for 5 to 40 ppm)[59] and a colorimetric method (applicable for 1 to 6 ppm).[60]

III. POLYACRYLAMIDE — PAAm

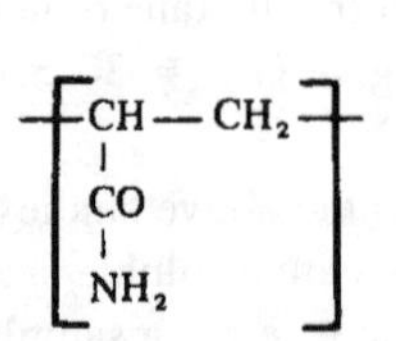

$M_0(C_3H_5NO) = 71.08$

Other chemical names, abbreviations, and trade names are poly(acrylamide); polyacrylic amide; poly(1-carbamoylethylene) (IUPAC); PAm; PAM; *Cyanamer* (American Cyanamid); *Purifloc* (Dow); *Separan* (Dow).

A. General Features

Polyacrylamide is distinguished by its highly hydrophilic character, which is greater than that of most other nonionic water-soluble polymers. This shows itself, e.g., by the ready absorption and tenacious retention of water by the bulk polymer, by the absence of phase separation in aqueous solution over the normal temperature range (which is in accord with the estimated value of the ideal temperature of this system $\Theta_{-} = -38°C$ — see Section III.E.1), and by its relatively high resistance to precipitation from aqueous solution by electrolytes (see Chapter 2 of Volume II). A further reflection of this character is that its solubility in nonaqueous liquids is limited to hydrophilic ones such as ethylene glycol, glycerol, and formamide, while it is insoluble in most other common organic liquids such as methanol, ethanol, acetone, diethyl ether, and aliphatic and aromatic hydrocarbons.

Paradoxically, however, despite this highly hydrophilic character, the main technical application of PAAm is for the flocculation of aqueous suspensions and allied purposes, where it is therefore required to have a relatively high affinity to the surface of the suspended solid if the necessary strong adsorption of the polymer is to take place (see Chapter 4 of Volume II); in practice, however, this adsorption is often enhanced by using modified (anionic or cationic) forms of PAAm.

The general properties and applications of PAAm have been discussed by Montgomery.[61] Technical and other data on commercial grades are available from the manufacturer's information bulletins (e.g., see References 62 to 64).

B. Production and Purification

Polyacrylamide may be produced from its monomer by free radical polymerization using a wide variety of initiation methods.[65,66]

The purely thermal polymerization at 140°C has been studied,[66] as well as the polymerization in aqueous solution initiated by free radical sources such as hydrogen peroxide,[65,66] persulfate,[67,68] and an azo compound (where the reaction was followed to high conversions).[69] The azo-initiated polymerization has been studied in benzene and diglyme (diethylene glycol dimethyl ether), in the latter case at temperatures of 60, 100, and 140°C;[66] similarly initiated polymerization in ethanol leads to the precipitation of the polymer during the course of the reaction.[4]

For the production of PAAm by one of these thermal methods, the initiator of choice would seem to be hydrogen peroxide, since it allows the reaction to be carried out at convenient temperatures (40 to 67°C) to give a high molecular weight polymer free of contaminants; if need be the molecular weight can be controlled (i.e., reduced) by including methanol in the reaction medium as a chain transfer agent.[65]

Redox systems have also been used as free radical initiators in aqueous solution, such as permanganate + tartaric acid,[70] chlorate + sulfite,[71] and persulfate + thiosulfate.[72]

The electro-initiated reaction in aqueous solution with perfluoroacetic acid and potassium perfluoroacetate leads to only relatively short-chain polymers (M = 5D3).[73]

Irradiation methods for polymerization in aqueous solution that have been studied include those using UV or visible light through the photodecomposition of hydrogen peroxide,[74] or via photosensitizers such as ferric ions[75] or riboflavin;[76] using ultrasonic irradiation;[77] or using X- or γ-rays.[78-81] In connection with the mechanisms of these processes, Collinson and co-workers[82,83] have studied the reactions that take place between polyacrylamide radicals and metal ions in aqueous solution in such irradiated systems.

The γ-ray-initiated polymerization of solid acrylamide has also been studied.[84]

Minsk and co-workers[66] have investigated suggestions that the apparent low nitrogen content of PAAm produced by the free radical mechanism arises from an imidization cross-linking during the course of the reaction:

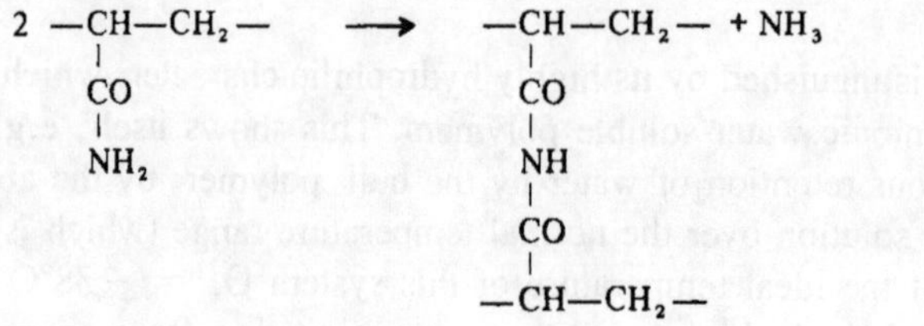

(1)

They showed, however, that this imidization is not likely to occur to any appreciable extent during the polymerization, and that the apparent low nitrogen content is more likely due to the presence of retained water or other solvent in the polymer finally isolated (see Section III.H.).[66]

In certain cases, polymerization of acrylamide does not lead to polyacrylamide; with Grignard reagents (in the presence of an inhibitor of free radical polymerization) or with alkali metal alkoxides, there is an anionic "transition polymerization" leading to poly(β-alanine) (nylon-3):[81,84]

$$CH_2{=}CH{-}CO{-}NH_2 \longrightarrow {-}\!\!\left[CH_2{-}CH_2{-}CO{-}NH\right]\!\!{-} \qquad (2)$$

Polyacrylamide is also available commercially, e.g., in the form of *Cyanamer* P-250 produced by the American Cyanamid Co. (other grades of *Cyanamer* such as A-370 or P-35 are modified forms of PAAm)[62,63] or of *Separan* and *Purifloc* produced by the Dow Chemical Co. (where again only certain grades are true unmodified PAAm).[64] Where such commercial products (or indeed, any other sample is though to have an appreciable carboxylic acid content due to hydrolysis of some of the amide groups, these may be reconverted back to amide groups by treatment with thionyl chloride and ammonia.[85]

The polymer may be purified by precipitating it from aqueous solution with methanol,[65,66,73,81] ethanol,[79] or acetone.[71,86]

Various grades of cross-linked polyacrylamide in bead form, with different degrees of cross-linking and different particle sizes, are available commercially (e.g., Reference 87) for use in gel permeation chromatography. The cross-linked polymer is also used in gel electrophoresis, where it is generally prepared *in situ* in water-swollen form using methylene-*bis*-acrylamide as the cross-linking agent.

C. Fractionation and Characterization

PAAm may be fractionated by conventional stepwise precipitation methods with the solvent/nonsolvent system water/acetone.[67] The molecular weight distributions of whole polymer samples have been studied by GPC.[69]

D. Bulk Polymer

The glass transition temperature, t_g, of PAAm has been given variously as 165°C (without further details)[38,88] and 188°C (by differential thermal analysis).[38] Whichever value is correct, this shows that up to these relatively high temperatures the polymer is in the glassy state and thus relatively resistant to the ingress or egress of small-molecule penetrants (such as retained solvent), and that it is necessary to use higher temperatures to put the polymer in the rubbery state where absorbed solvent molecules are sufficiently mobile to be desorbed.

The IR absorption spectrum of PAAm has been given for the range 200—4000 cm^{-1} by Chapiro and Perec,[81] for 625—4000 cm^{-1} by Nyquist,[89] and for 600—4000 cm^{-1} by Kulicke and Klein,[65] accompanied in the last case by an interpretation; the section over the more limited range 3,000—37,000 cm^{-1} has been given by Griot and Kitchener[85] in connection with the adsorption of the polymer from aqueous solution onto silica. The IR spectrum may be used to distinguish true PAAm (which has a primary amide band at 1655 cm^{-1}) from the alternative polymerization product poly(β-alanine) (which has instead a secondary amide band at 1555 cm^{-1}). (See Section III.B).[81]

E. Solution Behavior

1. Aqueous Solution

The results of various studies on the behavior of PAAm in aqueous solution are summarized in Table 3. The table also includes some studies in the presence of simple salts, such as 0.2 M NaCl (entry 4), 0.5 M NaCl (entry 37), and 1 M $NaNO_3$ (entries 16, 17, 19, and 28); these are included because the presence of salts even at these relatively high concentrations seems to have little effect on such properties as conformational size, while the MHS parameters for the polymer in 1 M $NaNO_3$ solution have been widely used for the estimation of molecular weight values (see below). Although no reason for adding salt has been given in any of these cases, it seems that this was done to suppress the effect of the ionization of any small proportion of carboxylic acid groups present in the polymer chain.

One complicating feature of any study of the behavior of aqueous solutions of PAAm is the "aging" effects that occur, particularly, with very high molecular weight samples, which lead to a progressive fall in the viscosity of the solution (see Section III.G).[86,93,95,97] From the practical viewpoint, two approaches can be adopted to deal with these effects. In the first approach, the systems are studied as soon as possible after the preparation of the solution, using where feasible an additive, such as 2% 2-propanol,[95] to slow down the process involved — this approach has the inevitable disadvantage that any studies must themselves take some time, while the presence of an additive can alter conformational and related properties of the polymer. Alternatively, the systems are stored until these effects have gone to completion,[97] where possible attempting to speed up the process — this has the disadvantage of being time consuming, and in the interval the solutions may undergo other changes due to hydrolysis, microbial attack, etc., while even heating does not seem to greatly speed up these changes.[86] Fortunately, these phenomena only seem to occur with very high molecular weight samples, but on the other hand these are just the types of materials that are important from the technical viewpoint (as flocculants, etc.).

The *partial specific volume* of the polymer in aqueous solution has been studied by three groups of workers, with the following disparate values for $\bar{v}_2^{\circ}$ (in cm^3 g^{-1}): 0.769 for M = 3D4 at 20°C, by Scholtan;[67] 0.656 and 0.669 for M = 2D5 and M = 6D6, probably at 25°C, by Shyluk and Stow;[93] 0.716 for M = 3D4 at 25°C, by Roy-Chowdhury and Kale.[4] It seems unlikely that the differences in the values (~16% between the extremes) arise from the differences in molecular weights of the samples or in the experimental temperatures. Differences in the concentration ranges that were used also seem to be an unlikely source, since the third group[4] showed that the apparent specific volume ψ_2 remained essentially constant (±0.003) from 0.1 up to 2%. Possibly they arise from retention of water by the samples used or the evaporates of their solutions, in which case the highest value, i.e., of Scholtan,[67] would be the preferred one.

For the *specific refractive index increment, $\tilde{n}$,* five values have been reported which fall into two groups, i.e. (with units of cm^3 g^{-1}), a "low" group, comprising the values 0.149,[1] 0.161,[69] and 0.163,[90] and a "high" group, comprising the values 0.186[62] and 0.187.[97] Some clarification of this situation is obtained from the fact that in the case of the last quoted value it was also shown that a typical "low" value of 0.161 was obtained if the polymer sample used was not completely dry.[97] On this basis, the two high concordant values are to be preferred.

Considering the *intrinsic viscosity-molecular weight* relations for PAAm in aqueous solution, the situation here is rather complex. In the first place, there are the "aging" effects

Table 3
STUDIES ON PAAm IN AQUEOUS SOLUTION

Entry	Year	Polymer samples: No.	Polymer samples: Range of $\overline{M}$	Techniques used	Temperature (t/°C)	Polymer concentration	Comments	Ref.
1	1952	1	5D6	LS	25	0.02—0.1%	436, 546 nm: → $\overline{M}_w$; A_2[a]	90
2	1952	1	5D6	RI	25?	1%	546 nm: → $\bar{n}$ = 0.163 cm³ g⁻¹	90
3	1952	1	5D6	VI	25?		→ [η]	90
4	1954	7	2D4—5D5	SS	28	0.2—1%	+ NaCl (0.2M): → $s°$	67
5	1954	1	~5D4?	SV	20		→ $\bar{v}_2°$ = 0.769 cm³ g⁻¹	67
6	1954	7	2D4—5D5	TD	20	0.4—1.5%	→ $D°$; $\overline{M}_{sD}$	67
7	1954	7	2D4—5D5	VI	25	0.1—1%	→ [η] (k';) MHS (0.80; 6.31d3)	67
8	1955	?	?	VI		10%	→ No negative thixotropy;	42
9	1957	?	~1D4—3D5?	VI	25?		(AAm → PAAm) → η (progress of reaction);	79
10	1957	24	~1D4—3D5	KI	25		(AAm → PAAm) → $\overline{M}_n$ (?)	80
11	1957	24	~1D4—3D5	VI	25?		→ [η]; MHS (0.66; 6.8d2)[b]	80
12	1957	5	~5D4—1D5	VI	25	15%?	→ η; [η] ?; $\overline{M}_v$?	74
13	1957	1	4D5	LS	22—65	0.05—0.9%	546 nm: → $\overline{M}_w$; A_2; thermodynamic parameters	1
14	1957	1	4D5	RI	20—65		546 nm: → $\bar{n}$ = 0.149 cm³ g⁻¹	1
15	1957	1	4D5	VI	22—62		→ [η]; d [η]/dt	1
16	1958	22?	1D6—3D6?	VI	25?		+ $NaNO_3$ (0,1M):$_c$ → [η] (H_2O) = [η](1M $NaNO_3$)$\overline{M}_w$	71
17	1960	13	7D4—2D5	VI	30		+ $NaNO_3$ (1M)?: → [η]; k'; $\overline{M}^c$	84
18	1964	3	4D5—6D6	VI	25	0.6—12%	Rheology; → η; [η]; $\overline{M}^c$	76
19	1966	2	2D5, 2D6	VI	30	0.07—0.8%	+ $NaNO_3$ (0,1M): → η; [η]; $\overline{M}^c$; "aging" effects	86
20	1969	1?	?	USA	25	0.01—28%	1—12 MHz	47
21	1969	2	2D5, 5D6	SE,SS	25—65	0.1—0.86%	→ $\overline{M}$; "aging" studies	93
22	1969	1	2D5	SV	25?		→ $\tilde{v}_2°$ = 0.669 cm³ g⁻¹	93
23	1969	1	5D6	SV	25?		→ $\tilde{v}_2°$ = 0.656 cm³ g⁻¹	93
24	1969	2	2D5, 5D6	VI	25	0.01—2.5%	→ η; "aging" studies	93

25	1970	1	2D4	SV	25	0.1—2%	$\rightarrow \tilde{v}_2^{\circ} = 0.716\ cm^3\ g^{-1}$	4
26	1970	1	2D4	USV	25	0.1—2%	5 MHz: $\rightarrow \tilde{\kappa}_2^{\circ}$	4
27	1971	?	?	RI	25		546 nm: $\rightarrow \tilde{n} = 0.186\ cm^3\ g^{-1}$ (footnote d)	62
28	1971	?	?	VI	30		$+NaNO_3$ (1M): $\rightarrow$ MHS (0.66; 3.73d2)[c]	62
29	1973	1	6D6	LS	r.t.	0.03—0.3%	546 nm: $\rightarrow \bar{M}_w\ (;A_2)$[a]	69
30	1973	?	?	RI			546 nm?: $\rightarrow \tilde{n} = 0.161\ cm^3\ g^{-1}$	69
31	1973	42	3D6—1D7	VI	25	0.02—0.1%	$\rightarrow [\eta]; \bar{M}$[c]	69
32	1975	1	Cross-linked	TD	25	$w_2 = 0.07$	$\rightarrow D\ (H_2O)$	94
33	1977	10	5D5—6D6	VI	25	0.001—5%	$\rightarrow \eta; [\eta]$; "aging" studies	95
34	1978	?	Cross-linked	NMR		0—20%	6 MHz: $\rightarrow T_1$ (^{1}H)	96
35	1978	12?	5D5—5D6?	RI	25		546 nm: $\rightarrow \tilde{n} = 0.187\ cm^3\ g^{-1}$	97
36	1978	12	5D5—5D6	VI	25		$\rightarrow [\eta]$; MHS (0.80; 4.9d3)	97
37	1978	12?	5D5—5D6?	VI	25		$+NaCl$ (0.5M): $\rightarrow [\eta]$; MHS (0.77; 7.19d3)	97
38	1978	9	5D5—5D6	VI	25	0.01—0.3%	$\rightarrow [\eta]; k'$; [e]$\bar{M}$	65
39	1979	10	?	NMR		1—50%	$\rightarrow T_1$ (^{1}H)	98
40	1979	8	Cross-linked	NMR		8—16%	$\rightarrow T_1$ (^{1}H)	98
41	1979	10	?	NMR	−35	1—70%	Line area: $\rightarrow$ % unfrozen water	98
42	1979	8	Cross-linked	NMR	−35	8—16%	Line area: $\rightarrow$ % unfrozen water	98

[a] Zimm plot: $c_2 = 0$ lines have double curvature.

[b] MHS parameters obtained using $\bar{M}_n$ (?) values from entry 10.

[c] The MHS parameters quoted in entry 28 have been used for estimation of $\bar{M}$ values in other published work through their apparently having been quoted in earlier versions of the American Cyanamid booklet, i.e., in the 1955 version (Reference 91) which is cited directly or indirectly in References 71, 84, 86, and 88; and in the 1967 version (Reference 92) cited in Reference 69.

[d] This value of $\tilde{n}$ was apparently quoted in the 1967 version of the booklet (Reference 92), as cited in Reference 69.

[e] Huggins plots ($\eta_c c$ vs. c_2) curved for $\bar{M} > 1.7D6$; Schulz-Blaschke plots (η_c vs. η_{sp}) linear for all $\bar{M}$ values studied.

referred to elsewhere (see above in this section and also Section III.G below), which complicate the procedures of obtaining the value of [η], and which also raise doubts as to the status of the values reported in the literature. Second, there are four different sets of MHS parameters reported, which however fall into two fairly closely similar pairs; the parameters (given as a; $K/\mathrm{cm^3\ g^{-1}}$) are as follows:

1. American Cyanamid Co.:[62] 0.66; 3.73d2 (30°C, 1 M $NaNO_3$)
2. Collinson et al.:[80] 0.66; 6.8d2 (25°C?)
3. Klein and Conrad:[97] 0.80; 4.9d3 (25°C)
4. Scholtan:[67] 0.80; 6.31d3 (25°C)

Despite the apparent fair concordance between the first two sets, they are both open to criticism. The parameters given by Collinson and co-workers[80] were obtained using samples of whole (i.e., unfractionated) polymers, with the molecular weights as $\overline{M}_n$ values calculated from a presumed mechanism for the reaction in which they were produced. The parameters given by the American Cyanamid Co.[62] are supported by no further details other than that the molecular weights used are $\overline{M}_w$ values.

By contrast, the parameters given by Klein and Conrad[97] and Scholtan[67] are free of such criticisms and are in quite good concordance with one another. Furthermore, the high value of a = 0.80, which is essentially the upper limit for a flexible chain nonionic polymer in a good solvent, is also in accord with the known goodness of water as a solvent for this polymer (i.e., as shown by the hygroscopic nature of the bulk polymer, its ideal temperature well below 0°C and its resistance to precipitation by electrolytes). On this basis, these latter two sets seem the most trustworthy, and we may take the average values as those which can be recommended for the system, i.e., 0.80; 5.6 (±0.7)d3.

2. *Nonaqueous Solution*

The limited number of nonaqueous liquids which are solvents for PAAm has correspondingly restricted studies on the behavior of the polymer in nonaqueous systems. Kulicke and Klein[65] have reported the following MHS parameters (given as before) for PAAm in formamide at 25°C: 0.59; 9.18d2 (for M = 5D4—5D6). They have also reported[65] for ethylene glycol the parameters 0.18; 21.2, which however are discordant with those reported for the same system by this group in a parallel publication[97] — 0.54; 0.136. Judging from data presented in other publications by the same group (e.g., Figure 12 in Reference 95), it seems that these last set of parameters relate also to formamide rather than to ethylene glycol. The extremely low value of a (0.18) reported by them[65] for ethylene glycol is also suspect, since it indicates such a poor solvent power that it is surprising the polymer dissolves at all.

In view of the absence of "aging" effects with PAAm in these two solvents,[95] they would seem more suitable media for molecular weight estimations than water, so long as their MHS parameters could be definitely established.

F. Phase Equilibria

Considering water vapor uptake by the bulk polymer, little if any quantitative information is available on the absorption isotherm, other than the frequently stated observation that the polymer has a high affinity for water which once absorbed is tenaciously held.[1,4,65,66,67]

The polymer shows no sign of phase separation in aqueous solution over normal temperature ranges, which is in accord with the ideal temperature value of Θ_{-} = −38°C estimated by Silberberg and co-workers from their light-scattering measurements.[1]

The partition of PAAm between phenol and water, or more precisely between the phenol-rich and the water-rich phases formed by the two components, has been studied by Dobry;[99] the respective equilibrium concentrations of PAAm in the two phases were 0.18 and 3.24%, giving a partition coefficient of 18 in favor of the water-rich phase.

G. Chemical Behavior

The thermal stability of the bulk polymer has been studied by thermal gravimetric analysis from room temperature up to approximately 700°C by Minsk and co-workers,[66] and up to 450°C by Kulicke and Klein;[65] with rigorously dried samples, only a small weight loss occurred below the initial decomposition temperature (about 185°C), which was shown by mass spectrometry to be water.[66]

With aqueous solutions of high molecular weight ($M \approx$ D6) there are "aging" effects as shown by steady reductions with time in the viscosity of the solutions.[86,93,95] These effects seem to be largely physical, relating either to the disentangling or the disaggregation of the polymer molecules, although there may also be chemical contributions from the breaking of "weak links" whose infrequency means that they only become significant in their effect with such long polymer chains.

The degradation of the polymer in aqueous solution by high-speed stirring has been studied by Nagashiro and co-workers.[100]

The most important chemical reaction of PAAm in aqueous solution is the hydrolysis of the amide groups. For neutral solutions, Silberberg and co-workers[1] found that appreciable decomposition of the polymer only set in above approximately 67°C. For the alkaline hydrolysis, Conix and co-workers[101,102] found that the reaction in 1 M NaOH followed second-order overall kinetics up to about 50% conversion and then slowed down abruptly, although still subsequently following the same kinetic dependence; the Arrhenius parameters for the two regions were A = 4.3D5 M^{-1} min^{-1}; E = 14.3 ± 0.2 kcal mol^{-1} (59.8 ± 0.8 kJ mol^{-1}), and A = 2.5D4 M^{-1} min^{-1}; E = 13.8 ± 0.2 kcal mol^{-1} (57.7 ± 0.8 kJ mol^{-1}). For this same system, Klein and Heitzmann[38] found that the reaction went only to a limit of about 70 mol conversion, and that an equivalent limit was reached starting from PAA-AAm copolymers, while with copolymers having a lower acrylamide content there was no appreciable hydrolysis. This seems to represent the formation of a stabilized structure, seen also with PMAAm (see Section V.G), where each residual amide group is bordered on either side by an ionized carboxyl group.

The amide groups in PAAm also undergo other reactions similar to those of small-molecule amides. Thus, on treatment with alkali and hypochlorite they undergo the Hofmann degradation giving basic vinylamine, [$-CH(NH_2)-CH_2$], groups along the chain.[62,63,103]

Similarly, reaction with formaldehyde in alkaline medium leads to partial methylolation of the amide groups, i.e., $-NH_2 \rightarrow -NHCH_2OH$. Further reaction with sulfite or bisulfite gives sulfomethylol groups, $-NH_2CH_2SO_3^-$; these represent a useful modification of the polymer since they are strong acid groups and hence less sensitive to changes in pH than the carboxylic groups produced by hydrolysis. Alternatively, treatment of the methylol groups with an amine, RNH_2, leads by the Mannich reaction to secondary amine groups, $-NH_2CH_2NHR$, which are stronger bases than the primary amine groups produced by the Hofmann degradation.[62,63,103]

H. Analysis

The analysis of PAAm, and indeed all quantitative work on the polymer, is made awkward by the highly hydrophilic character of the polymer, which frequently leads to a high water content for polymer samples used, and the tenacious retention of water by the evaporates from aqueous solutions.[1,4,65,66,97] It appears that this retained water can only be removed by prolonged vacuum drying of finely powdered samples, which ought to be done preferably at temperatures exceeding the glass transition temperature, i.e., approaching 200°C. Rather than drying down samples of solution to determine the polymer content, it might be preferable to precipitate the polymer out in a large volume of volatile, water-miscible nonsolvent (e.g., methanol, ethanol, or acetone); choosing this to be one that slightly swells the polymer would ensure that the residual water is still sufficiently mobile to be extracted, while a subsequent vacuum drying would remove the residual nonsolvent so long as it is sufficiently volatile.

In view of these difficulties, many workers have preferred to determine the PAAm content of the solid or of solutions by nitrogen analysis (the pure polymer contains 19.71% nitrogen), using the Dumas method either alone,[1,38] or in conjunction with the Kjeldahl method.[65,66] Nonspecific methods such as refractometry are also applicable to these analyses, although eventually they have to be calibrated or checked by one of the absolute methods discussed above. For assaying very dilute solutions (~0.001%), Griot and Kitchener[85] used alkaline hydrolysis (presumably to the limit AA-AAm copolymers discussed above in Section III.G), followed by turbidimetry on the colloidal suspension formed on the addition of a cationic surfactant, *Hyamine* 1622.

IV. POLY(METHACRYLIC ACID) — PMAA

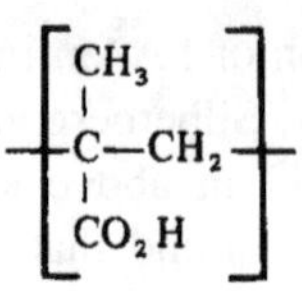

$M_0(C_4H_6O_2) = 86.09$

Other names and abbreviations are poly(1-carboxy-1-methylethylene) (IUPAC); polymethacrylic acid; HPMA (i.e., "hydrogen polymethacrylate"); PMA.

A. General Features

The general properties and applications of poly(methacrylic acid) have been reviewed, in association with those of poly(acrylic acid), by Glavis.[11]

Although poly(methacrylic acid) is not so important technically or commercially as others of this group such as PAA or PAAm, nevertheless it has been the subject of considerable scientific study. The main reason for this interest lies in the fact that nonionized PMAA exists in aqueous solution in a relatively compact conformation. This compactness shows itself, e.g., in a relatively low [η] value and, less directly, in a much more marked binding power for nonpolar small-molecule cosolutes than is seen with analogous polymers such as PAA (see Chapter 2 of Volume II). At the same time, PMAA shows inverse solubility-temperature behavior in aqueous solution, in contrast with the normal behavior shown by PAA, PAAm, and PMAAm.

Upon ionization, the molecules of PMAA undergo a fairly sharp conformational transition, which takes place over a relatively narrow range of degree of ionization, to the dimensions expected for a normal polyelectrolyte molecule; thus much of the information on the forces responsible for the production of this compact form comes from studies on the polyelectrolyte behavior of PMAA at intermediate degrees of ionization (see Chapter 1 of Volume II).

There is still much controversy as to the full explanation for the anomalous effects which are seen with PMAA. One contributing factor may be the attractive hydrophobic forces which must exist in aqueous solution between the substituent methyl groups; another factor could be the very marked stiffening effect on the chain of the presence of two substituent groups (CH_3 and CO_2H) on the alternate backbone carbon atoms — this effect is readily evident from an examination of molecular models.[104]

A particular incentive to these studies on the conformational behavior of PMAA is the parallel it provides with the transitions from compact to expanded forms shown by many proteins.

B. Production and Purification

A wide variety of methods have been used to prepare and to purify poly(methacrylic acid).

In certain cases these are aimed at the production of a particular stereoregular form; even when this is not the aim, the stereoregularity of the product is often markedly influenced by the route used. For these reasons, the methods used have been listed and discussed in some detail; correspondingly, the method which has been used for producing a particular sample needs to be kept in mind when interpreting the behavior of the sample.

Free radical polymerization of the monomer is quite naturally a commonly used route to prepare PMAA. Frequently this is done in aqueous solution, using peroxy initiators such as hydrogen peroxide,[1,16,32,105-109] or persulfate ions;[110-113] hydrogen peroxide has the advantage over many other initiators in that it does not leave any organic or ionic impurities in the system. Redox systems have also been used, such as persulfate + thiosulfate[114] and ferrous ions + hydrogen peroxide;[115-117] however, the latter system has the disadvantage that it leads to an appreciable iron content for the final polymer.[115] The polymerization may also be carried out in the absence of solvent, i.e., using the undiluted monomer,[51,118] necessitating an organic-soluble initiator such as benzoyl peroxide;[118] in this case high temperatures (i.e., above 80°C) should be avoided since the exothermic reaction may then become explosive.[118] Nonaqueous diluents have also been used, such as butanone (with azobisisobutyronitrile as initiator)[19-121] and dioxane (with benzoyl peroxide as initiator).[2,22]

Even with poly(methacrylic acid) produced by this free radical mechanism, the product seems to be somewhat syndiotactic rather than being purely atactic as might be expected. Thus with hydrogen peroxide, although Priel and Silberberg[109] state that this leads to an essentially atactic polymer, Mandel and co-workers find the product to be somewhat syndiotactic rich ($\sigma = 0.245$ — see Section IV.C).[107,122] The azo-initiated polymerization in butanone also leads to a product with a similar bias ($\sigma = 0.3$).[120] Use of a relatively high pH in aqueous solution polymerization leads not only to the expected reduction in the rate of polymerization,[112] but also increases the tendency to syndiotactic character in the product.[123] For these reasons, it is customary to refer to samples of PMAA produced by free radical mechanisms as being of the "conventional" type (represented by *cv*-PMAA), rather than as being more specifically atactic.

Polymerization of methacrylic acid may also be induced by irradiation with gamma rays. This is done most commonly with methanol as the solvent, either at room temperature,[124] or at low temperatures (-70 to -80°C);[104,110,120,125] in the latter case this leads to a product which is very markedly syndiotactic rich ($\sigma = 0.08$).[104,120,125] Lando and co-workers[126] have studied the effect of temperature and nature of the solvent (for a range of organic solvents) on the stereoregularity of the polymers so produced.

The other main route which has been used to produce PMAA is the hydrolysis of a poly(alkyl methacrylate). The hydrolysis occurs much less readily than with the corresponding acrylate esters; some of the kinetic features of the hydrolysis have been discussed by Platé.[127] Because the parent polymers can be produced fairly readily in specific stereoregular (or approximately stereoregular) forms, this has proved to be a useful method for producing the corresponding stereoregular forms of PMAA.

The most common parent polymer that is used for this purpose is the methyl ester, PMMA; the hydrolysis is normally carried out using concentrated sulfuric acid, in which case complete hydrolysis is indicated by complete dissolution of the starting polymer, and confirmed by IR or NMR examination of the product (see Section IV.D). With "conventional" PMMA, i.e., that produced by free radical polymerization, the hydrolysis occurs fairly easily to give the corresponding form of PMAA;[124] as with PMAA produced directly by free radical polymerization, this must be expected to have a syndiotactic bias corresponding to that in the parent polymer.[128] With isotactic PMMA the hydrolysis may be carried out either at room temperature,[121] or at 60°C,[110,125] or using each in succession to give essentially complete reaction.[104,129] With syndiotactic PMAA the conversion is more troublesome;[120,123] hydrolysis with 96% sulfuric acid at 45°C gives only 85% conversion, and repetition of the cycle is necessary to raise the conversion to greater than 98%.[130,131]

Other esters may give more ready conversion, particularly for specific stereoregular forms; the iso- and syndiotactic forms of the trimethylsilyl ester have been used to produce the corresponding forms of PMAA,[123] and the triphenylbenzyl ester has also been used for the syndiotactic form.[24] It should be noted that alkaline hydrolysis of the amide, i.e., polymethacrylamide, is not a suitable route for the production of PMAA since the reaction only goes maximally to about 60% conversion (see Section V.G).

Rather than using any of these methods to prepare their own samples of PMAA, some workers have chosen to use materials from commercial sources for their studies;[43,132] however, the polymer does not seem to be readily available from general chemical suppliers.

Breitenbach and Karlinger[133] have described the preparation and swelling behavior of a cross-linked sample of PMAA. The purification procedures used for PMAA, to free it of small-molecule contaminants, depend to some extent on the stereoregular type. With "conventional" PMAA a variety of solvent/nonsolvent precipitation systems have been used, most commonly methanol/diethyl ether,[2,22,43,51,107,110,111,132,149] but also water/hydrochloric acid,[1,16,105] water/butanone,[117] ethanol/diethyl ether,[118] and methanol/methyl isobutyl ketone.[134] Alternatively, or in addition, such contaminants may be removed by exhaustive dialysis[16,108-110,114,117,119] or by electrodialysis (which is naturally more efficient for removing ionic impurities).[113,115,116] The polymer may be finally isolated in a convenient form by freeze drying.[16,109,113,119,135]

With syndiotactic PMAA, similar solvent/nonsolvent systems have been used for purification, e.g., water/hydrochloric acid,[123,130] methanol/diethyl ether,[110] methanol/hexane.[120] Exhaustive dialysis has been carried out using either water[104] or dimethylformamide[129] as the solvent, and once again the final isolation may be carried out by freeze drying.[104,110]

With isotactic PMAA different purification methods must be used because the polymer is insoluble in water.[104,125] Precipitation may be carried out using concentrated sulfuric acid/water (i.e., directly from the hydrolysis reaction mixture),[104,125] or with alkaline aqueous solution/acid,[110,123] or with dimethylformamide/water.[121] Exhaustive dialysis may be carried out using as the solvent either alkaline aqueous solution[110] or dimethylformamide.[104,121] Freeze drying of an aqueous suspension has been used to finally isolate the polymer.[104]

In all these production and purification procedures, it is advisable to avoid contact of the polymer and its solutions with metals to prevent it from becoming contaminated with traces of polyvalent cations.[107] The special precautions required to prepare a sample of PMAA suitable for surface-activity studies have been described by Katchalsky and Miller.[108]

C. Fractionation and Characterization

The fractionation of conventional PMAA according to molecular weight has been carried out using a variety of solvent/nonsolvent precipitation systems: water/concentrated hydrochloric acid (for rough fractionation);[1,16,117] acidified water/butanone (or butanone + acetone for the highest molecular weights);[113,115,116] 2-propanol/hexane;[32] methanol/methyl isobutyl ketone;[134] methanol/diethyl ether;[22,31,107] or methanol/ethyl acetate + acetic acid (95:5 v/v).[135,136] These last two systems have inverse solubility-temperature behavior.

Aylward[123] has used preparative GPC for the fractionation of conventional and stereoregular samples of PMAA. Characterization of PMAA as regards its average molecular weight may be done by any of the standard solution methods (see Section IV.E). Characterization as regards stereoregularity is also important, not only in the case of deliberately produced stereoregular forms (where the extent of the stereoregularity needs to be quantified since samples are seldom if ever purely iso- or syndiotactic), but also for the "conventional", nominally atactic forms which in fact have a syndiotactic bias.[119-122,128]

The most direct quantitative estimates of stereoregularity are obtained from NMR spectra of solutions of either the parent alkyl ester or of a derived alkyl ester (generally methyl), which give signals at chemical shifts of δ = 0.91, 1.05, and 1.22 ppm (τ = 9.09, 8.95, and 8.78 ppm) from the α-methyl groups at the center, respectively, of syndio-, hetero-, and isotactic triads;[104,119,120,123,125,126,128-130] the relative areas of the three peaks are proportional to the amounts of the three types of placements, and where Bernoullian statistics prevail may be used to calculate the sigma (σ) value of Bovey and Tiers[128] which is a quantitative measure of the stereoregularity.* It is necessary to use the alkyl ester for these spectra since the corresponding peaks in the spectra of aqueous solutions of the acid itself or of its salts are not sufficiently well resolved for more than semiquantitative estimations.[119,123,129]

Estimates of stereoregularity can be obtained also from IR spectra of the bulk polymer (see Section IV.D), and also less precisely from such features as the crystallinity (crystallizability) of the bulk polymer, aqueous solubility (see Section IV.F.2), ease of hydrolysis of the parent alkyl esters[127] and of alkylation of the acid,[123] and binding ability towards cosolutes in aqueous solution (see also Chapter 2 of Volume II).[104,121,125]

D. Bulk Polymer

The dynamic mechanical properties of PMAA have been studied by Fitzgerald and Nielsen[137] over the temperature range -50 to $+250$°C, giving the shear modulus and damping (logarithmic decrement) for frequencies of about 1 Hz. The value of the glass transition temperature was found to be t_g = 185°C, but it was stated that this value may be too low since the polymer sample was believed to contain at least 5% water. Aylward[123] has studied an isotactic sample of PMAA by differential thermal analysis (DTA), observing endothermic transitions at 68 and 183°C; these were assigned by him to t_g and t_m (melting temperature), respectively, but in the light of the results of Fitzgerald and Nielsen,[137] this assignment should be reversed, i.e., giving t_m = 68°C.

X-ray diffraction examination by Barone and co-workers[110] of their sample of *st*-PMAA showed it to be crystalline; however, Aylward[123] was not able to obtain a sharp diffraction pattern from his sample of *it*-PMAA.

The infrared absorption spectra for conventional (*cv*-), *it*-, and *st*-samples of PMAA, taken over the region 500—2200 cm^{-1}, have been presented by Barone and co-workers.[110] With the *it*- and *st*-samples the spectra show clear differences in the 900—1000 and 1400—1500 cm^{-1} regions; in the former region, particularly, the *it*-form has only one peak while the *st*-form has two.[123] The spectrum of the *cv*-sample is intermediate in form but closer to that of the *st*-form, showing its syndiotactic bias.[110] Fitzgerald and Nielsen,[137] from their studies on PMAA and copolymers of methacrylic acid and styrene (PMAA-S), found that the hydrogen-bonded carboxyl group has a peak at 5.88 μm (1700 cm^{-1}) whereas that for the nonbonded form is at 5.74 μm (1740 cm^{-1}). Infrared spectra have also been used to detect "impurities" in the polymer chain; e.g., completeness of hydrolysis of a parent alkyl ester is demonstrated by the absence of absorption at 1730 cm^{-1},[104,125] while any anhydride groups produce a band at 1020 cm^{-1}.[110]

E. Solution Behavior

1. Aqueous Solution

The results of various studies on the behavior of nonionized PMAA in aqueous solution are summarized in Table 4. For such studies to relate to the nonionized form, it is clearly necessary to ensure that the ionization is effectively suppressed. With studies involving fairly high concentrations there is more or less self suppression of the ionization, e.g., at concentrations of ~7% used in the study of gelation by Silberberg and co-workers[139,143] the system has pH = 3 — i.e., degree of ionization i = 0.001. For the dilute solutions which are more usually studied the suppression is commonly accomplished by adding hydrochloric acid. To decide what is the minimum acid concentration required, we note that Braud and

* Values of σ range from 0 for purely syndiotactic polymer, through 0.5 for the atactic form, up to 1.0 for purely isotactic polymer.

Table 4
STUDIES ON NONIONIZED PMAA IN AQUEOUS SOLUTION[a]

Entry	Year	Polymer samples: No.	Polymer samples: Range of $\overline{M}$	Techniques used	Temperature (t/°C)	Polymer concentration	Comments	Ref.
1	1949	1	Cross-linked	SV	r.t.	13%	+HCl (10 mM): → Q	133
2	1951	7	9D4—9D5	LS	30	0.01—0.1%	+HCl (0—3 mM); 546 nm: → $\overline{M}_w$; $A_2 \approx 0$ for 2mM HCl)	22
3	1951	8?	9D4—8D5?	RI	30		+HCl (2 mM); 546 nm: → $\bar{n}$ = 0.138 $cm^3\ g^{-1}$	22
4	1951	8	9D4—8D5	VI	30		+HCl (2 mM): → [η]; MHS (0.50; 6.6d2 or 0.57; 2.7d2 — see text)	22
5	1952	2	4D4, 9D4	LS		0.1—1%	+HCl (pH 1.55); 436 nm: → $\overline{M}_w$; A_2	117
6	1952	2?	4D4, 9D4?	RI			+HCl (pH 1.55); 436 nm: → $\bar{n}$ = 0.159 $cm^3\ g^{-1}$	117
7	1952	2	4D4, 9D4	VI	25		+HCl (pH 1.55): → [η]	117
8	1952	3	4D4—2D5	VI	30		+HCl (2 mM): → [η]; $\overline{M}_v$	134
9	1955	5	2D5—4D5	LS	25		+HCl (0.1M); 546 nm: → $\overline{M}_w$; A_2	115
10	1955	6	9D4—7D5	OP	25		+HCl (0.1, 1M): → $\overline{M}_n$; A_2	115
11	1955	5	2D5—4D5	RI	25?	0.01—0.12%	+HCl (2 mM—1M); 546 nm?: → $\bar{n}$ = 0.155 $cm^3\ g^{-1}$ (0.1M HCl)	115
12	1955	2	2D5, 4D5	VI	25	'0.4%	+HCl (0—2M): → η_{sp}, [η]	115
13	1955	1?	9D4?	VI		5—15%	Effect of stirring: → shear hardening (negative thixotropy)	42,138
14	1957	1	6D5	LS	20—58	0.1—1%	+HCl (20 mM); 546 nm: → $\overline{M}_w$; A_2; TP	1
15	1957	1	6D5	RI	20—58		+HCl (20 mM); 546 nm: → $\bar{n}$ = 0.158 $cm^3\ g^{-1}$	1
16	1957	1	6D5	VI	8—58	0.2—1%	+HCl (20 mM): → [η]; k'; d [η]/dt	1
17	1959	3	7D3—9D4	LS		0.1—4.7%	+HCl (0.05—0.1M); 436 nm: → $\overline{M}_w$	16
18	1959	3	7D3—9D4	OP		1.2—4.6%	+HCl (0.05—0.1M); → $\overline{M}_n$; osmotic coefficients	16
19	1959	3?	7D3—9D4?	RI			+HCl?; 589 nm: → $\bar{n}$ = 0.159 $cm^3\ g^{-1}$	16
20	1962	4	2D5—9D5	VI	10—48	6—8%	pH ≈ 3.6 (no additive)[b]	139
21	1963	1	6D4	SV	21	0.5—6%	+HCl (0.01M); 1—4900 atm: → $\bar{v}_2^{\circ}$ = 0.712 $cm^3\ g^{-1}$ (1 atm)	139a
22	1963	1	7D5	SV	25	0.1—4%	+HOAc(0,0.1M): → $\bar{v}_2^{\circ}$ = 0.694, 0.707 $cm^3\ g^{-1}$	2
23	1963	1	7D5	VI	25?	0.1—1%	+HOAc (0,0.1M): → η_c (;[η]; k')	2
24	1963	1	7D5	USV	25	0.1—4%	+HOAc (0,0.1M); 5 MHz: → k_2°	2

25	1964	1	5D4	LS	20—60	0.05—0.15%	$+$NaCl (0.05, 0.5M); i = ?; 546 nm: $\rightarrow \overline{M}_w$; A_2; $\hat{L}$ ($\hat{R}$); Θ (= 68, 43°C) (footnote c)	140
26	1964	1	5D4	RI	20—60		$+$NaCl (0.05, 0.5M); i = ?; 546 nm: $\rightarrow \bar{n}$ = 0.168, 0.183 $cm^3\ g^{-1}$ (footnote c)	140
27	1964	1	5D4	VI		'0.3%	$+$NaCl (0, 0.05, 0.5M); i = ?: $\rightarrow \eta_{sp}$; $[\eta]^c$	140
28	1965	1	1D5	UV		0.03—1%	pH ≈ 3.5 (no additive); 190—240 nm: $\rightarrow A$; ϵ	43
29	1965	1	1D5	VI	25	0.4%	i = ? (No additive): $\rightarrow \eta_c$	43
30	1965	1	7D5	UV			i = ? (No additive): 182—250 nm	110
31	1965	1[b]	5D5[b]	UV			i = ? (No additive): spectra not obtainable due to contamination	110
32	1965	1[e]	3D4[e]	UV			i = ? (No additive); 182—250 nm: $\rightarrow \epsilon$	43a,110
33	1965	1	7D5	VI	27—52	0.2%	i = ? (No additive): $\rightarrow \eta_c$	43a,110
34	1965	1[d]	5D5[d]	VI	23—63	0.1, 0.3%	i = ? (No additive): $\rightarrow \eta_c$	110
35	1965	1[e]	3D4[e]	VI	27—63	0.2, 0.6%	i = ? (No additive: $\rightarrow \eta_c$	110
36	1967	1	2D5	IR			D_2O; i = ? (no additive); 1300—2000 cm^{-1}	45
37	1967	1	2D5	UV		0.02—0.5%	$+HClO_4$ (20 mM): 200—220 nm: $\rightarrow A$; ϵ	45
38	1967	1	1.7D6	RI		0.1—0.4%	pH 2.6: $\rightarrow$ intrinsic flow birefringence	141
39	1967	1	1.7D6	VI		0.1—0.4%	pH 2.6: $\rightarrow [\eta]$; k'	141
40	1968	1	1D6	VI	20	0.2—0.6%	$+$HCl (2 mM): $\rightarrow [\eta]$; k'	136
41	1969	4	3D4—2D5	USA	5—50	1.6—6%	$+$HCl (0, 0.1M); 1—90 MHz	47
42	1970	?[e]	~3D4[e]	OP	37		$+$HCl (5 mM): $\rightarrow \overline{M}_n$	123
43	1970	1	3D5	NMR	31	1—5%	D_2O; pH 1; 2—9 MHz: $\rightarrow T_1$ (^2H)	142
44	1970	1	3D5	VI	13—43	6.5%—8%	pH [g] 3 (i = 0.001)[b]	143
45	1972	4	4D4—4D5	SV	25	0.04—1.7%	i = ? (no additive): $\rightarrow \bar{v}_2^\circ$ = 0.690 $cm^3\ g^{-1}$	50
46	1974	1	3D5	VI	25	'0.02%	$+$HCl (0—5 mM): $\rightarrow \eta_c$; $[\eta]$; $\overline{M}_v$	51
47	1974	1	4D5	VI	30	0.15%	pH 3.7; $+$NaCl (5 mM); viscoelastic studies (0—125 kHz)	111
48	1976	1	2D4	NMR	26	4, 8%	D_2O; $+$ DCl (2 mM); 25 MHz: $\rightarrow \delta(^{13}C)$; T_1 (^{13}C); nuclear Overhauser enhancement	114
49	1976	1[f]	2D4[f]	NMR	30	5%	pH 2: $\rightarrow \delta(^2H)$	114
50	1976	1	4D4	NMR	35	3%	D_2O; no additive[g]: $\rightarrow \delta(^1H)$; line width	119

[a] Conventional PMAA (i.e., nominally atactic, but with syndiotactic bias), unless otherwise indicated.
[b] Viscoelastic studies on gelation; ionization self suppressed at these relatively high polymer concentrations.
[c] No acid added to suppress ionization of PMAA; ionization effects probably somewhat suppressed with added NaCl.
[d] Isotactic PMAA.
[e] Syndiotactic PMAA.
[f] Perdeutero PMAA.
[g] Ionization largely self suppressed at this polymer concentration.

co-workers[51] found that the value of η_{sp} for a 0.02% solution was markedly reduced when 1 mM HCl was added, but then fell only gradually for concentrations up to 5 mM; similarly Katchalsky and Eisenberg[22] found that the light-scattering intensity for 0.01 to 0.1% solutions of PMAA rose to a limiting value at this same HCl concentration of 1 mM. For the maximum concentration which may be used, Arnold and Caplan[115] found that η_{sp} of 0.4% solutions was fairly constant for HCl concentrations over the range 0.01—0.1M, but fell steadily above 0.1M (due, presumably, to "salting-out" effects). Thus any HCl concentration within the range 1 mM—100 mM would seem to be suitable. Omission of the precaution of adding this acid to suppress PMAA ionization vitiates or invalidates the results of several studies, but the results of these are still included in Table 4 for greater completeness and for the purpose of comparison.

The *partial specific volume* of PMAA in aqueous solution has been investigated by Andersson,[139a] Roy-Chowdhury,[2] and Tondre and Zana.[50] Unfortunately, only in the first of these was the solution made sufficiently acid (pH 2.0, with 0.01M HCl) to suppress the ionization over the full concentration range (0.5 to 2.5%); the value of the apparent specific volume ψ_2 was constant at 0.711 cm^3 g^{-1} over this range, and is thus also the value of $\bar{v}_2^{\circ}$.[139a] With the second studies, measurements were made in the absence of additive and in the presence of 0.1M acetic acid (pH = 3); only at concentrations above 2% were the values of ψ_2 essentially constant (the values showing a marked rise at lower concentrations even in the presence of the acetic acid), and for the range 2 to 4% the ψ_2 values (taken to be equal to $\bar{v}_2^{\circ}$) were 0.694 cm^3 g^{-1} and 0.707 cm^3 g^{-1}, respectively.[2] In the third studies, no additive was used; the values of ψ_2 were found to be essentially independent of polymer concentration (over the range 0.04 to 1.7%) at 0.690 ($\pm$0.003) cm^3 g^{-1}.[50] Thus on the basis of these studies we may take the value of $\bar{v}_2^{\circ}$ for nonionized PMAA to be 0.709 $\pm$ 0.002 cm^3 g^{-1}, using the mean of the first two sets.[2,72]

The *specific refractive index increment, $\tilde{n}$*, has been measured by five groups: Alexandrowicz,[16] Arnold and Caplan,[115] Kanevskaya and co-workers,[140] Oth and Doty,[117] and Silberberg and co-workers.[1] In all cases except the third group, the appropriate amount of HCl was added to suppress ionization; Arnold and Caplan[115] found that $\tilde{n}$ varied somewhat with HCl concentration from 2 mM up to 0.05 M, but was essentially constant from 0.1 to 1.0 M. The values of n (in cm^3 g^{-1}) for the remaining four groups (i.e., omitting the third) with wavelengths are 0.155 (546 nm); 0.1594 (436 nm); 0.158 (546 nm); 0.159 (589 nm). As usual, the values show no marked correlation with wavelength, and lead to an overall average for this wavelength range of 0.157 $\pm$ 0.002 cm^3 g^{-1}. The value seems to also be insensitive to temperature, since Silberberg and co-workers[1] found it to be essentially constant over the temperature range 20 to 58°C. The values of $\tilde{n}$ obtained by Kanevskaya and co-workers were as expected higher (0.183 cm^3 g^{-1} with 0.5 M NaCl; 0.168 cm^3 g^{-1} with 0.05 M NaCl), indicative of partial ionization of the polyacid (compare the value 0.243 cm^3 g^{-1} for i = 0.80 obtained in these same studies).[140]

The *intrinsic viscosity-molecular weight* relation has been determined by Katchalsky and Eisenberg[22] using 2 mM HCl at 30°C; they stated that the data may be fitted by the form of the MHS equation with a = 0.5. No value of the other parameter K was quoted, and for this purpose it is necessary to estimate the [η] values which were only presented graphically; with the above a value this procedure gives K = 6.64 ($\pm$0.48)d2 cm^3 g^{-1} (i.e., a 7% uncertainty, as estimated from the standard deviation). These parameters have been used subsequently quite widely for the estimation of molecular weights. However, from the graph presented (Figure 7 in Reference 22) the value a = 0.50 seems somewhat too low for the best fitting line, and reevaluation gives the best fit as a = 0.57; K = 2.71 ($\pm$0.14)d2 cm^3 g^{-1} (i.e., with the uncertainty reduced to $\pm$5%). This is also in better accord with the fact that Silberberg and co-workers[1] obtained the ideal temperature for PMAA in 20 mM HCl as Θ_+ = 56°C, for which system a = 0.50; the more dilute HCl solution and the lower temperature used by Katchalsky and Eisenberg[22] should then have led to it being a better solvent system, i.e., with $a > 0.5$, as is in fact obtained from the above reevaluation.

However, the values of [η] and M obtained by other workers for nonionized PMAA in aqueous solution do not seem to fit in very well with either of these two sets of parameters. Thus Silberberg and co-workers[1] obtained [η] = 34 cm³ g⁻¹ for a sample with $\overline{M}_w$ = 5.9D5 in 20 mM HCl at 30°C, as compared with the value of 53 cm³ g⁻¹ expected for this molecular weight from either of the two above sets of parameters. Also, Oth and Doty,[117] using 30 mM HCl at 25°C, with two samples having $\overline{M}_w$ = 4.31D4 and 9.03D4, obtained [η] = 26.5 and 45 cm³ g⁻¹, respectively, compared with 12 and 18 cm³ g⁻¹ calculated from the second set of parameters above. These discrepancies are much greater than would be expected from any differences in HCl concentration and temperature. On the other hand, Eliassaf and co-workers[106] obtained [η] = 28 cm³ g⁻¹ for a sample with $\overline{M}_w$ = 1.7D5 in 20 mM HCl at 30°C, which is in accord with the value 26 cm³ g⁻¹ expected from the second set of MHS parameters. In those cases, therefore, where molecular weights are to be estimated solely from intrinsic viscosity values, it is preferable to use other solvent systems (see below), in addition to dilute aqueous hydrochloric acid discussed above, to provide an independent check on the molecular weight value obtained.

Where the studies on PMAA are necessarily restricted to aqueous solutions for some reason, in place of the above MHS parameters for nonionized PMAA it is possible to use alkaline solutions and obtain molecular weights from the MHS parameters for the *ionized* polymer, as discussed in Chapter 1 of Volume II.

2. Nonaqueous Solution

Conventional PMAA is soluble in a variety of nonaqueous liquids, including methanol, ethanol, acetone, dioxane, ethylene glycol, and dimethylformamide, but it is insoluble in alkanes, diethyl ether, and the higher ketones such as butanone and methyl isobutyl ketone (2-methylpentan-4-one); *st*-PMAA is soluble in methanol and dimethylformamide but insoluble in hexane and diethyl ether, and *it*-PMAA (which is insoluble in water but soluble in concentrated sulfuric acid and in aqueous alkali) is soluble in dimethylformamide (see Section IV.B).

The results of various studies on PMAA in nonaqueous and mixed solvents are summarized in Table 5. The most extensively studied systems are those with methanol as solvent; the MHS parameters obtained for this system by Wiederhorn and Brown[134] have been widely used, as alternatives to those of Katchalsky and Eisenberg[22] applying to PMAA in acidified water, for the determination of molecular weights.

F. Phase Equilibria

1. Uptake of Water Vapor

The equilibrium uptake of water vapor by PMAA has been studied at 40°C by McLaren and co-workers;[145,146] the data presented for this system by Peterson[56] are simply a replotting of their values. The isotherm values are plotted in Figure 2 with the uptake U_1 (grams water/100 grams dry polymer) on a logarithmic scale. The initial section of the isotherm shows a "toe" (also evident when U_1 is plotted directly) equivalent to 1 to 2% of water; this probably represents the water tenaciously held by the polymer even after prolonged treatment in vacuo or freeze drying.[1,110,113] At the high water-activity end, the plot is essentially linear, and extrapolates to a finite uptake of 21 g/100 g polymer, which is coincidentally close to 1 mol H_2O/basemole MAA unit. The fact that this extrapolation gives a finite value, rather than indefinitely high uptake corresponding to dissolution of the polymer, seems to be a result of the relatively high temperature used (40°C), which is close to the ideal temperature value (Θ_+ = 56°C) for PMAA in water.[1]

Table 5
STUDIES ON PMAA IN NONAQUEOUS AND MIXED SOLVENTS

Entry	Year	Polymer samples: No.	Polymer samples: Range of $\overline{M}$	Techniques used	Temperatue (t/°C)	Polymer concentration	Comments	Ref.
					H_2O + CH_4O: Water + Methanol			
1	1967	1[a]	2D5[a]	IR			D_2O (ϕ = 0.6—0.9) + CH_3OD; 1300—2000 cm^{-1}	45
2	1967	1	2D5	UV	20	0.02—0.06%	$\phi(H_2O)$ = 0.6—0.8; 200—220 nm: → ϵ	45
3	1968	3	3D5—1D6	VI	20	0.1—0.8%	$\phi(H_2O)$ = 0.1—0.9: → $[\eta]$ (;k')	136
					CH_4O: Methanol (See Also H_2O + CH_4O)			
4	1952	3	1D5—2D5	OP	27		→ $\overline{M}_n$	134
5	1952	72	3D3—3D5	VI	26		→ $[\eta]$; k^1; MHS (0.51; 0.24)[b]	134
6	1964	1	5D5	RI	20—60		546 nm: → $\hat{n}$ = 0.134 cm^3 g^{-1}	140
7	1964	1	5D5	LS	20—40	0.1—0.3%	546 nm: → $\overline{M}_w$; A_2; $\hat{L}(\hat{R})$	140
8	1974	1	5D5	VI	20—50	~0.3%	→ η_{sp}; $[\eta]$	140
9	1967	1	2D5	UV	20		200—220 nm: → ϵ	45
10	1968	3	3D5—1D6	VI	20	~0.5%?	→ $[\eta]$	136
11	1968	3?	3D5—1D6?	VI	26		→ $[\eta]$; $\overline{M}_v$	136
					$C_2H_4Cl_2$: Dichloroethane			
12	1975	1	9D4	USA	−10 to +40	4%	7—65 MHz	144
					C_3H_6O: Acetone			
13	1975	2	6D4, 1D5	USA	−10 to +25	4, 6%	7—85 MHz	144
					C_7H_8: Toluene			
14	1975	1	1D5	USA	10—30	3%	7—75 MHz	144

[a] Polymer carboxylic acid groups deuterated.

[b] Viscosity plots (η_c vs. c_2) linear — hence no appreciable ionization; MHS parameters obtained with data from only six polymer samples having M = 4D4—2D5.

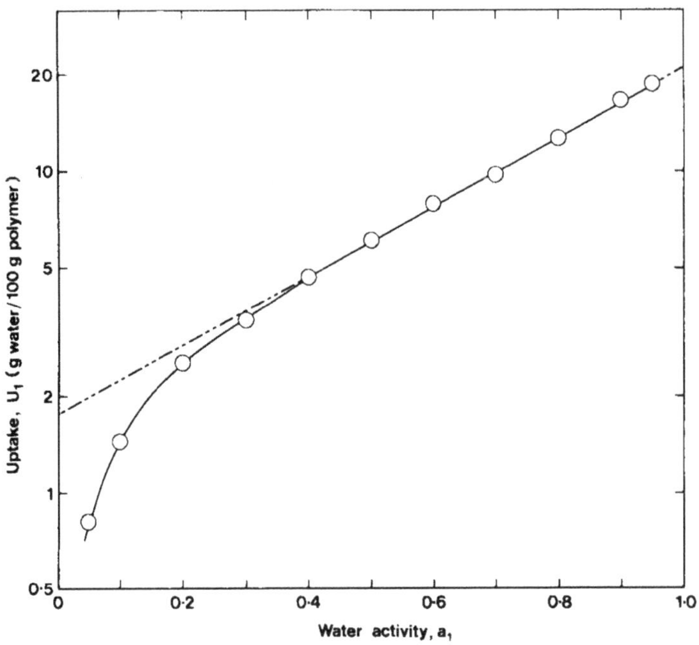

FIGURE 2. Equilibrium uptake of water, U_1 (logarithmic scale) by poly(methacrylic acid) at 40°C, vs. water activity, a_1.[145,146]

2. Aqueous Solution

Conventional PMAA is stable physically in water and in dilute acid at normal temperatures and below, but the solutions show phase separation at higher temperatures; thus a sample with $\overline{M}_w$ = 5.9D5 separates from dilute solution in 20 mM HCl at about 58°C,[1] while the phase diagram for a similar sample has its minimum (i.e., the critical temperature t_c) close to 50°C (at c_2 = 7%, which is an unexpectedly high value).[139] These observations are in line with the light-scattering data of Silberberg and co-workers[1] which gave the Θ_+ value as 56°C. These systems are also notable for forming gels at concentrations around 7%;[139,143] this gelation is induced by heating and also by stirring ("negative thixotropy"; "shear hardening"),[42,138] and may be viewed as a form of "homogeneous" phase separation by the production of a network of aggregated polymer chains.

Considering more specifically the stereoregular forms of PMAA, the *st*-form is stated to be sparingly soluble in water,[123] although it is apparently still sufficiently soluble (up to at least 0.6%) for various physical measurements to be made such as UV, VI, and OP (see Table 4).

The *it*-form is stated to be essentially insoluble in water,[121,123,125] but aqueous solutions can be prepared by dissolving in alkali and then treating with a cation-exchange resin in the H^+-form; the polymer precipitates sufficiently slowly from these so that it is possible for physical measurements to be made.[110] Even with alkali as solvent, the polymer only dissolves when the degree of ionization is at least 0.3, while higher degrees of ionization are required for dissolution in salt solutions.[104,121,125]

G. Chemical Behavior

Considering the chemical stability of PMAA, certain samples have been reported to degrade spontaneously even in the solid state.[109,115] Such effects, which can be monitored by viscosity measurements,[109] are possibly the result of the presence of "weak links" in the chains of these samples.

At higher temperatures (100°C and above), PMAA in the bulk state gradually loses water to form the cyclic anhydride.[57] This may be avoided (at least in the case of the conventional polymer) by drying below 50°C.[108] The reaction takes place most readily with the *it*-form,[57,110] as shown by the appearance of a band at 1020 cm^{-1} in the IR absorption spectrum.[110] The kinetics of the reaction have been discussed by Boucher[57] in the general context of these intramolecular reactions. The acid may be regenerated by warming (80°C) an alkaline solution of the polymer followed by reprecipitation of the acid (for the *st*-form),[130] or in the case of the (insoluble) *it*-form by treating the solid polymer with water at 80°C.[110]

PMAA in aqueous solution at normal temperatures is often found to undergo a slow degradation which may be due to bacterial attack[50] or to oxidative breakdown.[22,116] It has therefore been recommended that such solutions should either be used freshly prepared or stored in a refrigerator (5°C) under nitrogen.[22] However, Wiederhorn and Brown[134] found that UV light was the major degrading agent, in that when their solutions (apparently, either in water or in methanol) were kept in the dark or in yellow light (presumably at room temperature), they were stable for up to 2 weeks. Considering the effect of higher temperatures, dilute aqueous solutions of PMAA have been found to be stable for most experimental purposes up to about 58°C,[1] but in more concentrated solutions (~7%) significant thermal and bacterial attack (as well as reversible phase separation) occurs above approximately 47°C.[143]

Vink[147] has examined the effect on PMAA in aqueous solution at 25°C of the following reagents (studied in each case both in oxygen and in nitrogen): 1 M HCl; 1 M NaOH alone and with 10mM hydroxylamine; and 1 M NaCl alone and with various transition metal ions (Fe^{2+}, Fe^{3+}, Co^{2+}, Mn^{2+}, Ni^{2+}, Cr^{3+}) at 5 mM. Only in the case of the NaCl/Fe^{2+}/O_2 combination was there any appreciable degradation (with η_{sp} falling by 0.2%/min).

The most important chemical modification that is applied to PMAA is its esterification, most commonly methylation (i.e., to give PMMA). This last reaction is usually applied for the purpose of the estimation of stereoregularity (see Section IV.C), but it is also occasionally used for the determination of molecular weight via the measurement of the intrinsic viscosity of the PMMA formed.[2]

Katchalsky and Eisenberg[22] showed that this conversion may be carried out by treating a suspension of PMAA in benzene with diazomethane at room temperature; the reaction goes smoothly with *cv*-PMAA[2,22,120,126] and *it*-PMAA,[104,123,129] and generally also with *st*-PMAA,[104,110,120,126] although Aylward[123] found difficulties with the methylation of this latter form, possibly due to the high stereoregularity of his sample. It has been shown that the cycle of methylation and hydrolysis for PMAA (or vice versa for PMMA) does not lead to any appreciable degradation or change of stereoregularity.[22,104] The distribution statistics of the intermediate copolymers has been determined for the methyl ester,[130] the benzyl and diphenylmethyl esters,[131] and the mixed esters of each of the latter with methyl.[130]

Considering other routes to the alkyl esters, the reaction of PMAA with dimethyl sulfate in neutral solution has been used specifically to produce the intermediate copolymers (PMAA-MMA);[148] however, attempts to carry out methylation by treating the silver salt of PMAA with methyl iodide were not successful.[22]

H. Analysis

Although determination of the concentration of aqueous PMAA solutions has been carried out apparently successfully by means of evaporation to give the solids content,[50,143] in general this method gives high results because of the tenacious retention of between 4 and 10% solvent.[1,110,113] Attempts to remove the solvent by high temperature treatments may lead to chemical loss of water to give the anhydride (see Section IV.G), and possibly even to some decarboxylation.

For this reason, analysis of PMAA by means of the carboxyl group content is preferred, using titration with 0.1M carbonate-free NaOH, optionally with added salt (0.1 or 1 M NaCl) to sharpen the end point; the latter may be detected either with an indicator such as Alizarin Yellow R,[72,116] or thymolphthalein,[1,115] or more commonly now by potentiometry which has the advantage in that automatic methods may be applied. Conductimetric detection of the end point has also been reported,[116] and indeed presumably has certain advantages already discussed under PAA (see Section II.H).

V. POLYMETHACYYLAMIDE — PMAAm

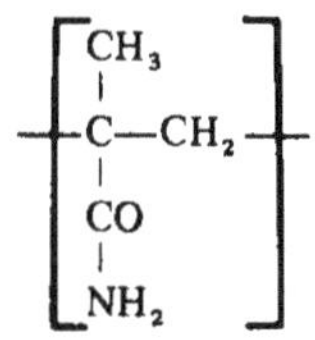

$M_0(C_4H_7NO) = 85.11$

Other names and abbreviations are poly(1-carbamoyl-1-methylethylene) (IUPAC); poly(methacrylamide); polymethacryl amide; polymethacrylic amide; PMAm.

A. General Features

This member of the acrylic group does not seem to be readily available commercially, neither does it seem to have any great technical importance or applications. In the present context its significance from the fundamental viewpoint lies in the fact that it represents the fourth member of the "acrylic quartet", although it has not been studied in anything like so much detail as the other three members have.

Examination of molecular models shows that, as with PMAA, the presence of the two substitutent groups (CH_3 and $CONH_2$) on alternate backbone carbon atoms of polymethacrylamide leads to a highly rigid chain with marked inhibition of rotation about the C-C bonds.[149]

Polymethacrylamide seems to have solubility characteristics similar to those of PAAm, since it is insoluble in many common water-miscible organic solvents such as methanol, ethanol, acetone, and dioxane.

B. Production and Purification

The monomer can be polymerized using conventional free radical initiators, for example in aqueous solution using hydrogen peroxide[1,150,151] or persulfate,[149] or in dioxane solution (from which the polymer precipitates) using benzoyl peroxide.[3] Purification may be effected by precipitating the polymer from aqueous solution with ethanol[1,3] or acetone,[149] or by exhaustive dialysis[150,151] followed by freeze drying.[150]

C. Fractionation and Characterization

It is likely that the polymer can be fractionated by stepwise precipitation from aqueous solution using miscible organic nonsolvents such as methanol, ethanol, acetone, or dioxane, or alternatively by taking advantage of the precipitation that occurs when aqueous solutions are cooled (see Section V.F). The polymer may be characterized in aqueous solution using the usual range of techniques (see Section V.E).

D. Bulk Polymer

On analogy with the glass-transition values for the other acrylic polymers (see Sections II.D, III.D, and IV.D), it is likely that the value of t_g for PMAA is fairly high, certainly above 100°C. Thus evaporated films would need to be heated to a correspondingly high temperature to give absorbed water molecules the required mobility to leave the polymer. The infrared absorption spectrum (650—4000 cm^{-1}) of the bulk polymer has been published by Nyquist.[89]

E. Solution Behavior

The results of various studies on PMAAm in aqueous solution are summarized in Table 6. Single estimate values are seen to be available for the partial specific volume $\bar{v}_2^\circ = 0.734$ $cm^3\ g^{-1}$ (entry 9) and the specific refractive index increment $\tilde{n} = 0.209\ cm^3\ g^{-1}$ (entry 3) in this solvent.

Considering the relation between intrinsic viscosity and molecular weight, there seems to be only two data points available, both for 30°C: $[\eta] = 11\ cm^3\ g^{-1}$ for $\overline{M}_w = 7.8D4$[150] and $[\eta] = 32\ cm^3\ g^{-1}$ (interpolated) for $M_w = 3.2D5$;[1] these give the MHS parameters as: $a = 0.76$; $K = 2.1d3\ cm^3\ g^{-1}$. Relying as they do solely upon two data points, these parameters should be used with caution, but may be applicable to interpolate values of $\overline{M}$ for samples with intermediate $[\eta]$ values,[3] and also for extrapolation immediately outside the range. The value of a is certainly very high for a polymer which is fairly close to its precipitation point (see Section V.F below).

F. Phase Equilibria

The only information on the equilibrium water-vapor uptake by PAAm is that, like other polymers of the acrylic group, it retains appreciable amounts (8 to 10%) of water tenaciously, even after vacuum drying.[1]

For aqueous solutions, it has been stated that a sample of the polymer ($\overline{M}_w = 7.8D4$) precipitated out of solution (concentration presumably ~1%) when it was cooled to 0°C.[150] This is in accord with the deduction that Silberberg and co-workers[1] made from their light-scattering measurements, that for this system the ideal temperature has the value $\Theta_- = 6°C$. Highly concentrated aqueous solutions of PMAAm set to gels, with the ease of setting decreased by rise in temperature.[1]

G. Chemical Behavior

Silberberg and co-workers[1] found that neutral aqueous solutions of PMAAm deteriorated above 65°C, presumably as a result of the progressive hydrolysis of the amide groups. The alkaline hydrolysis of PMAAm (and also PMAA-MAAm) has been studied by Pinner,[149] who found that even in boiling 4 M NaOH only about 62 mol % of the amide groups are hydrolyzed; the residual amide groups are apparently stabilized by being bordered on either side by an ionized carboxyl group. On the other hand, acid hydrolysis as used, for example, in the Kjeldahl method for nitrogen content, always leads to full conversion.[149] Polymethacrylamide also seems to be modifiable chemically in much the same manner as PAAm (see Section III.G).[103]

H. Analysis

For the analysis of fairly concentrated solutions, although evaporation and vacuum drying have apparently been used to analyze these solutions by means of the total solids content,[3] in other studies the polymer so obtained seemed still to retain 8 to 10% of tenaciously held water.[1] For this reason analysis by the Kjeldahl method for the total nitrogen content seems preferable.[1,149] For routine analysis, a rather more rapid but nonspecific method such as refractometry would be applicable.

VI. OTHER ACRYLIC HOMOPOLYMERS

In this section some other water-soluble acrylic homopolymers that have been studied from the fundamental viewpoint are considered. These are all N-substituted derivatives of

Table 6
STUDIES ON PMAAm IN AQUEOUS SOLUTION

Entry	Year	Polymer samples: No.	Polymer samples: Range of $\bar{M}$	Techniques used	Temperature (t/°C)	Polymer concentration	Comments	Ref.
1	1955	?	?	VI		~10%?	No negative thixotropy	42
2	1957	1	3D5	LS	22—56	0.1—1%	546 nm:$\rightarrow \bar{M}_w$; A_2; thermodynamic parameters	1
3	1957	1	3D5	RI	20—65?		546 nm: $\rightarrow \tilde{n} = 0.209$ cm^{3}g^{-1}	1
4	1957	1	3D5	SS			$\rightarrow s^o$ (footnote a)	1
5	1957	1	3D5	TD			$\rightarrow D^o$ (footnote a); $\bar{M}_{sD}$	1
6	1957	1	3D5	VI	22—62		$\rightarrow [\eta]$; k'; K_Θ	1
7	1970	1	8D4	LS			λ = ?: $\rightarrow M_w$; A_2	150
8	1960	1	8D4	VI	30	0.5—1.5%	$\rightarrow [\eta]$; k'	150
9	1969	1	~1D5	SV	25	0.1—2%	$\rightarrow \tilde{v}_2^o = 0.734$ cm^{3}g^{-1}	3
10	1969	1	~1D5	USV	25	0.1—2%	$\rightarrow x_2^o$	3
11	1969	1	~1D5	VI	25		$\rightarrow [\eta]$; $\bar{M}_v$(?)	3

[a] Value not stated.

either PAAm or PMAAm, in these cases produced directly from the corresponding monomer (rather than by chemical modification of the parent polymer). Of these, the N-monosubstituted derivatives have general significance in containing peptide groupings similar to those in proteins, although in the present case these groupings are in the side chains rather than in the backbone of the polymers. The N,N-disubstituted derivatives are also interesting because of their relation to polyvinylpyrrolidone, which possesses this same type of amide grouping and which is another very important water-soluble synthetic polymer (see Chapter IV).

A. Poly(N,N-Dimethylacrylamide)

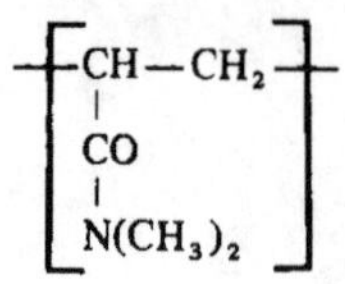

$M_0(C_5H_9NO) = 99.13$

The solution behavior of this polymer has been studied in some detail by Trossarelli and Meirone.[152] They prepared a number of samples by free radical polymerization of the monomer in dioxane, varying the reaction conditions to obtain polymers of a range of average molecular weights (from 5D4 to 1D6). The polymer was found to be soluble in water, methanol, and dioxane, but insoluble in diethyl ether. The polymer was studied at 25°C in water and in methanol by light scattering (giving $\overline{M}_w$ and A_2), refractometry ($\tilde{n}$), and viscometry ([η], also obtained for water at 40°C).

The kinetics of the alkaline hydrolysis of the related N,N-*diethyl* compound in 1 M NaOH have been studied by Conix and co-workers,[101,102] leading to the Arrhenius parameters: $A = 2D10\ M^{-1}\ s^{-1}$; $E = 22.8 \pm 0.2$ kcal mol^{-1}. It is interesting that, in contrast with the corresponding reaction with PAAm itself (Section III.G) and with PMAAm (Section V.G), there is here no mention of any anomalies in the reaction from the formation of stabilized forms of intermediate copolymers.

B. Poly(N-Isopropylacrylamide) — PIPAAm

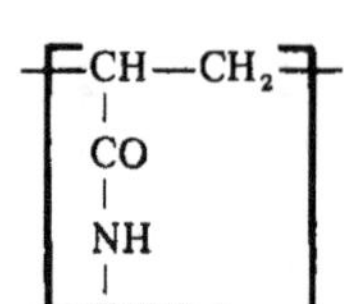

$M_0(C_6H_{11}NO) = 113.16$

A number of different aspects of the solution behavior of this polymer have been studied by Heskins and Guillet.[153] Their work was carried out using a single sample of fairly broad distribution ($\overline{M}_w/\overline{M}_n = 3.5$), and their studies in aqueous and in chloroform solutions are summarized in Table 7. The polymer showed inverse solubility-temperature behavior in aqueous solution; the phase diagram gave the critical solution temperature $t_{c+} = 31$°C ($\phi_{2c} = 0.16$); the heat of phase separation was also measured with a differential scanning calorimeter.

Table 7
STUDIES ON POLY(N-ISOPROPYLACRYLAMIDE) IN SOLUTION[a]

Entry	Techniques used	Temperature (t/°C)	Polymer concentration	Comments
		Water		
1	HS	20—30	0.03%	$\rightarrow \Delta H$
2	LS	33	0.3—1%	546 nm: $\rightarrow \overline{M}_w;(A_2);\hat{R}$
3	RI	33		546 nm: $\rightarrow \tilde{n} = 0.171\ cm^3 g^{-1}$
4	SS	20—33		$\rightarrow s^o$
5	SV	15—33		$\rightarrow \bar{v}_2^o$
6	VI	15—33		$\rightarrow [\eta]$
		Chloroform		
7	OP			$\rightarrow \overline{M}_n$
8	LS	25	0.15—0.4%	546 nm: $\rightarrow \overline{M}_w;(A_2);\hat{R}$ small
9	RI	25?		546 nm?: $\rightarrow \tilde{n} = 0.043\ cm^3\ g^{-1}$
10	VI	25		$\rightarrow [\eta]$

[a] Studies on single unfractionated sample, $\overline{M} \approx$ 1D6, by Heskins and Guillet.[153]

C. Poly(N-Acrylylglycinamide)* — PAG

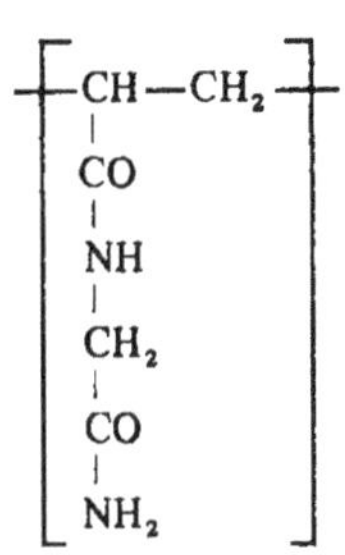

$$M_0(C_5H_8N_2O_2) = 128.13$$

This polymer has been investigated extensively by Haas and co-workers, the results being reported in a sequence of six papers.[154-159] The stated aim of these studies was to prepare a purely synthetic substitute for photographic gelatin; apparently, the rationale behind the choice of this structure was that it contains dipeptide side chains (mimicking the structure of gelatin) attached to a carbon chain backbone. In fulfilment of the original aim, PAG was found to produce thermally reversible gels in aqueous solution, with the melting point of the gel dependent on the molecular weight and concentration of the polymer. In addition to being soluble in water, the polymer dissolves in 2 M aqueous sodium thiocyanate (which disperses the aggregates found in plain aqueous solution) and also in dimethylsulfoxide (hot or cold); it is insoluble in methanol, ethanol, acetone, glacial acetic acid, and dimethylformamide, and only swollen in warm pyridine.

The sequence of six parts deals with the following topics: part I[154] gives an account of the preliminary results obtained; part II[155] deals with the synthesis of the polymer (by free radical polymerization of the monomer), its density and IR absorption spectrum in the bulk state, and viscometry on its solutions in water and in 2 M NaSCN, as well as its gelation behavior in aqueous solution; part III[156] reports the characterization of the polymer by osmotic

* Alternative names are poly(N-acetamidoacrylamide); poly(α-acrylamidoacetamide).

pressure measurements and viscometry, the role of methanol as a chain-transfer agent in the polymerization reaction, and further studies on gelation; part IV[157] deals with the kinetics of the polymerization initiated by persulfate; and parts V[158] and VI[159] report comparisons of the gelation and gel melting behavior of PAG with that of gelatin, while the last part also contains some comments on the storage stability and rheology of the polymer.

D. Poly(N-Methacrylylglycinamide)* — PMAG

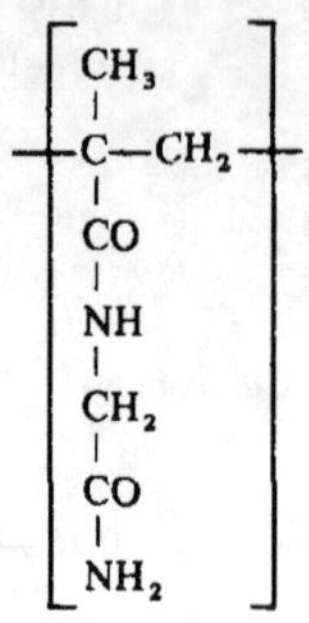

$M_0(C_6H_{10}N_2O_2) = 142.16$

This polymer has also been synthesized by Haas and co-workers, and its properties studied, although in less detail than those of PAG.[154,160] Like PAG, it forms thermally reversible gels in aqueous solution, but higher concentrations and/or molecular weights are required, and the gel melting points are lower than with PAG. The mode of preparation and the general properties of PMAG are also similar to PAG; it is soluble in water, 2 M aqueous sodium thiocyanate, hot glacial acetic acid, and hot dimethylsulfoxide, swollen in warm pyridine and *m*-cresol, and insoluble in the lower alcohols, acetone, tetrahydrofuran, and dimethylformamide.

VII. ACRYLIC COPOLYMERS

Further information on the factors that lead to a polymer being water soluble, and on the behavior of such polymers in aqueous solution, can be obtained by investigating copolymers. Most studies have been carried out on copolymers containing only two types of monomer unit, and for this reason and for the sake of simplicity, attention will be restricted to these. A selection of such copolymers which contain at least one monomer of the acrylic type, are water soluble even when not ionized, and have been studied from various scientific aspects, is listed in Table 8.

The production of copolymers from two monomers leads in principle to an almost indefinite range of possibilities for properties and behavior, since in the simplest case we would expect any feature to be a continuously varying function of composition between the extremes of the two homopolymers. Further possibilities still arise by varying the distribution of the two units within the structure, i.e., by producing random, alternating, block, and graft copolymers. It is for these reasons that such copolymers are also technically and commercially important.

In the present context of the acrylic group, particular interest arises with the interpolymers, where both units are of the acrylic type. Such interpolymers arise naturally in the interconversion reactions already discussed, especially PAA $\rightleftharpoons$ PMA, PAA $\rightleftharpoons$ PAAm, and the corresponding reactions with the methacrylates. In the case, e.g., of PAA-MA, the nominally random copolymer may be obtained in principle in three distinct ways, i.e., partial meth-

* Alternative names are poly(N-acetamidomethacrylamide); poly(N-methacrylamidoacetamide); poly(α-methacrylylglycinamide).

Table 8
SOME WATER-SOLUBLE NONIONIC ACRYLIC COPOLYMERS: POLY(A-B)

Entry	Monomer A	Monomer B	Ref.
1	Acrylamide	Acrylic acid	38, 97, 161, 162
2	Acrylamide	Acrylonitrile	161, 163
3	Acrylic acid	N-Acrylylglycinamide	155, 156
4	Acrylic acid	Ethyl acrylate	164
5	Acrylic acid	Methyl acrylate	37
6	Acrylic acid	Methylenebutyrolactam	102
7	N-Acrylylglycinamide	N-Isopropylacrylamide	159
8	Methacrylamide	Methacrylic acid	149
9	Methacrylic acid	Benzyl methacrylate	131
10	Methacrylic acid	Diphenylmethyl methacrylate	131
11	Methacrylic acid	Methyl methacrylate	130, 148
12	Methacrylic acid	Styrene	124, 137

ylation of PAA; partial saponification of PMA; and copolymerization of AA and MA. In the general case, these three different routes may be expected to lead to different distributions of chain units in the chains. Similarly, since in the course of any interconversion reaction each molecule must go through the complete range of copolymers between the two extremes, then the success of the reaction depends upon the reactivity of each of these successive copolymers.

POSTSCRIPT

II.A. Review of the properties and applications of PAA[166]

II.C. Fractionation (of *Carbopol* 940) by extraction with hydrochloric acid (6—0.5M)[167]

II.D. Infrared spectrum (KBr disc; 800—1500 cm^{-1})[168]

II.E.1. Fluorescence depolarization (dye-labeled *at*, *it*-, and *st*-PAA);[169] viscometry, (giving MHS parameters);[170] NMR relaxation [T_1 (^{17}O)][171]

II.E.2. Fluorescence depolarization (dye-labeled *at*-, *it*-, and *st*-PAA) in methanol, dimethylformamide, and tetrahydrofuran[169]

II.G. Effects of radiation;[172] wet oxidation[173]

III.A. Review of the properties and applications of PAAm;[174] further detailed studies (including aqueous solution) by Klein, Kulicke, and co-workers[175-180]

III.B. Review of methods of preparation[181]

III.D. Infrared spectrum (KCl disc; 1000—5000 cm^{-1})[182]

III.E.1. In addition to References 175 to 180: NMR;[183] fluorescence depolarization (fluorescein-labeled polymers);[184,185] heats of dilution and partial specific volume[186]

III. G. Hydrolysis (acid and alkaline);[187-190] effects of radiation;[172] wet oxidation[173]

III.H. Analysis of solution by hydrolysis and Nessler determination of NH_3 liberated,[191] or turbidimetry on the hydrolysate with *Hyamine* 1622,[191] or direct nephelometry with tannic acid[192]

IV.A. Review of the properties and applications of PMAA[166]

IV.B. Effects of pH during free radical polymerization on stereoregularity;[193] kinetics of the hydrolysis of PMAA-MMA for production of PMAA[194]

IV.D. Infrared spectrum (KBr disc; 625—4000 cm^{-1}) for *cv*- and *it*-PMAA[195]

IV.E.1. Osmotic pressure (0.4M HCl);[196] partial specific volume;[197] viscometry and fluorescence (with Auramine O);[198] fluorescence depolarization (dye-labeled *cv*-, *it*-, and *st*-PMAA);[169] NMR relaxation [T_1 (^{17}O)][171]

IV.E.2. Fluorescence depolarization (dye-labeled *cv*-, *it*-, and *st*-PMAA in methanol, dimethylformamide, and tetrahydrofuran)[169]

IV.G. Effects of radiation;[172] kinetics of dehydration (of *cv*- and *it*-PMAA) to cyclic anhydride[195]

V.A. High molecular weight ($M > 7D4$) samples of PMAAm may be insoluble in water at normal temperatures[199,200]

V.B. Kinetics of photosensitized polymerization of MAAm;[199,201] preparation of PMAAm with redox system ammonium persulfate + sodium metabisulfite,[200] or thermal polymerization with potassium persulfate;[183] purification by precipitation from water[199] or 0.45M aqueous magnesium perchlorate[200] with methanol

V.C. Fractionation with formamide/methanol;[200] characterization of water-insoluble (high molecular weight) samples in aqueous magnesium perchlorate (0.4 or 1.2M) or aqueous urea (2—8M)[200]

V.D. Infrared spectrum (KCl disc; 1000—5000 cm^{-1})[182]

V.E. NMR spectrum[183]

V.F. Dissolution temperatures in water of high molecular weight PMAAm[200]

V.G. Kinetics of alkaline and acid hydrolysis[187]

VI.A. Poly(N,N-dimethylacrylamide): kinetics of polymerization,[202] heat of dilution, and partial specific volume (aqueous solution);[186] kinetics of hydrolysis of poly(N,N-diethylacrylamide) and of poly(N,N-diethylmethacrylamide)[187]

VI.B. Poly(N-methylacrylamide): heat of dilution and partial specific volume (aqueous solution),[186] kinetics of alkaline, and acid hydrolysis[187]

VII. (Table 8): acrylamide-acrylic acid:[179,187,188,192] acrylamide-methacrylamide;[182] acrylic acid-vinylpyrrolidone;[187] methacrylic acid-methyl methacrylate.[194]

REFERENCES

1. **Silberberg, A., Eliassaf, J., and Katchalsky, A.,** Temperature-dependence of light scattering and intrinsic viscosity of hydrogen bonding polymers, *J. Polym. Sci.*, 23, 259, 1957.
2. **Roy-Chowdhury, P.,** Adiabatic compressibility of polyelectrolytes in aqueous solutions: poly(methacrylic acid), *J. Appl. Polym. Sci.*, 12, 751, 1968.
3. **Roy-Chowdhury, P.,** Ultrasonic velocity and adiabatic compressibility of polymethacrylamide in aqueous solution at 25°C, *Indian J. Chem.*, 7, 692, 1969.
4. **Roy-Chowdhury, P. and Kale, K. M.,** The adiabatic compressibility of poly(acrylic acid) and polyacrylamide in aqueous solution, *J. Appl. Polym. Sci.*, 14, 2937, 1970.
5. **Molyneux, P.,** Synthetic polymers, in *Water: A Comprehensive Treatise*, Vol. 4, Franks, F., Ed., Plenum Press, New York, 1975, chap. 7.
6. **Anon.,** *IARC Monographs on the Evaluation of the Carcinogenic Risk of Chemicals to Humans*, Vol. 19, World Health Organization/International Agency for Research on Cancer, Lyon, France, 1979.
7. **Anon.,** Acrylic acid, methyl acrylate, ethyl acrylate and polyacrylic acid, in *IARC Monographs on the Evaluation of the Carcinogenic Risk of Chemicals to Humans*, Vol. 19, World Health Organization/International Agency for Research on Cancer, Lyon, France, 1979, 47.
8. **Kern, W.,** Über heteropolare Molekülkolloide. I. Die Polyacrylsäure, ein Modell des Eiweisses, *Z. Phys. Chem. (Leipzig) Abt. A*, 181, 249, 1938.
9. **Kern, W.,** Über heteropolare Molekülkolloide. II. Die Viscosität von Lösungen der Polyacrylsäure und ihrer Salze, *Z. Phys. Chem. (Leipzig) Abt. A*, 181, 283, 1938.
10. **Miller, M. L.,** Acrylic acid polymers, in *Encyclopedia of Polymer Science and Technology: Plastics, Resins, Rubbers, Fibers*, Vol. 1, Mark, H. M., Gaylord, N. G., and Bikales, N. M., Eds., Interscience, New York, 1964, 197.
11. **Glavis, F. J.,** Poly(acrylic acid) and its homologs, in *Water-Soluble Resins*, 2nd ed., Davidson, R. L. and Sittig, M., Eds., Van Nostrand Reinhold, New York, 1968, chap. 8.
12. **Anon.,** *Polyacrylic Acids: Primal/Acrysol A-1, A-3, and A5*, Tech. Bull. RC-10, Rohm & Haas Co., Philadelphia, 1968.

13. **Anon.,** *Carbopol Water-Soluble Resins*, Service Bull. GC-36 (Revised), B. F. Goodrich Chemical Division, Cleveland, 1978.
14. **Breitenbach, J. W. and Kauffmann, H. F.,** Acrylic acid popcorn polymers, *Makromol. Chem.*, 175, 2597, 1974.
15. **Newman, S., Krigbaum, W. R., Laugier, C., and Flory, P. J.,** Molecular dimensions in relation to intrinsic viscosities, *J. Polym. Sci.*, 14, 451, 1954.
16. **Alexandrowicz, Z.,** The correlation between activities of polyelectrolytes, measured by light-scattering and osmotic methods, *J. Polym. Sci.*, 40, 91, 1959.
17. **Kay, P. J. and Trelor, F. E.,** The intrinsic viscosity of poly(acrylic acid) at different ionic strengths: random coil and rigid rod behavior, *Makromol. Chem.*, 175, 3207, 1974.
18. **Chapiro, A. and Sommerlatte, T.,** Polymérisation radiochimique de l'acide acrylique en phases liquide et solide I. Étude cinétique, *Eur. Polym. J.*, 5, 707, 1969.
19. **Chapiro, A. and Sommerlatte, T.,** Polymérisation radiochimique de l'acide acrylique en phases liquide et solide. II. Examen des polymères formés — discussion, *Eur. Polym. J.*, 5, 725, 1969.
20. **Monjol, P.,** Acides polyacryliques stéréoréguliers. Photopolymérisation en solution de l'acide acrylique, *C. R. Acad. Sci. Ser. C*, 265, 1426, 1967.
21. **Eastmond, G. C., Haigh, E., and Taylor, B.,** Solid-state polymerization of acrylic acid initiated by polarized ultraviolet radiation, *Trans. Faraday Soc.*, 65, 2497, 1969.
22. **Katchalsky, A. and Eisenberg, H.,** Molecular weight of polyacrylic and polymethacrylic acid, *J. Polym. Sci.*, 6, 145, 1951.
23. **Miller, M. L., Botty, M. C., and Rauhut, C. A.,** Crystalline polyacrylic acid, *J. Colloid Sci.*, 15, 83, 1960.
24. **Miller, M. L., O'Donnell, K., and Skogman, J.,** Crystalline polyacrylic acid, *J. Colloid Sci.*, 17, 649, 1962.
25. **Monjol, P.,** Préparation d'acide polyacrylique syndiotactique par hydrolyse du polyacrylate d'isopropyle. Propriétés cristallographiques des acides polyacryliques stéréoréguliers, *C. R. Acad. Sci. Ser. C*, 266, 81, 1968.
26. **Bevington, J. C., Eaves, D. E., and Vale, R. L.,** Tests on the hydrolysis of certain synthetic polymers, *J. Polym. Sci.*, 32, 317, 1958.
27. **Reynolds, J. E. F. and Prasad, A. B., Eds.,** *Martindale: The Extra Pharmacopoeia*, 28th ed., The Pharmaceutical Press, London, 1982.
28. **Barry, B. W. and Meyer, M. C.,** The rheological properties of carbopol gels. I. Continuous shear and creep properties of carbopol gels, *Int. J. Pharm.*, 2, 1, 1979.
29. **Barry, B. W. and Meyer, M. C.,** The rheological properties of carbopol gels. II. Oscillatory properties of carbopol gels, *Int. J. Pharm.*, 2, 27, 1979.
30. **Okamoto, H., Nakajima, H., and Wada, Y.,** Dynamic viscosity of dilute aqueous solutions of poly(acrylic acid) at frequencies of 2 to 500 kHz, *J. Polym. Sci. Polym. Phys. Ed.*, 12, 1035, 1974.
31. **Crescenzi, V., Delben, F., and Quadrifoglio, F.,** Calorimetric investigation of poly(methacrylic acid) and poly(acrylic acid) in aqueous solution, *J. Polym. Sci. Part A-2*, 10, 357, 1972.
32. **Noda, I., Tsuge, T., and Nagasawa, M.,** The intrinsic viscosity of polyelectrolytes, *J. Phys. Chem.*, 74, 710, 1970.
33. **Mandel, M.,** The potentiometric titration of weak polyacids, *Eur. Polym. J.*, 6, 807, 1970.
34. **Lleras, J., Ahouande, F., and Combet, S.,** Fractionnement de l'acide polyacrylique par démixtion par élévation de température, *C. R. Acad. Sci. Ser. C*, 278, 689, 1974.
35. **Monjol, P.,** Acides polyacryliques et polyacrylates stéréoréguliers. Détermination de la tacticité par résonance magnétique nucléaire et spectroscopie infrarouge, *C. R. Acad. Sci. Ser. C*, 267, 1021, 1968.
36. **Smith, K. L., Winslow, A. E., and Petersen, D. E.,** Association reactions for poly(alkylene oxides) and polymeric poly(carboxylic acids), *Ind. Eng. Chem.*, 51, 1361, 1959.
37. **Thompson Hughes, L. J. and Fordyce, D. B.,** Sorption of water vapor by water-soluble polymers: kinetic, equilibrium and glass transition data, *J. Polym. Sci.*, 22, 509, 1956.
38. **Klein, J. and Heitzmann, R.,** Preparation and characterization of poly(acrylamide-co-acrylic acid), *Makromol. Chem.*, 179, 1895, 1978.
39. **Bardet, L., Cassanas-Fabre, G., and Alain, M.,** Étude de la transition conformationelle de l'acide polyacrylique syndiotactique en solution aqueuse par spectroscopie de vibration, *J. Mol. Struct.*, 24, 153, 1975.
40. **Boyer-Kawenoki, F.,** Contribution à l'étude spectroscopique infrarouge du composé d'addition polyvinylpyrrolidone-acide polyacrylique, *C. R. Acad. Sci. Ser. C*, 263, 278, 1966.
41. **Miyamoto, T. and Cantow, H.-J.,** A nuclear magnetic resonance study on the conformation and molecular interaction of polyelectrolyte model compound in alcohol/water mixtures, *Makromol. Chem.*, 169, 211, 1973.
42. **Eliassaf, J., Silberberg, A., and Katchalsky, A.,** Negative thixotropy of aqueous solutions of polymethacrylic acid, *Nature (London)*, 176, 1119, 1955.

43. **Barone, G., Crescenzi, V., and Quadrifoglio, F.,** pH-Induced transition in aqueous solutions of poly(methacrylic) acid. Note I: conventional polyacid, *Ric. Sci. Parte 2 Sez. A,* 35, 393, 1965.

43a. **Liquori, A. M., Barone, G., Crescenzi, V., Quadrifoglio, F., and Vitagliano, V.,** Hydrophobic interactions in polyelectrolyte solutions, *J. Macromol. Chem.,* 1, 291, 1966.

44. **Porter, R. S. and Johnson, J. F.,** The entanglement concept in polymer systems, *Chem. Rev.,* 66, 1, 1966.

45. **Mandel, M., Leyte, J. C., and Stadhouder, M. G.,** The conformational transition of poly(methacrylic acid) in solution, *J. Phys. Chem.,* 71, 603, 1967.

46. **Ise, N. and Okubo, T.,** The partial molal volume of polyelectrolytes, *J. Am. Chem. Soc.,* 90, 4527, 1968.

47. **Michels, B. and Zana, R.,** Absorption ultrasonore de solutions aqueuses d'acides polyacrylique et polyméthacrylique non neutralisées, *Kolloid-Z. Z. Polym.,* 234, 1008, 1969.

48. **Friedman, S., Caillé, A., and Daoust, H.,** Apparent molal volume of poly(acrylic acid) in aqueous solution, *Macromolecules,* 3, 700, 1970.

49. **Williams, D. J. A. and Ottewill, R. H.,** The stability of silver iodide sols in the presence of polyacrylic acids of various molecular weights, *Kolloid-Z. Z. Polym.,* 243, 141, 1971.

50. **Tondre, C. and Zana, R.,** Apparent molal volumes of polyelectrolytes in aqueous solutions, *J. Phys. Chem.,* 76, 3451, 1972.

51. **Braud, C., Muller, G., Fenyo, J.-C., and Sélégny, E.,** Conformational transition of poly(methacrylic acid) in methanol-water mixtures, *J. Polym. Sci. Polym. Chem. Ed.,* 12, 2767, 1974.

52. **Bardet, L. and Alain, M.,** La gélification d'un haut polymère d'acide acrylique utilisé en pharmacie, *Ann. Pharm. Franc.,* 33, 651, 1975.

53. **Inzelt, Gy. and Gróf, P.,** Study of the hydration of macromolecules. III. Measurement of the self-diffusion of water in poly(acrylic acid) solutions, *Acta Chim. Acad. Sci. Hung.,* 93, 117, 1977.

54. **Bardet, L. and Cabrol, M.,** Étude de la diffusion de l'eau dans des sols d'acide polyacrylique, *Ann. Pharm. Franc.,* 36, 323, 1978.

55. **Flory, P. J. and Osterheld, J. E.,** Intrinsic viscosities of polyelectrolytes. Poly(acrylic acid), *J. Phys. Chem.,* 58, 653, 1954.

56. **Peterson, G. H.,** Sorption of water by vinyl polymers, *J. Polym. Sci.,* 28, 458, 1958.

57. **Boucher, E. A.,** Reaction kinetics of polymer substituents. Neighbouring-substituent effects in pairing reactions, *J. Chem. Soc. Faraday Trans. 1,* 68, 2281, 1972.

58. **Crisp, S., Lewis, B. G., and Wilson, A. D.,** Conductometric titration of aqueous solutions of poly(acrylic acid) and its copolymers, *J. Dent. Res.,* 54, 1238, 1975; *Chem. Abstr.,* 84, 49898r, 1976.

59. **Wimberley, J. W. and Jordan, D. E.,** Automated methods for the determination of low concentrations of polyelectrolytes, *Anal. Chim. Acta,* 56, 308, 1971.

60. **Sweet, F. and Rolfe, P. F.,** Estimation of traces of poly(acrylic acid) and other poly(carboxylic acids) in water and salt solutions by complexing with methylene blue, *Anal. Chem.,* 38, 1958, 1966.

61. **Montgomery, W. H.,** Polyacrylamide, in *Water-Soluble Resins,* 2nd ed., Davidson, R. L. and Sittig, M., Eds., Van Nostrand Reinhold, New York, 1968, chap. 9.

62. **Anon.,** *Cyanamer Polyacrylamides: Cyanamer A370 Modified Polyacrylamide, Cyanamer P26 Modified Polyacrylamide, Cyanamer P250 Polyacrylamide,* Tech. Bull. PRC 21, American Cyanamid Co., Wayne, N.J., 1971.

63. **Anon.,** *Cyanamer Polyacrylamides for the Processing Industries,* Tech. Bull. PRT-21C, American Cyanamid Co., Wayne, N.J., 1975.

64. **Anon.,** *Dow Products,* Dow Chemical Co., Midland, Mich., 1978.

65. **Kulicke, W.-M. and Klein, J.,** Zur Präparation und Charakterisierung hochmolekularer Polyacrylamide, *Angew. Makromol. Chem.,* 69, 169, 1978.

66. **Minsk, L. M., Kotlarchnik, C., Meyer, G. N., and Kenyon, W. O.,** Imidization during polymerization of acrylamide, *J. Polym. Sci. Polym. Chem. Ed.,* 12, 133, 1974.

67. **Scholtan, W.,** Molekulargewichtsbestimmung von Polyacrylamid mittels der Ultrazentrifuge, *Makromol. Chem.,* 14, 169, 1954.

68. **Riggs, J. P. and Rodriguez, F.,** Persulfate-initiated polymerization of acrylamide, *J. Polym. Sci. Part A-1,* 5, 3151, 1967.

69. **Ishige, T. and Hamielec, A. E.,** Solution polymerization of acrylamide to high conversion, *J. Appl. Polym. Sci.,* 17, 1479, 1973.

70. **Misra, G. S. and Rebello, J. J.,** Kinetics of the aqueous polymerization of acrylamide initiated by the permanganate-tartaric acid redox system, *Makromol. Chem.,* 175, 3117, 1974.

71. **Suen, T. J., Jen, Y., and Lockwood, J. V.,** The polymerization of acrylamide with chlorate-sulfite initiator, *J. Polym. Sci.,* 31, 481, 1958.

72. **Riggs, J. P. and Rodriguez, F.,** Polymerization of acrylamide initiated by the persulfate-thiosulfate couple, *J. Polym. Sci. Part A-1,* 5, 3167, 1967.

73. **Ogumi, Z., Tari, I., Takehara, Z., and Yoshizawa, S.,** The electro-initiated polymerization of acrylamide in an aqueous solution of trifluoroacetic acid and potassium trifluoroacetate, *Bull. Chem. Soc. Jpn.*, 47, 1843, 1974.
74. **Dainton, F. S. and Tordoff, M.,** The polymerization of acrylamide in aqueous solution. III. The hydrogen peroxide photosensitized reaction at 25°C, *Trans. Faraday Soc.*, 53, 499, 1957.
75. **Dainton, F. S. and Tordoff, M.,** The polymerization of acrylamide in aqueous solution. IV. The ferric ion photosensitized reaction at 25°C and the burial-emergence cycle, *Trans. Faraday Soc.*, 53, 666, 1957.
76. **Rodriguez, F. and Goettler, L. A.,** The flow of moderately concentrated polymer solutions in water, *Trans. Soc. Rheol.*, 8, 3, 1964.
77. **Henglein, A. and Schulz, R.,** Die Auslösung der Polymerization des Acrylamids durch Ultraschall, *Z. Naturforsch.*, 7b, 184, 1952.
78. **Collinson, E., Dainton, F. S., and McNaughton, G. S.,** The effect of acrylamide on the X- and γ-ray yields of hydrogen peroxide from de-aerated water, *Trans. Faraday Soc.*, 53, 357, 1957.
79. **Collinson, E., Dainton, F. S., and McNaughton, G. S.,** The polymerization of acrylamide in aqueous solution. I. The X- and γ-ray initiated reaction, *Trans. Faraday Soc.*, 53, 476, 1957.
80. **Collinson, E., Dainton, F. S., and McNaughton, G. S.,** The polymerization of acrylamide in aqueous solution. II. The effect of ferric perchlorate on the X- and γ-ray initiated reaction, *Trans. Faraday Soc.*, 53, 489, 1957.
81. **Chapiro, A. and Perec, L.,** Polymérisation radiochimique de l'acrylamide dans des milieux de nature differente, *Eur. Polym. J.*, 7, 1335, 1971.
82. **Collinson, E., Dainton, F. S., Smith, D. R., Trudel, G. J., and Tazuké, S.,** The oxidation and reduction of free radicals by metal ions in aqueous solution, *Discuss. Faraday Soc.*, 29, 188, 1960.
83. **Collinson, E., Dainton, F. S., Mile, B., Tazuke, S., and Smith, D. R.,** Thermal redox reactions between metal ions and radicals in aqueous solutions, *Nature (London)*, 198, 26, 1963.
84. **Ogata, N.,** The transition polymerization of acrylamide in the presence of Grignard reagents, *J. Polym. Sci.*, 46, 271, 1960.
85. **Griot, O. and Kitchener, J. A.,** Role of surface silanol groups in the flocculation of silica suspensions by polyacrylamide. I. Chemistry of the adsorption process, *Trans. Faraday Soc.*, 61, 1026, 1965.
86. **Narkis, N. and Rebhun, M.,** Ageing effects in measurements of polyacrylamide solution viscosities, *Polymer (London)*, 7, 507, 1966.
87. **Anon.,** *Gel Chromatography*, Bio-Rad Laboratories, Richmond, Calif., 1971.
88. **Rodriguez, F.,** *Principles of Polymer Systems*, Mcgraw-Hill, New York, 1970.
89. **Nyquist, R. A.,** *Infrared Spectra of Plastics (Polymers) and Resins*, 2nd ed., The Dow Chemical Co., Midland, Mich., 1961.
90. **Cantow, H.-J.,** Streulichtmessungen an wässrigen Polyacrylsäureamidlösungen, *Z. Naturforsch.*, 7b, 485, 1952.
91. **Anon.,** *Polyacrylamide*, New Product Bull. No. 34, American Cyanamid Co., New York, 1955.
92. **Anon.,** *Cyanamer Polyacrylamide*, Tech. Bull., American Cyanamid Co., Wayne, N.J., 1967.
93. **Shyluk, W. P. and Stow, F. S.,** Aging and loss of flocculation activity of aqueous polyacrylamide solutions, *J. Appl. Polym. Sci.*, 13, 1023, 1969.
94. **Brown, W. and Chitumbo, K.,** Solute diffusion in hydrated polymer networks. II. Polyacrylamide, hydroxyethylcellulose and cellulose gels, *J. Chem. Soc. Faraday Trans. I*, 71, 12, 1975.
95. **Kulicke, W.-M. and Klein, J.,** Zur Frage der Instabilität von Polyacrylamidlösungen, *Angew. Makromol. Chem.*, 69, 189, 1977.
96. **Katayama, S., Arata, Y., and Fujiwara, S.,** Proton magnetic relaxation time T_1 of water in acrylamide gel, *Bull. Chem. Soc. Jpn.*, 51, 1545, 1978.
97. **Klein, J. and Conrad, K.-D.,** Molecular weight determination of poly(acrylamide) and poly(acrylamide-co-sodium acrylate), *Makromol. Chem.*, 179, 1635, 1978.
98. **Katayama, S. and Fujiwara, S.,** NMR Study of the spatial effect of polyacrylamide gel upon the water molecules confined in it, *J. Am. Chem. Soc.*, 101, 4485, 1979.
99. **Dobry, A.,** Partage des substances macromoléculaires entre deux solvants, *Makromol. Chem.*, 18/19, 317, 1956.
100. **Nagashiro, W., Tsunoda, T., Tanaka, M., and Oikawa, M.,** Degradation of polyacrylamide molecules in aqueous solutions by high-speed stirring, *Bull. Chem. Soc. Jpn.*, 48, 2597, 1975.
101. **Conix, A. and Smets, G.,** Ring opening in lactam polymers, *J. Polym. Sci.*, 15, 221, 1955.
102. **Conix, A., Smets, G., and Moens, J.,** Alkaline hydrolysis of amide group containing polymers, *Ric. Sci. Suppl. Simposio Int. Chim. Macromol. Milan/Turin*, 1954, 25, 200, 1955; *Chem. Abstr.*, 54, 11545d, 1960.
103. **Schiller, A. M. and Suen, T. J.,** Ionic derivatives of polyacrylamide, *Ind. Eng. Chem.*, 48, 2132, 1956.
104. **O'Neill, J. J., Loebl, E. M., Kandanian, A. Y., and Morawetz, H.,** Dependence of the association of poly(methacrylic acid) with divalent cations on the stereoregularity of the polymers, *J. Polym. Sci. Part A*, 3, 4201, 1965.

105. **Alexandrowicz, Z.,** Results of osmotic and Donnan equilibria measurements in polymethacrylic acid-sodium bromide solutions. II, *J. Polym. Sci.*, 43, 336, 1960.
106. **Eliassaf, J., Eriksson, F., and Eirich, F. R.,** The interaction of poly(vinyl pyrrolidone) with cosolutes, *J. Polym. Sci.*, 47, 193, 1960.
107. **Leyte, J. C. and Mandel, M.,** Potentiometric behavior of polymethacrylic acid, *J. Polym. Sci. Part A*, 2, 1879, 1964.
108. **Katchalsky, A. and Miller, I.,** The surface activity of polymeric acids in aqueous solutions, *J. Phys. Chem.*, 55, 1182, 1951.
109. **Priel, Z. and Silberberg, A.,** Conformation of poly(methacrylic acid) in alcohol-water mixtures. I. Effect of concentration, temperature, pH and ionic strength, *J. Polym. Sci. Part A-2*, 8, 689, 1970.
110. **Barone, G., Crescenzi, V., and Quadrifoglio, F.,** pH-Induced transition of polymethacrylic acid in aqueous solution. Note II: stereoregular polyacids, *Ric. Sci. Parte 2, Sez. A*, 35, 1069, 1965.
111. **Okamoto, H. and Wada, Y.,** Viscoelastic study of the conformational transition of poly(methacrylic acid) in dilute aqueous solution, *J. Polym. Sci. Polym. Phys. Ed.*, 12, 2413, 1974.
112. **Pinner, S. H.,** Polymerization rate of methacrylic acid, *J. Polym. Sci.*, 9, 282, 1952.
113. **Van der Drift, W. P. J. T. and Overbeek, J. Th. G.,** Conductivity and relaxation effect of alkali polymethacrylates in alkali bromide solutions, *Recl. Trav. Chim. Pays-Bas*, 98, 81, 1979.
114. **Cutnell, J. D. and Glasel, J. A.,** ^{13}C Nuclear spin-lattice relaxation and nuclear Overhauser enhancement in aqueous solutions of poly(methacrylic acid), *Macromolecules*, 9, 71, 1976.
115. **Arnold, R. and Caplan, S. R.,** Solutions of polymethacrylic acid. I. Molecular weights and second virial coefficients of the undissociated acid, *Trans. Faraday Soc.*, 51, 857, 1955.
116. **Arnold, R. and Overbeek, J. Th. G.,** The dissociation and specific viscosity of polymethacrylic acid, *Recl. Trav. Chim. Pays-Bas*, 69, 192, 1950.
117. **Oth, A. and Doty, P.,** Macro-ions. II. Polymethacrylic acid, *J. Phys. Chem.*, 56, 43, 1952.
118. **Katchalsky, A. and Spitnik, P.,** Potentiometric titrations of polymethacrylic acid, *J. Polym. Sci.*, 2, 432 and 487, 1947.
119. **Kay, P. J., Kelly, D. P., Milgate, G. I., and Treloar, F. E.,** The conformational transition in poly(methacrylic acid) and butyl vinyl ether/maleic anhydride copolymers studied by ^{1}H NMR linewidth measurements, *Makromol. Chem.*, 177, 885, 1976.
120. **Loebl, E. M. and O'Neill, J. J.,** Preparation of syndiotactic polymethacrylic acid, *J. Polym. Sci. Part B (Polym.s30Lett.)*, 1, 27, 1963,
121. **Loebl, E. M. and O'Neill, J. J.,** Solution properties of isotactic polymethacrylic acid, *J. Polym. Sci.*, 45, 538, 1960.
122. **Stork, W. H. J., De Hasseth, P. L., Schippers, W. B., Körmeling, C. M., and Mandel, M.,** Interaction between Crystal Violet and poly(methacrylic acid) in aqueous solutions. I. Results from spectroscopic measurements and dialysis, *J. Phys. Chem.*, 77, 1772, 1973.
123. **Aylward, N. N.,** Stereoregular poly(methacrylic acids), *J. Polym. Sci. Part A*, 8, 319, 1970.
124. **Conio, G., Patrone, E., Russo, S., and Trefiletti V.,** Hydrophobic polyelectrolytes: preparation and conformational properties of poly(styrene-co-methacrylic acid), *Makromol. Chem.*, 177, 49, 1976.
125. **Morawetz, H. and Kandanian, A. Y.,** The association of bispyridinium cations with polycarboxylic acids, *J. Phys. Chem.*, 70, 2995, 1966.
126. **Lando, J. B., Litt, M., Kumar, N. G., and Shimko, T. M.,** Specific solvent interactions in the stereoregular polymerization of methacrylic acid, *J. Polym. Sci. Polym. Symp.*, 44, 203, 1974.
127. **Platé, N. A.,** Problems of polymer modification and the reactivity of functional groups of macromolecules, *Pure Appl. Chem.*, 46, 49, 1976.
128. **Bovey, F. A. and Tiers, G. V. D.,** Polymer NSR spectroscopy. II. The high resolution spectra of methyl methacrylate polymers prepared from free radical and anionic initiators, *J. Polym. Sci.*, 44, 173, 1960.
128a. **Bovey, F. A.,** The high resolution NMR spectroscopy of polymers, in ***Progress in Polymer Science***, Vol. 3, Jenkins, A. D., Ed., Pergamon Press, Oxford, 1971, chap. 1.
129. **Leyte, J. C., Arbow-van der Veen, H. M. R., and Zuiderweg, L. H.,** Irreversible potentiometric behavior of isotactic poly(methacrylic acid), *J. Phys. Chem.*, 76, 2559, 1972.
130. **Klesper, E., Strassila, D., and Regel, W.,** Copolymer statistics during esterification of syndiotactic poly(methacrylic acid) with diazomethane, *Makromol. Chem.*, 175, 523, 1974.
131. **Strasilla, D. and Klesper, E.,** Copolymer statistics during esterification of syndiotactic poly(methacrylic acid) with diazotoluene and diazodiphenylmethane, *Makromol. Chem.*, 175, 535, 1974.
132. **Vitagliano, V., Costantino, L., and Sartorio, R.,** Effect of chain tacticity on dye binding to polyelectrolytes, *J. Phys. Chem.*, 80, 959, 1976.
133. **Breitenbach, J. W. and Karlinger, H.,** Über Quelling von vernetzter Polymethacrylsäure, *Monats. Chem.*, 80, 312, 1949.
134. **Wiederhorn, N. M. and Brown, A. R.,** Viscosity-molecular weight relation for polymethacrylic acid in methanol, *J. Polym. Sci.*, 8, 651, 1952.

135. **Priel, Z. and Silberberg, A.,** Conformation of poly(methacrylic acid) in alcohol-water mixtures. III. Dependence on molecular weight, *J. Polym. Sci. Part A-2,* 8, 713, 1970.
136. **Anufrieva, E. V., Birshtein, T. M., Nekrasova, T. N., Ptitsyn, O. B., and Sheveleva, T. V.,** The models of the denaturation of globular proteins. II. Hydrophobic interactions and conformational transition in polymethacrylic acid, *J. Polym. Sci. Part C,* 16, 3519, 1968.
137. **Fitzgerald, W. E. and Nielsen, L. E.,** Viscoelastic properties of the salts of some polymeric acids, *Proc. R. Soc., Ser. A,* 282, 137, 1964.
138. **Silberberg, A., Eliassaf, J., and Katchalsky, A.,** Shear hardening of aqueous polymethacrylic acid solutions, *Ric. Sci. Suppl. Simposio Int. Chim. Macromol. Milan/Turin,* 1954, 25, 950, 1955.
139. **Eliassaf, J. and Silberberg, A.,** The gelation of aqueous solutions of polymethacrylic acid, *Polymer (London),* 3, 555, 1962.
139a. **Andersson, G. R.,** A study of the pressure dependence of the partial specific volume of macromolecules in solution by compression measurements in the range 1 to 8000 atm, *Arkiv Kemi,* 20, 513, 1963.
140. **Kanevskaya, Ye. A., Zubov, P. I., Ivanova, L. V., and Lipatov, Yu. S.,** The temperature dependence of the light scattering and viscosity of solution of poly(methacrylic acid), *Vysokomol. Soedin.,* 6, 981, 1964; *Polym. Sci. (U.S.S.R.),* 6, 1080, 1964.
141. **Zakharova, E. N. and Tsvetkov, V. N.,** Flow birefringence and viscosity of poly(methacrylic acid) in mixed solvents of methanol-water, *Vestn. Leningrad Univ. Fiz. Khim.,* 22, 41, 1967; *Chem. Abstr.,* 68, 40200a, 1968.
142. **Glasel, J. A.,** Participation of water in conformational changes of biopolymers as studied by deuteron magnetic relaxation, *J. Am. Chem. Soc.,* 92, 375, 1970.
143. **Silberberg, A. and Mijnlieff, P. F.,** Study of reversible gelation of partially neutralized poly(methacrylic acid) by viscoelastic measurements, *J. Polym. Sci. Part A-2,* 8, 1089, 1970.
144. **Nomura, H., Kato, S., and Miyahara, Y.,** Ultrasonic absorption in polymer solutions, *Mem. Fac. Eng. Nagoya Univ.,* 27, 72, 1975.
145. **Katchman, B. and McLaren, A. D.,** Sorption of water vapor by proteins and polymers. IV, *J. Am. Chem. Soc.,* 73, 2124, 1951.
146. **McLaren, A. D. and Rowen, J. W.,** Sorption of water vapor by proteins and polymers: a review, *J. Polym. Sci.,* 27, 289, 1951.
147. **Vink, H.,** Degradation of some polymers in aqueous solution, *Makromol. Chem.,* 67, 105, 1963.
148. **Mandel, M. and Stadhouder, M. G.,** The conformational transition in partially esterified poly(methacrylic acid), *Makromol. Chem.,* 80, 141, 1964.
149. **Pinner, S. H.,** Methacrylamide polymers and copolymers and their alkaline hydrolysis, *J. Polym. Sci.,* 10, 379, 1953.
150. **Eliassaf, J.,** The effect of hydrogen bonding addends on the dilute solution viscosity of poly(methacrylamide), *J. Appl. Polym. Sci.,* 3, 372, 1960.
151. **Eliassaf, J.,** The interaction of sodium dodecyl sulfate with vinyl polymers in aqueous solutions, *J. Appl. Polym. Sci.,* 7, S3, 1963.
152. **Trossarelli, L. and Meirone, M.,** Solution properties of poly(N,N-dimethylacrylamide), *J. Polym. Sci.,* 57, 445, 1962.
153. **Heskins, M. and Guillet, J. E.,** Solution properties of poly(N-isopropylacrylamide), *J. Macromol. Sci. - Chem.,* A2, 1441, 1968.
154. **Haas, H. C. and Schuler, N. W.,** Thermally reversible homopolymer gel systems, *J. Polym. Sci. Part B (Polym. Lett.),* 2, 1095, 1964.
155. **Haas, H. C., Moreau, R. D., and Schuler, N. W.,** Synthetic thermally reversible gel systems. II, *J. Polym. Sci. Part A-2,* 5, 915, 1967.
156. **Haas, H. C., Chiklis, C. K., and Moreau, R. D.,** Synthetic thermally reversible gel systems. III, *J. Polym. Sci. Part A-1,* 8, 1131, 1970.
157. **Haas, H. C., MacDonald, R. L., and Schuler, A. N.,** Synthetic thermally reversible gel systems. IV, *J. Polym. Sci. Part A-1,* 8, 1213, 1970.
158. **Haas, H. C., Manning, M. J., and Mach, M. H.,** Synthetic thermally reversible gel systems. V, *J. Polym. Sci. Part A-1,* 8, 1725, 1970.
159. **Haas, H. C., MacDonald, R. L., and Schuler, A. N.,** Synthetic thermally reversible gel systems. VI, *J. Polym. Sci. Part A-1,* 8, 3405, 1970.
160. **Haas, H. C., MacDonald, R. L., and Schuler, A. N.,** Synthetic thermally reversible gel systems. VII, *J. Polym. Sci. Part A-1,* 9, 959, 1971.
161. **Miller, M. L.,** Block and graft copolymers. II. Block polymers from acrylamide and acrylonitrile and acrylamide and acrylic acid, *Can. J. Chem.,* 36, 309, 1958.
162. **Miller, M. L. and Rauhut, C. E.,** Block and graft copolymers. III. Apparent dissociation constants of copolymers of acrylic acid and acrylamide, *J. Colloid Sci.,* 14, 524, 1959.
163. **Miller, M. L.,** Block and graft copolymers. I. Graft copolymers from acrylamide and acrylonitrile, *Can. J. Chem.,* 36, 303, 1958.

164. **Tan, J. S. and Schneider, R. L.,** Dye binding and its relation to polyelectrolyte conformation, *J. Phys. Chem.*, 79, 1380, 1975.
165. **Davidson, R. L., Ed.,** *Handbook of Water-Soluble Gums and Resins*, McGraw-Hill, New York, 1980.
166. **Greenwald, H. L., and Luskin, L. S.,** Poly(acrylic acid) and its homologs, in *Handbook of Water-Soluble Gums and Resins*, Davidson, R. L., Ed., McGraw-Hill, New York, 1980, chap. 17.
167. **Bardet, L. and Alain, M.,** Charactérisation physicochimique d'un haut polymère d'acide acrylique utilisé en pharmacie. III. La polydispersité, *Trav. Soc. Pharm. (Montpellier)*, 35, 273, 1975.
168. **Bardet, L. and Alain, M.,** Charactérisation physicochimique d'un haut polymère d'acide acrylique utilise en pharmacie. I. Détermination de la tacticité, *Trav. Soc. Pharm. (Montpellier)*, 35, 257, 1975.
169. **Valeur, B., Noël, C., Monjol, P., and Monnerie, L.,** Étude par polarisation de fluorescence de la flexibilité et de transitions conformationelles des acides polyacryliques et polymethacryliques atactiques et stéréoréguliers, *J. Chim. Phys.*, 68, 97, 1971.
170. **Bardet, L. and Alain, M.,** Charactérisation physicochimique d'un haut polymère d'acide acrylique utilisé en pharmacie. II. Détermination de la masse moléculaire, *Trav. Soc. Pharm. (Montpellier)*, 35, 263, 1975.
171. **Halle, B. and Piculell, L.,** Water oxygen-17 magnetic relaxation in polyelectrolyte solutions, *J. Chem. Soc. Faraday Trans. I*, 78, 225, 1982.
172. **Alexander, P. and Charlesby, A.,** Effect of X-rays and γ-rays on synthetic polymers in aqueous solution, *J. Polym. Sci.*, 23, 355, 1957.
173. **Imamura, S.-I., Tonomura, Y., Kawabata, N., and Kitao, T.,** Wet oxidation of water-soluble polymers, *Bull. Chem. Soc. Jpn.*, 54, 1548, 1981.
174. **Volk, H. and Friedrich, R. E.,** Polyacrylamide, in *Handbook of Water-Soluble Gums and Resins*, Davidson, R. L., Ed., McGraw-Hill, New York, 1980, chap. 16.
175. **Klein, J. and Conrad, K.-D.,** Characterisation of poly(acrylamide) in solution, *Makromol. Chem.*, 181, 227, 1980.
176. **Klein, J., Hannemann, G., and Kulicke, W.-M.,** Zum Fliessverhalten niedermolekularer Polyacrylamide in einen weiten Konzentrationsbereich, *Colloid Polym. Sci.*, 258, 719, 1980.
177. **Kulicke, W.-M. and Klare, J.,** Eine η-c-$\overline{M}$-Beziehung für Polymerlösungen, *Angew. Makromol. Chem.*, 84, 67, 1980.
178. **Kulicke, W.-M., Klein, J., and Porter, R. S.,** Viscoelastic properties of linear chain macromolecular gels. Growth of stress at the onset of steady shear flow, *Angew. Makromol. Chem.*, 76/77, 191, 1979.
179. **Kulicke, W.-M., Kniewske, R., and Hörl, H.-H.,** Untersuchung zur Flockungsfähigkeit von Polyacrylamid-co-acrylaten (0-100 mol-% Acrylamid), *Angew. Makromol. Chem.*, 87, 195, 1980.
180. **Kulicke, W.-M. and Kniewske, R.,** Langfristige Konformationsänderung von Makromolekülen in Losung. 1. Polyacrylamide, *Makromol. Chem.*, 181, 823, 1980.
181. **Bikales, N. M.,** Preparation of acrylamide polymers, in *Water-Soluble Polymers (Polymer Science and Technology, Vol. 2)*, Bikales, N. M., Ed., Plenum Press, New York, 1973, 213.
182. **Dainton, F. S. and Sisley, W. D.,** Polymerization of methacrylamide in aqueous solution. III. Copolymerization with acrylamide, *Trans. Faraday Soc.*, 58, 1385, 1962.
183. **Bovey, F. A. and Tiers, G. V. D.,** Polymer NMR spectroscopy. IX. Polyacrylamide and polymethacrylamide in aqueous solution, *J. Polym. Sci. Part A*, 1, 849, 1963.
184. **Nishijima, Y., Teramoto, A., Yamamoto, M., and Hiratsuka, S.,** Studies of micro-Brownian motion of a polymer chain by the fluorescence polarization method. I. Fluorescent conjugates of polyacrylamide, *J. Polym. Sci. Part A-2*, 5, 23, 1967.
185. **Teramoto, A., Hiratsuka, S., and Nishijima, Y.,** Studies of micro-Brownian motion of a polymer chain by the fluorscence polarization method. II. Segmental motions in concentrated polymer solutions, *J. Polym. Sci. Part A-2*, 5, 37, 1967.
186. **Day, J. C. and Robb, I. D.,** Thermodynamic parameters of polyacrylamides in water, *Polymer (London)*, 22, 1530, 1981.
187. **Moens, J. and Smets, G.,** Alkaline and acid hydrolysis of polyvinylamides, *J. Polym. Sci.*, 23, 931, 1957.
188. **Smets, G. and Hesbain, A. M.,** Hydrolysis of polyacrylamide and acrylic acid-acrylamide copolymers, *J. Polym. Sci.*, 40, 217, 1959.
189. **Nagase, K. and Sakaguchi, K.,** Alkaline hydrolysis of polyacrylamide, *J. Polym. Sci. Part A*, 3, 2475, 1965.
190. **Mukhopadhyay, S., Mitra, B. C., and Palit, S. R.,** Kinetics of alkaline hydrolysis of polyacrylamide, *Indian J. Chem.*, 7, 903, 1969.
191. **Crummett, W. B. and Hummel, R. A.,** The determination of traces of polyacrylamides in water, *J. Am. Water Works Assoc.*, 55, 209, 1963.
192. **Attia, Y. A. and Rubio, J.,** Determination of very low concentrations of polyacrylamide and polyethyleneoxide flocculants by nephelometry, *Br. Polym. J.*, 7, 135, 1975.
193. **Bovey, F. A.,** Polymer NMR spectroscopy. VIII. The influence of the pH of the polymerization medium on the stereochemical configuration of polymethacrylic acid, *J. Polym. Sci. Part A*, 1, 843, 1963.

194. **De Loecker, W. and Smets, G.,** Hydrolysis of methacrylic acid-methyl methacrylate copolymers, *J. Polym. Sci.,* 40, 203, 1959.
195. **Geuskens, G., Hellinckx, E., and David, C.,** Kinetic study of anhydride formation during thermal degradation of isotactic and atactic polymethacrylic acid, *Eur. Polym. J.,* 7, 561, 1971.
196. **Eley, D. D., Saunders, J., and Sparks, A. H.,** The entropy of dilution of polymethacrylic acid, *Trans. Faraday Soc.,* 48, 758, 1952.
197. **Conway, B. E., Desnoyers, J. E., and Smith, A. C.,** On the hydration of simple ions and polyions, *Phil. Trans. R. Soc. Ser. A,* 256, 389, 1964.
198. **Birshtein, T. M., Anufrieva, E. V., Nekrasova, T. N., Ptitsyn, O. B., and Sheveleva, T. V.,** Hydrophobic interaction and conformational transitions in polymethacrylic acid, *Vysokomol. Soedin.,* 7, 372, 1965.
199. **Dainton, F. S. and Sisley, W. D.,** Polymerization of methacrylamide in aqueous solution. I. Hydrogen-peroxide photosensitized reaction, *Trans. Faraday Soc.,* 58, 1369, 1962.
200. **Titkova, L. V., Prokopová, E., Sedláček, B., Petrus, V., Dušek, K., and Bohdanecký, M.,** Solution properties of poly(methacrylamide), *Eur. Polym. J.,* 14, 145, 1978.
201. **Dainton, F. S. and Sisley, W. D.,** Polymerization of methacrylamide in aqueous solution. II. The ferric-ion-photosensitized reaction, *Trans. Faraday Soc.,* 58, 1377, 1962.
202. **North, A. M. and Scallan, A. M.,** The free radical polymerization of N,N-dimethylacrylamide, *Polymer (London),* 5, 447, 1964.

Chapter 4

NONIONIC POLYMERS — THE VINYL GROUP

"In vino veritas"
Pliny: *Historia Naturalis*

I. INTRODUCTION

The group of water-soluble nonionic vinyl polymers to be considered in this chapter comprises two main polymers, poly(viny alcohol) (together with the closely associated vinyl acetate/vinyl alcohol copolymers) (Section II), and polyvinylpyrrolidone (Section III), together with the two less important examples, polyvinyloxazolidone and polyvinylmethyloxazolidone (Section IV). Although, strictly speaking, poly(vinyl methyl ether) falls within this same category, it has been found more convenient to deal with it alongside the structurally related polyoxides in Chapter 2.

These two main polymers are each very important technically and scientifically, but they are strongly contrasted in their properties and behavior. Poly(vinyl alcohol), although it is apparently among the simplest of polymers from the structural viewpoint, shows many complex features which arise from the crystallizability of even the common atactic form and variable degrees of its crystallinity in the solid state, while additional complexities arise from the presence in many commercial samples of so-called "poly(vinyl alcohol)" of residual acetate groups (from the parent poly(vinyl acetate) (see Section II). It is for this last reason, coupled with their water solubility, that the vinyl acetate/vinyl alcohol copolymers are also considered in Section II in conjunction with true poly(vinyl alcohol).

With polyvinylpyrrolidone, on the other hand, the monomer unit is more complex in structure; indeed there is a strongly amphiphilic character to it, arising from the joint presence of polar (hydrophilic) amide groups, and of nonpolar (hydrophobic) methylene and methine groups. This results in the polymer showing many interesting and useful features, e.g., being soluble in a wide range of organic liquids as well as in water, and also in having strong "adsorptive" properties, both towards small-molecule solutes (cosolutes) and at interfaces.

As with the two preceding chapters, the present chapter deals (with a few exceptions) only with the general features of the polymers, together with their behavior alone in solution; three other specific aspects are left until Volume II, i.e., interactions with small-molecule cosolutes (Chapter 2), interactions with other polymers (Chapter 3), and interactions with interfaces (Chapter 4).

II. POLY(VINYL ALCOHOL) — PVAL; POLY(VINYL ACETATE) — PVAC; THEIR COPOLYMERS — PVAA; AND RELATED POLYMERS

Poly(vinyl alcohol) – PVAL*

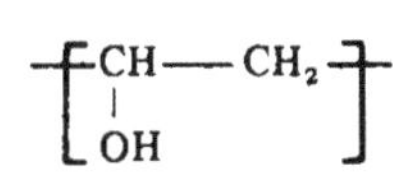

$M_0(C_2H_4O) = 44.05$

* The abbreviation PVAL is also used here for PVAA with less than 2 mol % acetate groups (see Section II.A).

Other chemical names and abbreviations are* ethenol homopolymer *(Chemical Abstracts)*; poly(1-hydroxyethylene) (IUPAC); polyvinyl alcohol; PVAl; PVA-OH; PVOH; PV-OH.

Poly(vinyl acetate) – PVAC

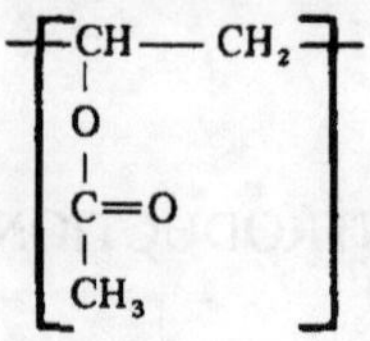

$$M_0(C_4H_6O_2) = 86.09$$

Other chemical names and abbreviations are* acetic acid ethenyl ester homopolymer *(Chemical Abstracts)*; poly(1-acetoxyethylene) (IUPAC); polyvinyl acetate; PVAc; PV-OAc.

Poly(vinyl acetate-*co*-vinyl alcohol)-PVAA**

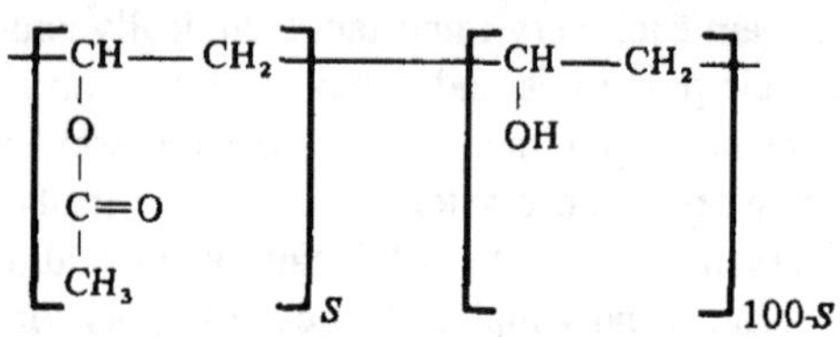

$$M_0 = 44.05 + 42.04S/100$$

Other chemical names and abbreviations are partially hydrolyzed poly(vinyl acetate); partially acetylated poly(vinyl alcohol); PVAC-VAL; PVA-A; PVAc-A.

Trade names for PVAA and PVAL are* *Alcotex* (Revertex); *Elvanol* (Du Pont); *Gelvatol* (Shawinigan); *Gohsenol* (Nippon Gohsei); *Lemol* (Borden); *Moviol* (Hoechst); *Polyviol* (Wacker); *Poval* (Kuraray); *Rhodoviol* (Rhône-Poulenc); *Vinol* (Airco Chemical).

A. Nomenclature

To clarify and simplify the subsequent consideration of these polymers, it is essential to start with a brief discussion of the nomenclature and abbreviations that will be used for them.

As listed above, the abbreviation to be used here for pure poly(vinyl alcohol) is PVAL, and that for poly(vinyl acetate) is PVAC. This represents a departure from the practice applied in these volumes and elsewhere of using capital letters for the initials of names and small letters otherwise (as in, e.g., PAAm for polyacrylamide); it is used here principally because it is the notation recommended specifically for these two polymers by IUPAC,[5,6] and at the same time serves to differentiate these two related and associated polymers more clearly.

For their copolymers, the systematic abbreviation would be PVAC-VAL, but in view of the importance of these materials it is simpler to use the shorter form PVAA; this, indeed, may also be used as a general form to include PVAC and PVAL as special cases.

However, in all considerations of these polymers, it is necessary to be aware of the ambiguities that arise with those commercial grades of "poly(vinyl alcohol)" which on the one hand are stated by the manufacturers to be "fully hydrolyzed", but which on the other

* See References 2 and 4 for more extensive lists.

** The abbreviation PVAL is also used here for PVAA with less than 2 mol % acetate groups (see Section II.A).

hand because of practical difficulties in ensuring the removal of the last few acetate groups are also stated (by the same manufacturers) to have an actual degree of hydrolysis between 98 and 100%, i.e. with possibly up to 2 mol % VAC. In addition, even in many scientific studies the polymer is often referred to simply as "poly(vinyl alcohol)" without any specific mention of the VAC content (if any). Because of the possibility that all such materials even when referred to as "poly(vinyl alcohol)" are likely to contain residual acetate groups, this would therefore lead to *all* such samples having to be referred to as PVAA. To avoid this situation, inasmuch as a VAC content of up to 2 mol % makes little difference to many (but not all) of the properties of the polymer, it is therefore convenient to still refer to polymers with this minor VAC content simply as PVAL.

To summarize, PVAC is used as the abbreviation for poly(vinyl acetate); PVAA is used as a shorter form of PVAC-VAL, for vinyl acetate/vinyl alcohol copolymers; and PVAL is used both for pure poly(vinyl alcohol) and for PVAA with 2 mol % or less of acetate groups.

B. General Features

Because of the industrial importance of PVAA and PVAL, these polymers have been the subject of an extensive technical and scientific literature. Much of the original published work is in Japanese, and some in Russian, which presents problems in the interpretation and evaluation of the work. The general properties and applications of the polymers have been dealt with in a diversity of reviews, including two encyclopedia articles;[7,8] Argana and Czerwin[9] have discussed the applications of commercially available materials, while Pritchard[10] has dealt succinctly with various scientific and general aspects of pure PVAL. A symposium on the joint subject of PVAA and PVAL was held in 1967 under the auspices of the Society of Chemical Industry (London), and the proceedings were edited by Finch and published in 1968;[11] the scope and contents of the book are reflected in the titles of the individual contributions.[12-21] Subsequently, a more extensive treatment has been published, again edited by Finch, and is in many respects an expansion of the earlier publication;[3] once again, the titles of the individual contributions indicate their scope and contents.[4,22-45] These reviews will be drawn upon fairly extensively in the subsequent discussion, along with a consideration of various other and more recent scientific studies.

In view of the normal way in which papers of PVAA and PVAL are scattered throughout the scientific and technical literature, it is useful to note a cluster of such papers on these and allied topics in a single 1966 volume of the *Journal of Polymer Science*.[46-51]

Technical bulletins are available giving the characteristics of the commercial grades from the respective manufacturers of the materials (e.g., References 52 to 54); Toyoshima[4] has given a very useful grid on which the main commercial grades are displayed, plotted in terms of the degree of hydrolysis and the viscosity of a 4% aqueous solution at 20°C (which is used in specifying commercial grades since it is also to some extent a measure of the molecular weight) as the two coordinates. Further incidental information on certain commercial grades may be found in various papers concerned with these polymers (e.g., References 55 to 57).

C. Structural Features and Structural Irregularities

From the structural viewpoint, poly(vinyl alcohol) is apparently one of the simplest polymers, and correspondingly it might be expected to show a very simple pattern of properties and behavior. Its monomer unit is isomeric with that of PEO, with only the additional feature that one of the hydrogen atoms (i.e., that on the hydroxyl group) of PVAL can enter into hydrogen bonds, although this ability (and that of the oxygen atoms for accepting such bonds) can be "neutralized" by the formation of intramolecular hydrogen bonds (i.e., between adjacent hydroxyl groups on the chains). However, in practice this expectation of simplicity is not fulfilled, in part because the methods used to produce the polymer lead to the occurrence of various types of structural irregularity, which includes residual vinyl acetate (or other ester groups) originating from the parent polymer.

1. Stereoregularity and Crystallizability

Poly(vinyl alcohol) cannot be produced by the direct polymerization of its monomer, because vinyl alcohol is unstable, with attempts at its synthesis leading to its isomer acetaldehyde. The polymer is therefore produced either by hydrolysis (using the term in its widest sense) of a poly(vinyl ester), which in commercial practice is most commonly the acetate, or alternatively from a poly(vinyl alkyl ether) by removal of the alkyl group. Thus any consideration of the behavior and properties of poly(vinyl alcohol) must inevitably also involve at least some consideration both of the mode of production of the parent polymer, and of the method used to carry out the conversion. For example, all these polymers will have chiral centers, and in the absence of racemization during the conversion reaction, the tacticity of the poly(vinyl alcohol) will be the same as that of the parent polymer. By and large, samples of common PVAL are found to be essentially atactic; the methods for the characterization of PVAL according to its tacticity are discussed in Section II.E.

With most chiral polymers, the property of crystallinity (or at least, crystallizability) is necessarily associated with marked stereoregularity, the atactic polymer thus being amorphous. However, with PVAL, as a result of the small size of the hydroxyl substituent group even the atactic form can crystallize, with the corresponding X-ray diffraction pattern indicating a statistical distribution of the hydroxyl groups in the crystal lattice.[58,59]

Furthermore, the degree of crystallinity of a sample (which among other things determines its solubility) depends upon its history, in that heat treatment of a sample prepared at room temperature leads to the progressive increase in crystallinity in what was initially a largely amorphous sample (see Section II.F.8).

2. Branching

If chain transfer to polymer occurs during the synthesis of the parent polymer, then this will lead to branches in the structure which may also appear in the final PVAL. In the common case where PVAC is the parent polymer, the chain transfer may either involve attack on the hydrogen atoms on the backbone of the polymer, in which case the links formed are resistant to hydrolysis, or alternatively it may involve attack on hydrogen atoms on the ester methyl group in which case the branches would be lost in the hydrolysis reaction.[60] Correspondingly, *intramolecular* chain transfer onto the ester methyl group would lead to a linear polymer but with ester groups in the main chain which would be broken in the hydrolysis reaction, resulting in a fall in $\overline{M}$.[61] Thus the structure and degree of polymerization of the PVAL produced would not necessarily be expected to be the same as that of the parent polymer, although the reconversion (i.e., reesterification) process should not lead to a further degradation (at least in the absence of other "weak links", or of other reactions such as oxidation).

3. Residual Acetate Groups

In the immediately preceding discussion we have been concerned specifically with true pure PVAL. In practice, however, samples of so-called "poly(vinyl alcohol)" (particularly those from commercial sources) almost invariably contain a number of types of structural irregularities in the chains; the possibilities for these have been listed and discussed by Zwick and van Bochove[56] and Pritchard.[62] The most important of these irregularities is residual acetate groups from the parent poly(vinyl acetate). There may be, e.g., an appreciable proportion deliberately left in (commonly, 12 mol %, i.e., the hydrolysis reaction is stopped short at this point) to confer cold-water solubility on the product, or there may be a relatively small proportion (commonly, 2 mol % or less) which remains because of the difficulty of carrying the reaction to completion without more drastic treatment. This often gives rise to problems in the interpretation and evaluation of published work, since not uncommonly no indication is given of the acetate content (if any) of the "poly(vinyl alcohol)" studied; in other cases the numerical value of this content is quoted without stating whether this is on a weight- or a mole-percent basis. (In the present discussion such values are taken to be on a mole-percent basis unless stated otherwise in the original source.)

Even when this content is clearly defined, there still remain various possibilities for the *distribution* of these units on the chain. At one extreme they may be distributed essentially randomly; at the other extreme they may be clustered together giving a block copolymer structure, or in general the distribution may be only "blocky", i.e., something in between. This hinges upon the manner in which the conversion from the parent polymer is carried out. When aqueous alkali is used to hydrolyze the PVAC there is a kinetic cooperative effect in that the presence of a hydroxyl group makes an adjacent acetate group much more readily attacked; this therefore leads to a block distribution of the residual acetate groups. When methanolic or ethanolic alkali is used, this kinetic effect is much less marked, so that the distribution is only somewhat "blocky". By contrast, if the polymer is completely hydrolyzed and then partially reacetylated, the acetate groups so introduced are distributed essentially randomly.[16,25,63]

4. 1,2-Diol (Glycol) Groups

An additional type of structural irregularity which may be present in PVAL and PVAA is 1,2-diol groups; these arise from the occurrence of head-to-head addition (rather than the normal head-to-tail placement) in the propagation reaction during the synthesis of the parent polymer, thus giving sequences in the PVAL of the form:

$$\text{—}\underset{\text{OH}}{\underset{|}{\text{CH}}}\text{—CH}_2\text{—CH}_2\text{—}\underset{\text{OH}}{\underset{|}{\text{CH}}}\text{—}\underset{\text{OH}}{\underset{|}{\text{CH}}}\text{—CH}_2\text{—}$$

Each such 1,2-diol group must therefore, as shown, also be associated with a 1,4-diol group which contains two adjacent methylene groups, i.e., a relatively *hydrophobic* feature. The detection and determination of these 1,2-diol groups were studied initially by Flory and Leutner,[64] subsequently by Harris and Pritchard,[65] and more recently still by Pritchard and Lan Chun Fung.[66] The groups occur to between 1 and 2 mol % in common samples of PVAL, and to a corresponding extent in PVAA.

5. Other Structural Irregularities[56,62]

Other structural irregularities which occur in PVAL and PVAA chains include carbonyl groups (formed by oxidation of hydroxyl groups) — while derived from these by condensation with hydroxyl groups there may be 1,5-hemiacetal rings and 1,3-dioxane rings (in the last case involving a cross-link); carbon-carbon double bonds, either isolated or conjugated with one another, and also possibly conjugated with the carbonyl groups; ether cross-links (formed by a dehydration reaction between pairs of hydroxyl groups); terminal carboxyl groups, along with terminal 1,5-lactone rings derived therefrom; and various other terminal groups derived from the initiator or from chain transfer reactions.

D. Production and Purification

As already stated (see Section II.C), PVAA and PVAL are produced commercially from PVAC. Accounts have been given of industrial methods used in the synthesis of this parent polymer,[13,24] and the conversion to PVAA and PVAL has been discussed both from the chemical[25] and technical aspects.[12,25] The manufacturing and engineering aspects of these processes have also been discussed.[26] Various laboratory procedures for obtaining PVAA and PVAL from PVAC have also been described,[25,67,68] while Pritchard[62] has outlined the procedure to be adopted to prepare "a reasonably pure specimen" of PVAL starting from VAC monomer. The progress of the conversion of PVAC to PVAL with alcoholic alkali has been followed by Beresniewicz;[67] its effect on the degree of polymerization has been studied by Bravar and co-workers.[60]

Samples of PVAA may also be obtained from PVAL by controlled reacetylation; the distribution of acetate groups is then essentially random in contrast with the more "blocky" distribution of those in the PVAA prepared directly from PVAC (see Section II.C.3).

The production of PVAL from polymers other than PVAC has been reviewed by Noro,[27] and also by Finch[31] more specifically in relation to the stereoregularity of the product. Various vinyl haloacetates have been studied as sources of PVAL by Cooper and co-workers,[69] and these and other vinyl esters similarly studied by Nozakura and co-workers;[70] with the trifluoroacetate, in particular, Matsuzawa and co-workers found that free radical polymerization at 60°C followed by hydrolysis gives a PVAL which is only slightly syndiotactic rich (σ = 0.45*),[71] whereas initiation induced by photodecomposition of an azo compound at −78°C gives a somewhat more syndiotactic product (σ = 0.36).[72]

The vinyl alkyl ethers provide an alternative starting point for the production of PVAL; by using bulky alkyl groups in the ethers, different stereoregular forms of the final product can be obtained. With the trimethylsilyl group (not strictly an *alkyl* group), use of different polymerization conditions can lead either to a PVAL which is isotactic rich (σ = 0.9) or to one which is syndiotactic rich (σ = 0.3).[27,31]

By using polymers obtained from vinyloxydialkylaluminium (with either ethyl or isobutyl as the alkyl group) it is possible to produce even more highly syndiotactic-rich products (σ = 0.1 to 0.3).[73]

The production of vinyl alcohol copolymers has been discussed by Warson[14] and by Noro.[28]

Samples of PVAL may be purified (i.e., freed of low molecular weight contaminants) by precipitation from aqueous solution with nonsolvents such as acetone,[64,74-77] acetone + diethyl ether (1:1),[68] or methanol.[70,78] Residual salts such as sodium acetate may be removed by exhaustive dialysis, or more efficiently by electrodialysis.[74] Foam fractionation has been used to purify PVAA for surface tension studies.[79]

Samples of PVAA with about 30 mol % acetate groups may be purified by using the precipitation of the polymer which takes place when an aqueous solution is heated (see Section II.H.2).[80]

E. Fractionation and Characterization

PVAA and PVAL may be fractionated by stepwise precipitation from aqueous solution with a variety of nonsolvents, most commonly using either 1-propanol[60,67,81-84] or acetone,[76,82,85-88] although methanol + acetone (1:1),[68] methanol + acetone + diethyl ether (1:1:2),[68] and dioxane[64] have also been used. Levi and co-workers[82] state that 1-propanol is preferable to acetone for this purpose, since with the latter there is local formation of precipitated polymer which is difficult to redissolve. It should be noted that in these types of systems, the fractionation of PVAA is according to both molecular weight and acetate content.[67,88] Alternatively, for PVAA with vinyl acetate contents between 10 and 20 mol % the fractionation may be carried out by raising the temperature of the aqueous solution, in which case the separation is largely according to vinyl acetate group content.[89]

Gel permeation chromatography has also been applied to the fractionation of PVAA, where it has the advantage that the separation is almost entirely by molecular weight;[90] however, the technique is said to be in general less convenient, less simple, and less rapid than fractional precipitation.[88]

* See Section II.E for significance of σ values.

Fractional extraction procedures have also been tried with PVAL, although not with uniformly satisfactory results because of the crystallinity of the polymer.[81,91]

An alternative approach is to fractionate the PVAC or other parent polymer from which the PVAA or PVAL is to be prepared;[82,87,92] however, if there is any degradation in the conversion reaction, through the breaking of hydrolyzable links in the main chain, then this will naturally lead to broadening of the molecular weight distribution.[60]

PVAL and PVAA may be characterized as to molecular weight by most of the standard solution methods, although light-scattering methods are objectionable because of their sensitivity to the aggregation effects that take place so readily in aqueous solutions of these polymers (see Section II.G.1).

For the characterization of PVAA or PVAL according to tacticity (stereoregularity), as already indicated crystallinity or even crystallizability is not an index of stereoregularity because even atactic PVAL will crystallize. Tacticity is therefore determined mainly either by NMR or IR measurements.

The application of NMR to the determination of the tacticity of PVAA or PVAL has been discussed by Pritchard[87] and Finch.[31] Although the polymer itself can be studied directly, better results are obtained by converting it first to PVAC; the three types of triad environments for a particular ester group then give rise to slightly different chemical shifts for the methyl protons in the NMR spectrum — e.g., for solutions of PVAC in chloroform at 60°C studied at 100 MHz, the peaks are located at: isotactic, τ = 7.98 ppm (δ = 2.02 ppm); heterotactic, τ = 8.00 ppm (δ = 2.00 ppm); syndiotactic, τ = 8.02 ppm (δ = 1.98 ppm). The areas under each of the three peaks then give the relative amounts of the three triads, and where Bernoullian statistics hold these can be used to calculate the sigma (σ) value of Bovey and Tiers[93,94] which ranges from 0 for purely syndiotactic, through 0.5 for atactic, up to 1.0 for purely isotactic polymer.

Infrared absorption measurements on solid PVAL (dispersed in KBr discs, or as films — see Section II.F.5) may be used as a more rapid method of estimating the tacticity since the spectroscopic features have been calibrated against the tacticity values derived from NMR.[31,87] The syndiotactic polymer has a peak at 916 cm^{-1} which is absent from the isotactic form, so that the ratio of optical densities (absorbances) D at 916 and at 849 cm^{-1}, represented D_{916}/D_{849}, may be used as a direct measure of tacticity (the measurement at the 849 cm^{-1} peak is used as a "reference" to adjust for differences in polymer content of the KBr disc or thickness of film), and also converted into the corresponding σ value using the calibration previously obtained.[31,87]

F. Bulk Polymer

The general features of the properties of PVAA and PVAL is the bulk state have been discussed by several authors;[29,31,35,39,59] these features are important in relation to such aspects of behavior as the solubility of the polymers, their ease of dissolution, and even their properties in solution (where aggregation that takes place has been likened to bulk state crystallization). For clarity of discussion it is useful to distinguish between, on the one hand, those properties which may be directly observed (such as density; refractive index; thermal, mechanical and spectroscopic properties; and X-ray diffraction patterns), and on the other hand the deduced features (such as crystallinity; glass-transition temperature; and melting point). Nevertheless, these two types of property are closely interconnected, and indeed often the interpretation of a set of experimental measurements for one property involves the indirect use of experimental results for another observed property through deduced features (e.g., the interpretation of density values requires a knowledge of the extent of crystallinity, which in its turn is put on an absolute basis by X-ray diffraction measurements).

1. Density

The density, ρ, of common PVAL often differs from one sample to another, although

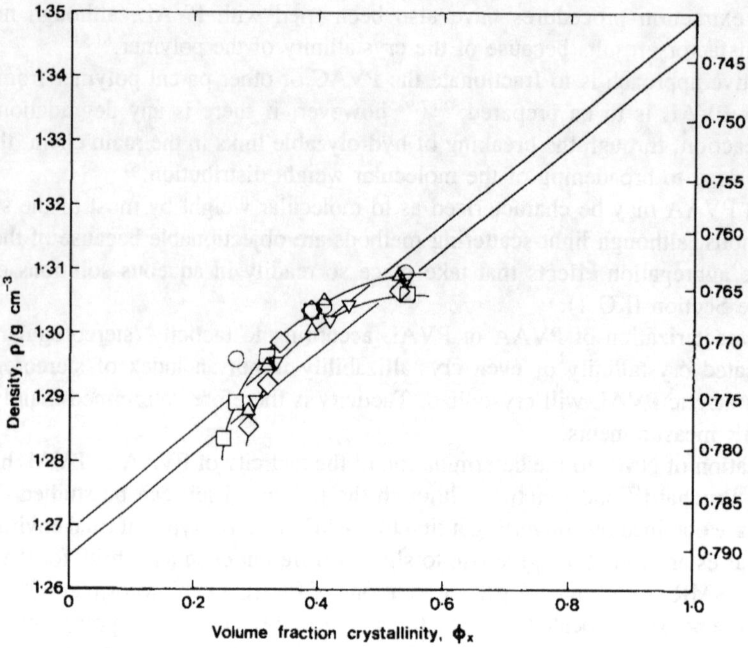

FIGURE 1. Density, ρ, vs. volume fraction crystallinity, ϕ_x, for five samples of poly(vinyl alcohol) of different molecular weight, heat treated at various temperatures to give various degrees of crystallinity. Molecular weights: (○) 1.4D4; (□) 3.1D4; (◇) 5.7D4; (△) 1.0D5; (▽) 2.0D5.[95,99]

generally it will be found to lie within the limits of 1.28 to 1.31 g cm^{-3} (References 95 and 96) and quite frequently within the narrower limits of 1.29 to 1.30 g cm^{-3} (References 39, 50, 53, 97, and 98). The main source of this variability seems to be differences in the degree of crystallinity of the sample (density naturally increasing with increase in crystallinity).

In the course of detailed pioneering investigations on the correlation between density and crystallinity for PVAL, Sakurada and co-workers[95,99] prepared low crystallinity films from five samples of the polymer with molecular weights ranging from 1.4D4 to 2.0D5. Each of these films was then heat treated at various temperatures (40 to 200°C) to develop the crystallinity to different extents. For the 23 final samples, the density was determined by flotation, and the volume fraction crystallinity ϕ_x determined by X-ray diffraction measurements (see Section II.F.8); the results obtained are plotted in Figure 1.

In fitting a correlation line to these data, it is normally assumed that the limiting density value for $\phi_x = 1$ (i.e., the pure crystalline state) must be near to the value deduced from the crystal lattice parameters obtained by X-ray diffraction, i.e., 1.350 ± 0.002 g cm^{-3} (see Section II.F.7). Two forms of correlation line have been fitted to these data.[59] First, a linear relation between density and ϕ_x:

$$\rho/\text{g cm}^{-3} = 1.265 + 0.084\,\phi_x \qquad (1)$$

where the lower and upper limiting values of ρ are 1.265 cm^3 g^{-1} ($\phi_x = 0$, i.e., pure amorphous polymer) and 1.349 cm^3 g^{-1} ($\phi_x = 1$, i.e., pure crystalline polymer). Second, a linear relation between the reciprocal of density (i.e., specific volume) and ϕ_x:

$$\rho^{-1}/\text{cm}^3\,\text{g}^{-1} = 1/1.269 - 0.0445\,\phi_x \qquad (2)$$

where the corresponding limiting values are 1.269 and 1.345 g cm^{-3}. In point of fact, the maximum span of the density values (i.e., between the lower and upper limiting values) is relatively small so that these two lines differ little from one another in their trends, as is evident from the similarities of the lower and the upper values in each case, and they can be jointly represented by the linear correlation band between density and ϕ_x:

$$\rho/\text{g cm}^{-3} = (1.267 \pm 0.002) + (0.084 \pm 0.004)\,\phi_x \qquad (3)$$

which is plotted as the linear band in Figure 1.

These correlation lines (Equations 1 and 2) are important in the present context because they have been used widely as a rapid means of estimating crystallinity of samples of PVAL. However, it should be clear from Figure 1 that because of the narrow range of the crystallinity values that can be obtained even from such careful work (i.e., ϕ_x = 0.25 to 0.55), and because of inevitable scatter in the data, any attempt to fit such correlation lines must be somewhat arbitrary. Indeed, if approached without any assumptions as to the form of the correlation or the limiting values, the best correlation is probably the *curved* band which has been drawn in Figure 1 so as to include the majority of the data points. Inasmuch as this curved correlation band relates directly to the original experimental data and furthermore covers the range of commonly encountered crystallinities, it would seem preferable to use it rather than the linear correlation band, which is quite obviously not such a good fit over the middle range of the experimental data. These uncertainties of interpretation emphasize the need for samples of PVAL of much lower, and much higher, crystallinity.

For PVAA, the value of the density falls off with increasing acetate content, the decrease being linear up to 15 mol % at which point there is an abrupt change to a second straight line of lower slope leading to the value of 1.175 g cm^{-3} for pure PVAC.[100]

2. Mechanical Properties

Elastic modulus measurements for PVAL have been discussed by Pritchard;[59] the modulus value falls fairly sharply to zero over the temperature range 220 to 240°C, i.e., the melting range of the polymer.[59] Measurements of the temperature variation of shear modulus at lower temperatures have been used by Thompson Hughes and Fordyce[101] to estimate the glass-transition temperature.

3. Calorimetric and Thermal Properties

The measurements of the calorimetric and thermal properties of PVAL have been reviewed by Pritchard[59] and Tubbs and Wu;[29] such measurements lead to estimates of the melting-point (Section II.F.10) and glass-transition temperatures and other second-order transition temperatures (Section II.F.9).

4. Refractive Index and Related Properties

The value of refractive index value of PVAL has been given as: n_D^{25} = 1.51 by Haas,[40] who has also discussed other optical properties of films of PVAL and derived materials. The refractive index-temperature correlation has been used to obtain the t_g value for PVAL (as well as to study the effect of absorbed water on this value) by Packter and Nerurkar.[96] The dielectric properties of PVAL have been discussed by Pritchard,[59] who quotes a value of about 40 as the dielectric constant at room temperature.

5. Infrared Absorption Spectra

The IR absorption spectrum of PVAL has been published for the range 625 to 4000 cm^{-1} by Nyquist[101a] (who has also given the corresponding spectra for PVAA and PVAC), for 770 to 5000 cm^{-1} (together with that of the polymer after deuteration with D_2O vapor) and 950 to 1350 cm^{-1} by Tadokoro and co-workers,[102,103] and of the iso- and sydiotactic-rich forms for 650 to 5000 cm^{-1} by Finch.[31]

Samples may be studied either as cast films (optimum thickness 5 to 10 μm[87]) or dispersed in halide discs. These spectra have been interpreted and discussed by Finch[31] and Pritchard.[87]

IR spectra have three main applications with PVAL and PVAA. First, they may be used for the estimation of stereoregularity, using the optical density ratio D_{916}/D_{849}, as discussed above in Section II.E.

Second, the IR spectra may also be used to estimate the degree of crystallinity using the optical density (absorbance) ratio D_{1144}/D_{2940}, where the band at 1144 cm^{-1} (8.74 μm) is crystallinity sensitive and that at 2940 cm^{-1} is used as a "reference" to adjust for differences in film thickness or polymer content of the halide disc;[84,86,103] alternatively, the band at 1409 cm^{-1} (7.1 μm) may be used as the reference.[49] These techniques have also been applied to the estimation of crystalline contents of PVAL gels.[104] Also, less directly, the IR study of films equilibrated with D_2O vapor shows the extent of deuteration, which is related to the amorphous content (since the crystalline regions are presumed impermeable to penetrants such as D_2O).[102]

Finally, IR spectra may also be used to detect the presence of impurities and structural irregularities in these polymers, such as acetate, carbonyl, peroxide, or carboxylic acid groups.[105] In the case of the vinyl acetate groups in PVAA, IR may be used to give information not only on the content but also on the distribution of these groups, since an acetate group bordered on either side by hydroxyl groups has a C = O stretching frequency of 1715 cm^{-1}, whereas for one bordered by other acetate groups the band is at 1734 cm^{-1} (Reference 63).

6. *Ultraviolet and Visible Absorption Spectra*

Pure PVAL should of course show no absorption either in the visible region of the spectrum or in the ultraviolet above approximately 200 nm. Dehydration or oxidation (e.g., by heat treatment) of the polymer leads to unsaturated groups such as C = C or C = O, which with increasing extent of conjugation give rise to increasing absorption at wavelengths above 200 nm.[87,103] This absorption is observed not only in the solid state but also in aqueous solutions of the polymer.[106]

7. *X-Ray Diffraction and Crystal Structure*

X-ray diffraction has been used widely to study both the structure of the crystalline regions and the degree of crystallinity of PVAL.[39,59,107] The crystal structure commonly accepted at present is essentially still that proposed more than 30 years ago by Bunn,[58] comprising a monoclinic (pseudoorthorhombic) lattice having two $-CH_2-CHOH-$ groups in the unit cell, with unit cell parameters: a = 7.805 Å, b = 2.533 Å, c = 5.485 Å, β = 92.2 ($\pm$0.3)°, giving a density of ρ = 1.350 ($\pm$0.002) g cm^{-3}.[59] However, more recently Colvin[107] has claimed to have detected two types of crystal structure within stretched fibers, i.e., a denser monoclinic one with ρ = 1.400 g cm^{-3} and 24 monomer units per unit cell, and a less dense orthorhombic form with ρ = 1.322 g cm^{-3} and 18 monomer units per unit cell. If this finding is confirmed, it may require changes in, e.g., the interpretation of density data (Section II.F.1).

Sakurada and co-workers[95,99,108-114] used X-ray diffraction measurements to determine the degree of crystallinity of various heat-treated samples of PVAL fibers and films (see Section II.F.1 and Figure 1). Crystallinity studies by X-ray diffraction have been made subsequently on films,[49] compressed powder,[69] and single crystal platelets.[50]

8. *Crystallinity*

In considering the crystallinity of a polymer such as PVAL, we adopt (whether implicitly or explicitly) the "two-phase model" used widely for polymers, in which the material is assumed to be composed of closely intermingled crystalline and amorphous regions which, despite this intermingling, retain many of their individual features (such as their density).

The degree of crystallinity as measured by X-ray diffraction is on a volume basis, generally expressed either as volume fraction, ϕ_x (e.g., $\phi_x = 0.4$) or as a percentage (i.e., 40% crystallinity); the amorphous content follows correspondingly by difference from 1 or from 100%, i.e., $\phi_a = 0.6$, or 60% amorphous content, in the example given. Samples of PVAL have degrees of crystallinity varying over a relatively narrow range; the minimum attainable seems to be about 20%, for films cast from solution at room temperature,[115] while the maximum attainable seems to be about 55%, obtained by heat treatment at 200°C, i.e., close to the melting point;[95,99] even single crystal platelets apparently have degrees of crystallinity (estimated from density measurements) of only about 42%.[50] Indeed, one of the problems of determining the degree of crystallinity of PVAL samples is that it does not seem possible to obtain the samples either of wholly crystalline or of wholly amorphous material which are required to properly calibrate the methods used.

Among the methods used for estimating crystallinity, that of X-ray diffraction provides the most nearly absolute values, as exemplified by the studies carried out by Sakurada and co-workers (see Sections II.F.1 and II.F.7).

One common, rapid method is to use density measurements, as already discussed in Section II.F.1.

Crystallinity may also be estimated from IR absorption spectrum using the optical density ratio D_{1144}/D_{2940} (see Section II.F.5).

A further method of determining crystallinity of PVAL is via studies on the extent to which the hydroxyl hydrogens of the polymer are exchanged for deuterium when the polymer is equilibrated with D_2O vapor, since only the hydroxyl hydrogens in the amorphous regions should be accessible to the D_2O molecules; the extent of this deuteration may be estimated from IR absorption spectra.[102] The actual equilibrium uptake of water (H_2O or D_2O) vapor by the polymer has also been used for crystallinity estimation (see Section II.H.1).[103,116]

The degree of crystallinity of a particular sample of PVAL depends upon its history, particularly its method of preparation and any subsequent treatment that it has received. Packter and Nerurkar[96] have investigated the effect of rate of evaporation and temperature on the properties of PVAL films prepared by the evaporation of aqueous solutions. The subsequent heat treatment of such a film leads to an increase in crystallinity, with an ultimate (equilibrium) degree of crystallinity being attained at each temperature below the melting point (~200 to 250°C) which increases with increase in the heat-treatment temperature.[95,99]

9. *Glass-Transition Temperature and Other Second-Order Transitions*

The values of the glass-transition temperature, t_g, which have been quoted for PVAL show some variation which seems to be the result of differences in the experimental methods used and in the molecular weights, degree of structural irregularity and water contents of the samples; however, most published values seem to lie within the range 75 ± 10°C,[59] although the rather higher value of 96 ± 3°C was obtained by Thompson Hughes and Fordyce[101] from shear modulus measurements.

Thus at normal temperatures the amorphous regions of PVAL are in the glassy state and relatively impermeable to the ingress or egress of small-molecule pentetrants, and the polymer needs to be heated above t_g in any drying procedure to give absorbed water molecules the requisite mobility for complete removal.

PVAL also shows a number of other minor second-order transitions which have been discussed by Pritchard.[59]

10. *Melting Behavior and Melting-Point Ranges*

The melting behavior of PVAL has been discussed by Pritchard[59] and Tubbs and Wu.[29] As usual with polymers, melting of PVAL takes place over a range of temperatures, and the melting range observed also depends somewhat upon the experimental method used.

The melting-point range also shows some dependence upon stereoregularity, the quoted values being in the sequence: isotactic, 212 to 235°C; atactic, 228 to 240°C; syndiotactic, 230 to 267°C.[29]

With PVAA, the melting point naturally falls with increasing acetate content (since the limiting polymer, PVAC, is completely amorphous), but in a manner that depends markedly upon the mode of production. At one extreme, if the polymer has been made by hydrolysis of PVAC with aqueous alkali the dependence upon acetate content is linear, by about 0.25°/mole % acetate, whereas at the other extreme with partially reacetylated PVAL the dependence is still linear but by about 8°/mol % acetate, i.e., about 32 times greater; the melting points of PVAA produced from PVAC by treatment with methanolic alkali lie in between. This behavior is a result of differences in the distributions of acetate groups on the three types of materials (see Section II.C.3).[16,63]

G. Solution Behavior

As in the case of the other nonionic polymers considered, it is convenient to deal with the behavior of PVAA and PVAL in aqueous solution separately from that in nonaqueous and mixed solvents. In considering each type of solution, the behavior of PVAL is dealt with first, followed by that of PVAA in relation to its vinyl acetate group content.

1. Aqueous Solution

The results of various studies on the behavior of PVAL and PVAA in aqueous solution are summarized in Table 1. There are certain general aspects to the preparation and stability of these solutions that also have to be borne in mind when considering such studies.[44]

First, with PVAL, as a result of its appreciable crystalline content the polymer does not dissolve in water at normal temperatures but has to be heated to a fairly well-defined dissolution temperature (generally, at least 80°C) and then kept at this temperature for full molecular dispersion to take place (see Section II.H.2).[15,92] Correspondingly, when cooled back down to normal temperatures the solutions are often prone to such phenomena as gelation, aggregation, or even crystallization of the polymer; the last may take place simply on standing[15] and is enhanced by stirring or other forms of shearing.[72,144,147] These aggregation phenomena may be studied as usual by light-scattering methods,[15,68,83,86,126] but at the same time the occurrence of these phenomena means that such methods cannot always be safely applied to the estimation of such features as the molecular weight of the polymer or its second virial coefficient.

For PVAA with less than about 25 mol % acetate groups, the polymer is soluble at normal temperatures, or at most requires only mild heating to dissolve it, but it may then precipitate from solution when it is heated up (see Section II.H.2). This behavior naturally narrows the range of temperatures over which PVAA can be studied in aqueous solution, while light-scattering measurements of molecular weight and other properties are somewhat restricted by the aggregation that takes place at temperatures approaching the precipitation point.[89,118]

The *partial specific volume* of PVAL in aqueous solution has been studied by eight groups as listed in Table 1: Dialer and co-workers (entry 11);[85] Dieu (entry 22);[68] Sakurada and co-workers (whose parallel studies on PVAA are discussed below) (entry 30);[125] Martin and co-workers (entry 42);[133] Kawai (entry 44);[74] Nakanishi and Kurata (entry 54);[137] Andersson (entry 62);[139] and Inzelt and Gróf (entry 83).[145] The studies have been done at various temperatures ranging from 20 to 35°C (mostly, at 30°C), and with sample molecular weights ranging from 8D3 to 1D5. To adjust those values of $\bar{v}_2^o$ not measured at 30°C to this temperature, we use the data obtained by Inzelt and Gróf[145] for 25 and 35°C which give the temperature coefficient as 3.4d3 deg^{-1}. Correspondingly, Nakanishi and Kurata[137] found $\bar{v}_2^o$ to be a linear function of reciprocal molecular weight (the value falling by 0.006 cm^3 g^{-1} in covering their range of M from 8.4D3 to 7.0D4), so that this dependence may be

Table 1
STUDIES ON PVAA AND PVAL IN AQUEOUS SOLUTION[a]

Entry	Year	Polymer samples No.	Polymer samples Range of $\bar{M}$	Techniques used	Temperature (t/°C)	Polymer concentration	Comments	Ref.
1	1940	6[b]	4D4—1D5	OP	27	0.2—2%	$\rightarrow \bar{M}_n$; (A_2)	117
2	1940	9[b]	5D4—1D5	VI	20	0.1—0.3%	$\rightarrow \eta_c$; ([η]; k')	117
3	1948	6	6D3—2D4	OP	30	0.25—2%	$\rightarrow \bar{M}_n$	64
4	1948	6	6D3—2D4	VI	25		$\rightarrow$[η]; MHS (0.76; 2.0d2)[c]	64
5	1949	6[d]	3D4—1D5	OP	30	0.1—1.4%	$\rightarrow \bar{M}_n$; (A_2)	98
6	1949	6[d]	3D4—1D5	VI	30	0.05—1.5%	$\rightarrow$[η]; (k')MHS (0.64; 6.66d2)	98
7	1949	1[e]	?	LS	30—50	0.05—1%	436 nm: $\rightarrow \bar{M}_w$ (apparent); A_2; temperature-dependent aggregation (PVAA)	118
8	1951	4[f]	1D5—2D5	LS	14—50	0.01—0.5%	436 nm: $\rightarrow$temperature-dependent aggregation (PVAA)	89
9	1951	3	1D5—2D5	LS	20—48	0.01—0.5%	436nm: $\rightarrow \bar{M}_w$; A_2; no aggregation (PVAL)	89
10	1952	4[d]	8D3—2D5	SS	28	0.2—1.5%	$\rightarrow s^0$	85
11	1952	1[d]	8D3	SV	25	0.5%	$\rightarrow \tilde{v}_2^0 = 0.748 \pm 0.005$ cm^{3}g^{-1}	85
12	1952	4[d]	8D3—2D5	TD	25	0.5—1.5%	$\rightarrow D^0$	85
13	1952	21[d]	700—5D4	VI	25	0.25—1%	$\rightarrow$[η]; MHS (0.5; 0.3)[g]	85
14	1952	10[d]	5D3—9D4	TD	25	1—10%	$\rightarrow D^0$	119
15	1952	13[d]	3D3—9D4	VI	25, 30	0.2—2%	$\rightarrow$[η]; $\bar{M}_v$; (k')	119
16	1952	6[d]	2D4—1D5	VI	20—80	0.1—1.5%	$\rightarrow$[η]; $\bar{M}_v$; MHS (see text)	120
17	1952		Heat of swelling measurements		17			121
18	1954	2	3D4, 9D4	LS	20	0.01—0.5%	436nm: $\rightarrow$ aggregation effects?	68
19	1954	2	7D4, 9D4	OP	20?		$\rightarrow \bar{M}_n$	68
20	1954	4?	1D4—9D4?	RI	20?		436nm: $\rightarrow \tilde{n} = 0.144$ cm^{3}g^{-1}	68
21	1954	4	1D4—9D4	SS	20	0.1—1%	$\rightarrow s^0$	68
22	1954	4?	1D4—9D4	SV	20?		$\rightarrow \tilde{v}_2^0 = 0.765$ cm^{3}g^{-1}	68
23	1954	4	1D4—9D4	TD		0.1—0.9%	$\rightarrow D^0$; $\bar{M}_{sD}$	68
24	1954	4	1D4—9D4	VI	0.5—78	0.1—1%	$\rightarrow$[η]; MHS (25°C:0.60; 0.14); TP	68
25	1954	1	8D4	OP			$\rightarrow \bar{M}_n$	122
26	1954	1	8D4	VI	22—30	0.1—1%	$\rightarrow$[η]	122
27	1955	14[h]	7D3—6D4	HS	30	0.8—1.1%	$\rightarrow \Delta H$	123
28	1955	15[i]	7D3, 6D4	HS	30		$\rightarrow \Delta H$ (swelling)	124

Table 1 (continued)
STUDIES ON PVAA AND PVAL IN AQUEOUS SOLUTION[a]

Entry	Year	Polymer samples: No.	Polymer samples: Range of $\bar{M}$	Techniques used	Temperature (t/°C)	Polymer concentration	Comments	Ref.
29	1955	7[j]	7D4	OP	30	0.1—0.3%	$\rightarrow \bar{M}_n$; A_2; thermodynamic parameters	125
30	1955	7[j]	7D4	SV	30		$\rightarrow \tilde{v}_2^\circ = 0.773—0.812$ cm³g⁻¹ (footnote j)	125
31	1955	7[j]	7D4	VI	15—60		$\rightarrow [\eta]$; k'; thermodynamic parameters	125
32	1955	1	1.3D5	LS		0.05—0.3%	436nm: → aggregation effects	126
33	1955	1	1.3D5	SS	20	0.08—0.8%	$\rightarrow s^\circ$; aggregation effects	126
34	1955	1	1.3D5	VI	25	0.2—0.8%	$\rightarrow [\eta]$; aggregation effects	126
35	1956	2	2D4, 3D4	HD	30	$\phi_2 = 0.04—0.135$	$\rightarrow \Delta H$	127
36	1956	3[k]	5D4	HD	30	$\phi_2 = 0.03—0.08$	$\rightarrow \Delta H$	128
37	1956	?[d]	?	HS	20—85		$\rightarrow \Delta H$ (swelling)	129
38	1956	4[e]	6D4—9D4	USV	20	~0.04%	$\rightarrow \kappa$; "bound water"	130, 131
39	1957	37	1D4—D5	VI	30	0.05—0.7%	$\rightarrow [\eta]$; k' (not constant)	92
40	1958	1	2D4	SS	(20)	0.4—1%	$\rightarrow s^\circ$; MWD	132
41	1958	1	2D4	TD	20	0.2—0.8%	$\rightarrow D^\circ$; MWD; $\bar{M}_{sD}$	132
42	1956	1[d]	?	SS	25	0.1—0.5%	H_2O or D_2O; +NaCl (0.5M): $\rightarrow s^\circ$; $\tilde{v}_2^\circ = 0.79 \pm 0.01$ cm³g⁻¹	133
43	1958	1[d]	7D4	FPD	0 to 4	$\phi_2 = 0.08—0.25$	$\rightarrow \Delta T$; anomalous effects	74
44	1958	1?[d]	7D4?	SV			$\rightarrow \tilde{v}_2^\circ = 0.748$ cm³g⁻¹	74
45	1958	11	9D3—1.8D5	RI	30	0.02—0.27%	436, 546, 578nm: $\rightarrow \tilde{n} = 0.168, 0.164, 0.164$ cm³g⁻¹ (footnote m)	134
46	1958	?[d]	?	LS, VI	25		→ MHS (0.63; 5.95d2)[n]	81
47	1958	22	1D4—3D5	VI	25		$\rightarrow [\eta]$; k'; $\bar{M}_v$	67, 81
48	1959	7[o]	8D4—3D5	LS		0.06—0.5%	546 nm?: $\rightarrow \bar{M}_w$; $\hat{R}$; A_2	67
49	1959	7[o]	8D4—3D5	RI			546 nm: $\rightarrow \tilde{n} = 0.155\ (\pm 0.003)$ cm³g⁻¹	67
50	1959	32[p]	7D4—4D5	VI	25	(4%)	$\rightarrow \eta(c_2 = 4\%)$; $[\eta]$; k'; MHS[q]	67
51	1959	3	5D4—2D5	VI	15—65		$\rightarrow [\eta]$	135
52	1959	1[c]	3D4	UV			220—380 nm: $\rightarrow \epsilon$; effect of hydrogenation	106
53	1960	15	1D4—7D5	VI	30	0.02—0.1%	$\rightarrow [\eta]$; MHS (0.62; 4.53d2)	136
54	1960	3	8D3—7D4	SV	30	0.05—0.9%	$\rightarrow \tilde{v}_2^\circ = 0.767—0.756$ cm³g⁻¹	137
55	1962	~20[d]	3D4—5D5	VI	25		$\rightarrow [\eta]$; MHS (0.616; 7.31d2)[r]	138

56	1962	5[c]	9D4—3D5	LS	30	0.3—1%	546 nm: →$\overline{M}_w$; A_2; aggregation effects	86
57	1962	5?[c]	9D4—3D5?	RI	25		546 nm: →$\widetilde{n}$ = 0.170 cm^3g^{-1}	86
58	1962	5	9D4—3D5	VI	30		→[η]; k'; $\overline{M}_v$	86
59	1962	8	1D5—5D5	LS	30—80	0.05—0.9%	436, 546 nm: →$\overline{M}_w$; A_2	83
60	1962	10?	1D5—5D5?	RI	60		436, 546 nm: →$\widetilde{n}$ = 0.1695, 0.1645 cm^3g^{-1}	83
61	1962	10	1D5—5D5	VI	30, 80		→[η]; k'; MHS (80°C:O.56;4d2)[g]	83
61a	1962	Reevaluation of the VI data of Matsumoto and Ohyanagi[136] (entry 53)						83
62	1963	1[d]	2D4	SV	21.5	0—3%	1-4536 atm →$\widetilde{v}_2^o$ = 0.750 cm^3g^{-1} (1 atm)	139
63	1966	Review of VI behavior in relation to entanglement						140
64	1968	Review of solution behavior						15
65	1970	Reviews of solution behavior						87, 141
66	1971	1	4D4	LS	25, 50, 150	0.1—4%	→$\overline{M}_w$; A_2	76
67	1971	1	4D4	VI	25, 50	0.1—1%	→[η]; k'	76
68	1972	19[t]	1D4—1D5	VI			→[η]; k'; $\overline{M}_v$ (footnote u)	55
69	1973	Review of solution behavior						4
70	1973	Review of solution behavior						29
71	1974	7[e]	8D3—6D4	SE	25	0.8%	→$\overline{M}_w$	90
72	1974	4[e]	3D4—7D4	SS	25		→s^o; R_h	90
73	1974	13[e]	4D3—7D4	VI	25		→[η]; k'; MHS (0.71; 2.7d2)[v]	90
74	1974	5[w]	3D4—1D5	LS	~25	0.3—0.9%	436, 546 nm: → aggregation effects	142
75	1974	5[w]	3D4—1D5	RI			436, 546 nm: → $\widetilde{n}$ = 0.166 (± 0.004) cm^3g^{-1}	142
76	1974	5[w]	3D4—1D5	VI	25		→[η]; k'	142
77	1974	1	8D4	LS	20—130	0.2—1%	436 nm: → $\overline{M}_w$; A_2	143
78	1974	1	8D4	RI	10—80		436, 546 nm: → $\widetilde{n}$ = 0.162—0.152 cm^3g^{-1}	143
79	1974	1	8D4	VI	25		→ [η]; $\overline{M}_v$	143
80	1974	4[x]	3D4—1D5	LS	20—100	0.1—10%	λ = ?: → "supermolecular order"	84
81	1974	4[x]	3D4—1D5	VI			→[η]	84
81a	1974	2[y]	1D5	—	30	0.05—1%	→ Flow-induced crystallization	144
82	1975	1[z]	9D4	LS	23	2.5—15%	500—800 nm: →transmission studies on frozen and thawed solutions	104, 105
83	1976	1[aa]	1D5	SV	25, 35	4—10%	→ $\widetilde{v}_2^o$ = 0.737, 0.763 cm^3g^{-1}	145
84	1976	1[aa]	1D5	TD	25, 35	4—10%	→ D (H_2O)	145
85	1976	1[aa]	1D5	VI	25—35	4—10%	→η	145
86	1976	1[aa]	1D5	EC	25—35	3—10%	20 kHz: →σ	146
87	1976	1[aa]	1D5	EP	25—35	3—10%	2—4 GHz: →ε; "hydration number"	146
88	1976	2[bb]	7D4, 1D5	SV	49—90	5%	→ΔV (gelation); gelation kinetics	72
89	1976	3[y]	1.0D5—1.2D5		30	~1%	→Flow-induced crystallization	147

Table 1 (continued)
STUDIES ON PVAA AND PVAL IN AQUEOUS SOLUTION[a]

Entry	Year	Polymer samples: No.	Polymer samples: Range of $\bar{M}$	Techniques used	Temperature (t/°C)	Polymer concentration	Comments	Ref.
90	1978	Reevaluation of the VI data of Garvey et al.[90] (entry 73)					→MHS (35°C: 0.70; 3.06d2)	88
91	1978	6[cc]	1D4—6D4	VI	25—50		→[η]; k'; MHS[dd]	88
92	1979	4[ee]	~2D5	UV			250—360 nm	148
93	1979	3[y]	1.0D5—1.4D5	—	18, 30	1—4%	→Flow-induced crystallization	71
94	1979	3[y]	1.0D5—1.4D5	VI	30	1.5—4%	→η; aging effects	71

[a] Entries refer to PVAL, i.e., with less than 2 mol % VAC, unless otherwise indicated.
[b] 1 to 7 mol % VAC.
[c] MHS parameters combining data of Flory and Leutner[64] with those of Staudinger and Warth (entries 1 and 2).[117]
[d] VAC content (if any) not stated.
[e] ~12 mol % VAC.
[f] 16 to 20 mol % VAC.
[g] MHS parameters from data for only four samples (fractions), $\bar{M}$ = 8D3—2D5.
[h] 0 to 35 mol % VAC.
[i] Same two PVAL samples, variously heat treated (25 to 218°C).
[j] Same parent polymer — 0 to 16 mol % VAC.
[k] 0.4 to 19 mol % VAC.
[j] 0 to 29 mol % VAC.
[m] Breaks in the refractive index plots (see text).
[n] Source of parameters given as "G.E. Munn, private communication" (Reference 81).
[o] 4 to 16 mol % VAC.
[p] 0 to 25 mol % VAC.
[q] MHS parameters for different VAC contents (mole %): 3.6%:0.61,6.9d2; 6.5%:0.60,7.4d2; 13.2%:0.58,8.0d4.
[r] MHS parameters using the combined data from Hosono and Sakurada,[119] (whose M values in fact were derived from the MHS relation of Nakajima and Furatate (see entry 6),[98] Dieu (see entry 24),[68] and Dialer et al. (see entry 13),[85] together with additional data from "M. Schonberg (TH, Hannover) and F. Patat (Farbwerke Hoechst AG)".
[s] MHS parameters for only six samples with $\bar{M}$ = 1D5—5D5; reevaluation, using same a (0.56), gives K = 8.0 (±0.6) d2 $cm^3\ g^{-1}$.
[t] 1 and 12 mol % VAC (commercial samples).

[u] $\overline{M}_v$ values obtained using the MHS parameters (source not disclosed): 2% VAC, 0.61, 8.8d2;12% VAC, 0.59, 9.9d2.

[v] MHS parameters for only seven samples with $\overline{M}$ = 8D3—6D4.

[w] "25 to 30% syndiotacticity".

[x] 1 to 8 mol % VAC.

[y] 55 mol % syndiotactic triads.

[z] *Elvanol* grade 73-125 G: ">70% atactic".

[aa] *Rhodoviol* 16/20 (~ 2 mol % VAC), partly fractionated.

[bb] VAC content not stated; 64% syndiotactic diads.

[cc] 9 to 16 mol % VAC.

[dd] MHS parameters quoted: 37°C, 0.67, 3.52d2; 50°C, 0.67, 3.04d2 — but these parameters are somewhat doubtful since the VAC contents of the samples vary with their molecular weights.

[ee] 1 to 25 mol % VAC.

used to adjust all the values to the standard condition of infinite molecular weight. The overall average value of $\bar{v}_2^\circ$ so obtained is 0.766 ± 0.016 cm³ g⁻¹ (30°C; $M \rightarrow \infty$). The spread in the values is evidently much greater than the uncertainty claimed by the individual groups (generally, ±0.001 cm³ g⁻¹), and seems to suggest the occurrence either of undetected systematic errors, or a sensitivity to structural irregularities or to small differences in stereoregularity; the effect of any differences in residual acetate content (see below) is too small to be an important contributing factor. The reciprocal of $\bar{v}_2^\circ$ is 1.306 (±0.027) g cm⁻³, which is thus the "effective density" of PVAL in aqueous solution; this value is in fact within the range of values obtained for the density of the (partly crystalline) solid polymer (see Section II.F.1) and thus it is definitely greater than that of the purely amorphous polymer, which is an indication of the compacting effect of the polymer on the solvent.

The *partial specific volume* of PVAA in aqueous solution in relation to its acetate content has been studied by Sakurada and co-workers,[125] who used seven samples with acetate contents ranging from 0 (i.e., PVAL) up to 16 mol %, all with M = 7D4. The values of $\bar{v}_2^\circ$ rose essentially linearly with acetate content, by 0.003 cm³ g⁻¹/mole %; within the limits of scatter of the data (±0.004 cm³ g⁻¹) there was no significant difference between the values for randomly acetylated and those for selectively acetylated PVAA.

The *specific refractive index increment,* $\tilde{n}$, for PVAL in aqueous solution has been determined by five groups as listed in Table 1: Dieu (entry 20);[68] Matsumoto and Ohyanagi (entry 45);[134] Matsuo and Inagaki (entries 57 and 60);[83,86] Gruber and co-workers (entry 75);[142] and Klenin and co-workers (entry 78).[143] The values for samples of PVAA of acetate contents from 4 to 16 mol % have been determined by Beresniewicz (entry 49).[67]

For PVAL, Matsumoto and Ohyanagi[134] reported that the plots of solution refractive index n vs. c_2 consist in each case of two parallel straight lines, the first one starting from the origin but then having a break at a certain critical concentration, c_2^*, beyond which the value of n remains constant for a small span of concentration after which the linear rise is resumed with the same slope again. The values of c_2^* ranged from 0.06 to 0.3%, and were found to be inversely proportional to $M^{1/2}$ and linearly related to $1/[\eta]$; this behavior therefore seems to represent the attainment of effective overlap of the polymer molecules at the break point. These effects have not been reported by any of the other groups (nor apparently for any other polymer in solution), but in any case they would have passed undetected if the concentration ranges that were used (also not reported by any of the other groups) were much higher than those quoted above; correspondingly, in such cases the measured $\tilde{n}$ values would also not be influenced if this effect had been present, so that it may be disregarded for the present purposes.

The measurements of Klenin and co-workers[143] were carried out over the temperature range 10 to 80°C, and showed the temperature coefficient of $\tilde{n}$ of PVAL in aqueous solution to be 9d4 deg⁻¹; this value may be used to adjust to 25°C the value of $\tilde{n}$ obtained at 60°C by Matsuo and Inagaki.[83] The collected values of $\tilde{n}$ for the range 20 to 30°C for PVAL and PVAA are plotted against mole percent acetate content in Figure 2, the data sources being given in the corresponding caption. The value of 0.144 cm³ g⁻¹ reported for $\tilde{n}$ for PVAL by Dieu[68] is well outside the range of the rest of the values, and it has accordingly been omitted. For PVAL the overall averages (including the value extrapolated to zero acetate content for PVAA samples from Beresniewicz[67]) for the two wavelengths are 436 nm, $\tilde{n}$ = 0.165 (±0.004) cm³ g⁻¹; 546 nm, $\tilde{n}$ = 0.164 (±0.004) cm³ g⁻¹, i.e., essentially the same values. The spread of the values is clearly much greater than the precision claimed for the individual values, and there are indications (particularly from the results of Gruber and co-workers[142]) that $\tilde{n}$ is sensitive to small differences in the structure of the PVAL samples.

For PVAA, the refractive index data of Beresniewicz[67] plotted in Figure 2 show that there is a linear fall of $\tilde{n}$ with acetate content, amounting to 8d4 cm³ g⁻¹/mol %. It is notable, in view of the scatter seen in the values of $\tilde{n}$ for PVAL samples in the same figure, that

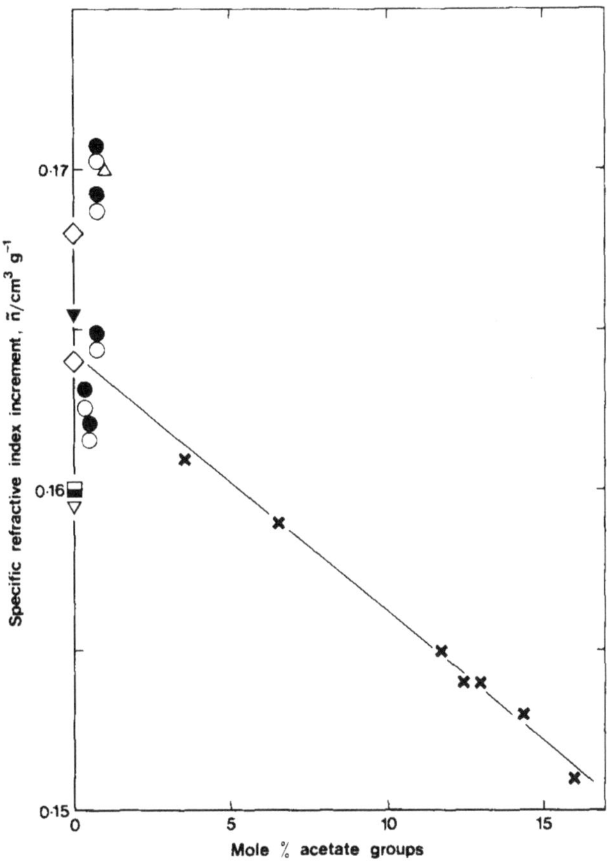

FIGURE 2. Specific refractive index increment, $\bar{n}$, of poly(vinyl alcohol) and poly(vinyl acetate-*co*-vinyl alcohol) in aqueous solution at 25 ($\pm$5)°C, vs. mole percent acetate content. Key (with wavelength in nm): (●, ○) Gruber et al. (436, 546);[142] (⬓) Klenin et al. (436, 546);[143] (♦, ◇) Matsumoto and Ohyanagi (436, 546);[134] (△) Matsuo and Inagaki (546);[86] (▼, ▽) Matsuo and Inagaki (436, 546 — $\bar{n}$ values adjusted to 25°C from 60°C);[83] (×) Beresniewicz (546).[67]

despite the wide range of molecular weights of these samples of PVAA (8D3—3D4) the scatter of the points about the linear correlation is within the limits of precision claimed for individual values ($\pm$0.001 cm^3 g^{-1}) by Beresniewicz.[67]

The *intrinsic viscosity-molecular weight* correlations for PVAL and PVAA in aqueous solution have been studied extensively, and this has correspondingly led to a rather complex situation of the type commonly encountered in such circumstances. In view of the technical and scientific importance of these polymers, a critical examination of the published data for these systems is necessary both to provide a sound basis for the determination of molecular weights from viscosity measurements (particular groups of workers tend to have a "favorite" set of MHS parameters which they use to the exclusion of other, possibly equally valid published sets), and also because the values of the two MHS parameters (particularly, that of the exponent *a*) indicate the solvent power of water for the polymer.

Considering, first, PVAL at normal temperatures, Figure 3 is a double logarithmic plot of the published data for intrinsic viscosity (at 20 to 30°C) vs. molecular weight from the sources listed in the corresponding caption. Examination of the data shows that there are four groups of workers whose sets of MHS parameters (i.e., *a* and K/cm^3 g^{-1}) are apparently equally trustworthy: Beresniewicz, 0.63 and 5.38d2;[81] Elias, 0.616 and 7.31d2;[138] Matsumoto

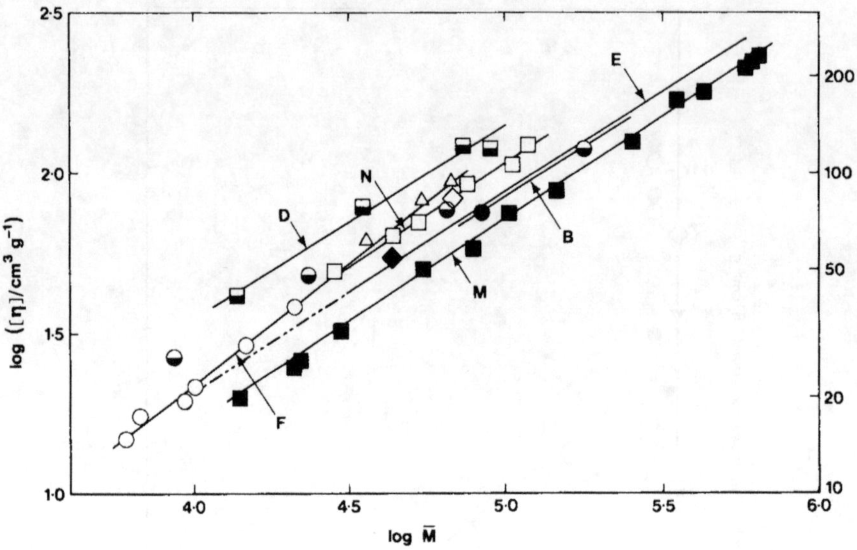

FIGURE 3. Double logarithmic plot of intrinsic viscosity vs. molecular weight for poly(vinyl alcohol) in aqueous solution at 25 (±5)°C. Key to data symbols and line labels: (B) Beresniewicz;[81] (◒) Dialer et al.;[85] (⬓, D) Dieu;[68] (E) Elias;[138] (○, F) Flory and Leutner;[64] (●) Klenin et al.;[143] (■, M) Matsumoto and Ohyanagi;[136] (□, N) Nakajima and Furatate;[98] (◇) Sakurada et al.;[125] (△) Staudinger and Warth;[117] (◆) Tager et al.[76] Type of molecular weight average indicated by shading of the symbols: open symbols, $\overline{M}_n$; half-filled symbols, $\overline{M}_{sD}$ (i.e., from sedimentation velocity and translation diffusion); filled symbols, $\overline{M}_w$.

and Ohyanagi, 0.64 and 4.35d2;[136] and Nakajima and Furatate, 0.64 and 6.7d2.[98] There is evidently little doubt about the value of *a*, since the quoted values all lie within the range 0.63 ± 0.01. For *K* the spread is much greater, presumably as a consequence of the different types of molecular weight average involved. However, although the relatively high value of *K* (i.e., 6.7d2) obtained by Nakajima and Furatate[98] may have been the result of their use of $\overline{M}_n$ values, any such effect should be minimized by the fact that the samples they used were fractions. The low value of *K* (i.e., 4.35d2) reported by Matsumoto and Ohyanagi[136] is even less easily understandable because they had corrected this parameter for any inhomogeneity in their samples. The best that can be done in these circumstances is to take a simple average of the four values of *K* reported. This therefore gives the recommended MHS parameters for this system as: $a = 0.63 \pm 0.01$; $K = 5.8\ (\pm 1.5)$d2 cm³ g⁻¹.

It has been stated[92] that a similar set of MHS parameters — $a = 0.62$; $K = 8.0$d2 cm³ g⁻¹ (at 30°C) — are used in the Japanese industry for molecular weight determinations of PVAL, and also of PVAA via viscometry on the PVAL obtained from it by the hydrolysis of the acetate groups; however, it appears from the details of the method as given by Finch[45] that these parameters are applied not to the intrinsic viscosity, [η], but rather to the inherent viscosity $(\ln \eta_r)/c_2$ (where η_r is the relative viscosity) measured at the polymer concentration of $c_2 = 1\%$, in which case the somewhat higher value of *K* which has to be used is understandable.

The concentration dependence of viscosity for dilute solutions of PVAL has been examined by numerous groups in connection with the determination of [η] (see, e.g., entries 2, 6, and 15 in Table 1), but the most detailed studies specifically on this dependence are those of Matsumoto and Imai (entry 39),[92] who showed that for the molecular weight range from 1D4 to 2D5 the Huggins parameter *k′* is not a constant but is dependent on [η], since the slopes of the plots of η_c vs. c_2 are proportional to $[\eta]^{2.28}$, rather than to $[\eta]^{2.0}$ as would be the case if *k′* were truly constant.

The effect of temperature on [η] and on the MHS parameters for PVAL in water has been studied over the range 20 to 80°C by Kuroiwa,[120] while some studies have also been carried out by Matsuo and Inagaki[83] at the single relatively high temperature of 80°C. In the first case the values of $\overline{M}$ were apparently obtained from [η] values measured at 30°C using the parameters derived by Nakajima and Furatate (see above);[98] the value of [η] decreases steadily with rise in temperature, while the MHS parameters (a and K/cm^3 g^{-1}) which may be calculated from the original data[120] vary smoothly from 0.650 and 6.25d2 at 20°C to 0.61 and 7.4d2 at 80°C. The fall in the value of a and the rise in that of K is further evidence for a reduction in the solvent power of water towards PVAL with rise in temperature. Matsuo and Inagaki[83] report rather different values of these parameters for 80°C, i.e., 0.56 and 9.4d2, but the range of their molecular weights is too small and the scatter of their data too great for these to be trustworthy.

Considering the intrinsic viscosity values of PVAA in water, the value of [η] falls with increasing acetate content, presumably as a result of the occurrence of hydrophobic interactions between segments in the polymer coil leading to an increase in the number of the segment-segment contacts.[67,125] The variation of the MHS parameters with acetate content has been investigated by Beresniewicz,[67] and by Lankveld and Lyklema;[55] both groups showed that, as expected, the value of a falls and that of K rises with increasing acetate content. In the first of these studies, which involved samples with acetate contents of 0, 4, 6, and 13 mol %, a rather extended extrapolation of MHS exponent a against acetate content suggests that a = 0.50 for PVAA with 50 mol % acetate groups in water at 25°C, i.e., this is an ideal system.[67] The variation of [η] with temperature (15 to 60°C) for PVAA samples of acetate content up to 15 mol % has been examined by Sakurada and co-workers.[125]

The dependence upon acetate content of the value of the Huggins parameter k' for PVAA in water has been studied by several groups, notably Beresniewicz[67] and Sakurada and co-workers.[125] The value of k' rises with acetate content, as would be expected for systems where the solvent is becoming an increasingly poor one, but the extent of the rise seems to depend very markedly on the distribution of the acetate groups within the polymer chain.

The value of the viscosity of a 4% solution of PVAA, which is widely used in the specifications for commercial samples, has been determined as a function of acetate content for samples of unfractionated polymers by Beresniewicz.[67] The value falls initially with increasing acetate content but then reaches a minimum at about 13 mol %; the subsequent rise is such that a sample with 25 mol % acetate has a 4% viscosity only somewhat less than that of PVAL itself. From the viewpoint of the dilute solution behavior, this is the result of the rise in the Huggins slope parameter k' eventually outweighing the accompanying fall in the value of [η].

2. *Nonaqueous and Mixed Solvents*

PVAA and PVAL are soluble in only a relatively limited range of organic liquids when compared with many other water-soluble synthetic polymers such as PEO and PVP; even the organic liquids which do dissolve PVAL, such as ethylene glycol, glycerol, and dimethylsulfoxide, are rather inconvenient ones from such practical aspects as their high viscosity.[9] Most solution studies on these polymers (other, i.e., than those in purely aqueous solution) have therefore been carried out on mixed solvent systems, particularly mixtures of water and an organic *nonsolvent;* the behavior of these mixtures, of course, overlaps with and merges into that of the water + cosolute systems considered more specifically in Chapter 2 of Volume II, but is it opportune to consider their behavior at this earlier stage because of their actual or potential use, e.g., in molecular weight determination, and also because they comprise systems where the amount of "cosolute" becomes comparable to that of the solvent. However, the gelation behavior of aqueous systems containing glycerol is conveniently left until Chapter 2 of Volume II.

The differential sorption of solvent components by PVAL samples with the mixtures of water + acetone (w = 0.03 to 0.38) and water + 1-propanol (w = 0.02 to 0.45) has been studied by Levi and co-workers;[82] for the first case there appeared to be no sorption or swelling above w(acetone) = 0.15, while in the second case the sorption of water decreased monotonically with increasing propanol concentration. The viscosity behavior of the corresponding homogeneous solutions with w(acetone) = 0.05 to 0.20 and w (propanol) = 0.5 to 0.33 has also been studied by the same group;[135] the addition of the acetone seemed to have no appreciable effect on $[\eta]$, but caused a marked decrease in the Huggins parameter k', while with propanol the value of $[\eta]$ increased up to w = 0.20 of the additive and thereafter fell off again sharply. The temperature variation (15 to 65°C) of $[\eta]$ for several of these mixtures has also been studied.[135]

The viscosity behavior of PVAL in water + dioxane mixtures with ϕ(dioxane) = 0.10 and 0.33 has been studied by Dieu,[68] in parallel with corresponding studies in water itself, for the overall temperature range 0 to 78°C. At the lowest end of the temperature range the addition of the dioxane makes little difference to the value of $[\eta]$, but with increasing temperature the value is correspondingly increased by such addition.

Similar viscosity measurements on PVAL in mixtures of water with methanol, ethanol, propanol, acetone, and dioxane have been reported by Naito and Kominami.[122]

Matsumoto and Imai[92] have studied the dilute solution viscosity behavior of PVAL at 30°C in aqueous phenol solution with w(phenol) = 0.85; the value of the intrinsic viscosity $[\eta]_p$ of the polymer in this solvent was found to be related to that in water, $[\eta]_w$ (where both values are in cm³ g⁻¹), by:

$$[\eta]_p = 0.75\,[\eta]_w^{1.25} \tag{4}$$

(the first numerical factor was given incorrectly as 4.2 in the original reference); using the MHS parameters previously recommended for PVAL in aqueous solution (see Section II.G.1) gives the corresponding parameters for this system as a = 0.79; K = 2.1d2 $cm^3\,g^{-1}$. Thus this system is a very good solvent for PVAL, and indeed the a value is essentially at the upper limit (i.e., 0.8) for a nonionic flexible chain polymer in solution; however, there are unfortunately obvious practical difficulties in the routine use of this system (particularly, its high viscosity and its caustic character).

The viscosity behavior of PVAL, PVAA, and also PVAC in dimethylsulfoxide has been studied by Naito;[149,149a] using molecular weights determined via viscosity measurements in water and the MHS parameters of Nakajima and Furatate,[98] the MHS parameters for PVAL in this solvent were deduced[149] to be a = 0.84; K = 1.6d2 $cm^3\,g^{-1}$. Again, as in the case of aqueous phenol discussed immediately previously, the value of a is at the upper limit for this type of polymer, showing this to be a very good solvent; the value of K is also closely similar to that in aqueous phenol.

H. Phase Equilibria

1. Water Vapor Uptake[141]

Literature data for the equilibrium uptake, U_1, of water vapor at various water activities, a_1, by PVAL are plotted in Figure 4, the sources being as given in the accompanying caption. These data refer to samples of PVAL of low crystallinity (20 to 30%), e.g., for films cast at room temperature and not subjected to higher temperatures subsequently. The uptake values are plotted on a logarithmic scale both because of their wide span and because this gives an essentially linear plot over the middle section. The open symbols correspond to the temperature range 25 to 30°C; there is good concordance between the different sources for $a_1 < 0.2$ and $a_1 > 0.9$, but some spread of values in between, presumably the consequence of differences in degree of crystallinity and residual acetate content. The filled symbols

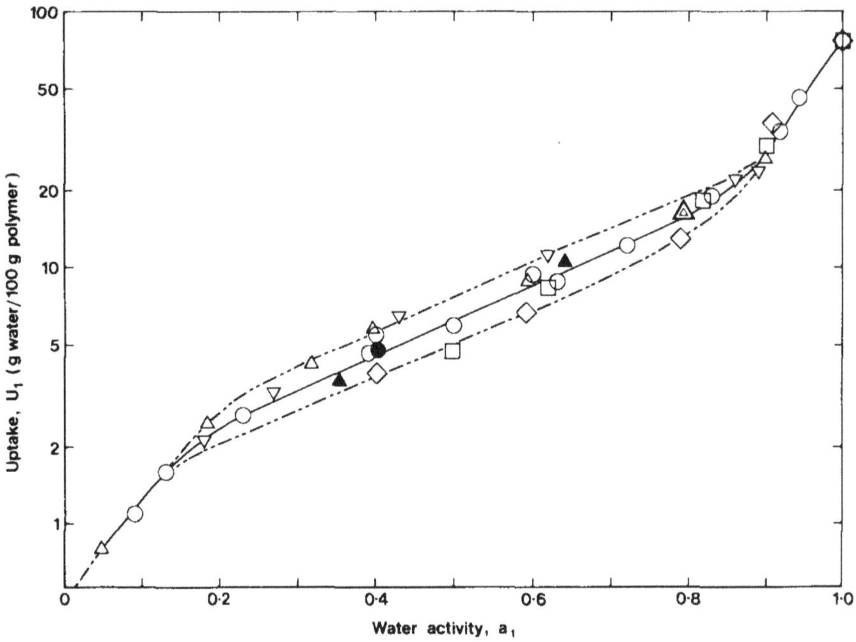

FIGURE 4. Equilibrium uptake of water, U_1 (logarithmic scale) by low crystallinity ($\phi_x \approx 0.25$) samples of poly(vinyl alcohol) vs. water activity, a_1. (○, ●) McLaren et al. (25, 36°C);[97,150] (▲) Long and Thompson (40°C);[75] (□, ◇) Sakurada et al. (30°C; samples F-2, F-6);[115] (△) Tadokoro et al. (25°C);[103] (▽) Tager et al. (25°C).[76]

correspond to somewhat higher temperatures (36 to 40°C) which nevertheless seems to have little effect on the uptake. The dependence of uptake on temperature has been studied more specifically by Dole and McLaren[151] and Tager and co-workers;[76] heat of swelling measurements,[124,129] since they relate to the limiting uptake at $a_1 = 1$, should also give information on the temperature dependence of this limiting uptake.

Heat treatment of PVAL reduces the subsequent water uptake, because of the resulting increase in crystalline content and thus the accompanying decrease in the amorphous content.[103]

Myers and co-workers[152] have reported uptake data for a plasticized PVAL film; the nature of the plasticizer was not revealed, but it was presumably glycerol or a similarly hydrophilic additive. Comparison of their data with that of unplasticized films (Figure 4) shows a marked increase on plasticization, although it is not clear whether this is only the result of the uptake by the plasticizer itself or whether there is also a contribution from an increase in the amorphous content of the PVAL.

Inasmuch as water vapor uptake by PVAL is so largely controlled by the amorphous content of the polymer (since the crystalline regions are presumably impermeable even to small-molecule penetrants such as water), then uptake data may be used as a method of estimating the relative amorphous and crystalline contents.[103] The uptake behavior of the polymer could be greatly clarified if data were available for purely amorphous PVAL, but these data are not directly obtainable because as already discussed (see Section II.F.8) even when prepared at room temperature and not heat treated subsequently, the samples have at least 20% crystallinity. One method to obtain such data might be to extrapolate data obtained for samples of a range of crystallinities (obtained by progressive heat treatment of a PVAL sample of low initial crystallinity) to zero crystallinity; however, application of this method to the data of this type obtained by Tadokoro and co-workers[103] gives plots which are too curved (particularly at the higher water activity values) for the extrapolation to be trustworthy. Alternatively, Valentine[116] has suggested that values for amorphous PVAL could be obtained by extrapolation of uptake data for (amorphous) PVAA to zero acetate content; however, once again, application of this method to the uptake data of this type obtained by Yamamura and Hamano[153] for both pure PVAC and for PVAA samples with 54 to 95 mol % vinyl acetate shows that the dependence is too markedly nonlinear and the extrapolation gap too great to give trustworthy results.

The interpretation of water vapor uptake by PVAL from the thermodynamic viewpoint is also complicated by the fact that the crystalline regions do not act simply as an inert diluent, but also function as cross-links for the absorbing amorphous regions.

The effects of absorbed water on the density of PVAL have been studied by Tadokoro and co-workers,[103,154] and those on the mechanical properties and the glass-transition temperature have been discussed by Toyoshima.[35]

The kinetic aspects of sorption of water vapor by PVAL have been studied by Hauser and McLaren,[97] Long and Thompson,[75] Sakurada and co-workers,[115] and Spencer and co-workers.[154a]

The water vapor uptake of PVAA in relation to its vinyl acetate content does not seem to have been reported in any detail, except for the studies of Yamamura and Hamano[153] (already mentioned above) on water-insoluble, acetate-rich samples (54 to 95 mol % vinyl acetate).

2. *Solubility Equilibria with Water*

There is a marked contrast between the solubility behavior of PVAL and that of PVAA; put into the simplest terms, PVAL is insoluble in cold water but soluble in hot water, whereas PVAA is soluble in cold water but insoluble in hot water.[15] This simple differentiation does indeed provide a useful guide to many of the practical applications of the polymers, but it needs to be qualified with many reservations if the real, more complex behavior is to be revealed; thus, as discussed below, the behavior of a sample of PVAL depends markedly upon its crystallinity, while that of a sample of PVAA correspondingly depends markedly upon its VAC content.

Considering PVAL, the insolubility in cold water is evidently related to the appreciable crystalline content; this is demonstrated most clearly by the facts that a sample of the polymer swells when equilibrated with cold water and that the extent of this swelling is directly related to the amorphous content, which may be reduced (and the crystalline content correspondingly increased) by heat treatment (see Section II.F.8). Thus, typically, a sample prepared at room temperature with 80% amorphous content has a tenfold equilibrium degree of swelling, but if the sample is heat treated to give it only 50% amorphous content this swelling is generally much less than twofold.[35,95,99]

When PVAL is heated in contact with water, the degree of swelling increases with rise in temperature until finally the polymer completely disperses at a fairly well-defined dissolution temperature.[4,5] Pritchard[141] has related this behavior with the need to melt the gel formed from the amorphous regions, and also with the need to disperse the crystalline regions. For samples of common PVAL the dissolution temperature is usually in the range 70 to 90°C (with the higher values associated with higher molecular weights);[4] previous heat treatment of the polymer sample increases the dissolution temperature because of the consequent higher crystalline content.[4,35] The dissolution temperature also depends on the steregularity of the sample;[31] thus, e.g., highly syndiotactic samples of PVAL have dissolution temperatures of 150 to 160°C.[70]

The homogeneous solutions of PVAL so formed seem to be unstable, being subject to aggregation and crystallization effects which take place particularly at the air/solution interface and enhanced by stirring.

Gelation (which may be viewed as a "homogeneous phase separation") occurs in the more concentrated solutions, with an ease dependent on tacticity.[31,72,141] The gelation of

PVAL in aqueous solution is greatly enhanced by the presence of dyes such as Congo Red, and of other cosolutes such as boric acid; these cosolute effects are considered more specifically in Chapter 2 of Volume II.

Tager and co-workers[76] found that PVAL solutions would separate into two phases on cooling if the concentration were greater than 6% (at lower concentrations no separation was observed even on prolonged cooling at 0°C); the critical temperature was about 80°C, with a critical concentration of about 35%.[76]

Considering the water solubility of PVAA, the substitution of PVAL with a small proportion of acetate groups makes the polymer more water sensitive at room temperature, the optimum solubility occurring at 12 mol % (which is the VAC content of the most common commercially available grade) where the dissolution temperature is in the range 25 to 35°C.[4,15] However, the solubility behavior also depends upon the distribution of the acetate groups.[16] Random copolymers can be dissolved readily; "blocky" copolymers dissolve in cold water if they have 5 to 10 mol % VAC, but higher dissolution temperatures are required for samples with only 1 to 2 mol % VAC; on the other hand, true block copolymers will not dissolve in cold water if they have more than 2 mol % VAC.[16]

This solubility behavior of sparsely acetylated PVAA may be explained by the disruptive effect of the acetate groups on the crystallizability of the remaining sequences of hydroxyl groups; this disruptive effect is clearly greatest with randomly distributed acetate groups, and least with those sequenced in blocks.

With higher VAC content the PVAA becomes less readily soluble at room temperature; thus Saito[155] found that PVAA samples with 13 and 20 mol % VAC were soluble at room temperature, one with 30 mol % was insoluble at 25°C but soluble at 15°C, while those with 54 and 69 mol % VAC were completely insoluble. The viscosity and light-scattering studies of Beresniewicz[67] indicate that water at 25°C is a theta solvent for PVAA with approximately 50 mol % VAC groups.

Samples of PVAA with moderate VAC contents (especially, between 12 and 35 mol %) show inverse solubility-temperature behavior in aqueous solution, i.e., the polymer precipitates on heating the solution, with the precipitation temperature decreasing with increasing VAC content.[4,15] This behavior has been investigated quantitatively by Sakurada and co-workers,[156] Saito and co-workers,[80,157-160] and Garvey and co-workers.[90] It has been used as a means to purify a PVAA sample with 30 mol % VAC,[80] and also to fractionate PVAA (where the separation is mainly according to VAC content and only to a lesser extent according to molecular weight).[89] Light-scattering studies on PVAA solutions at temperatures below the precipitation point show aggregation of the polymer molecules takes place, which becomes more extensive as the precipitation point is approached, but that this aggregation disappears if the PVAA is fully hydrolyzed to PVAL.[89,118]

Nord and co-workers[89] have shown that these systems also possess an upper critical temperature in that when the PVAA solutions were heated in sealed tubes, the turbidity disappeared at a temperature around 130°C, and reappeared on cooling.

Finally, considering the effects of electrolytes on the solubility of PVAL and PVAA in water, we note that Napper[161] has studied the phase separation of "poly(vinyl alcohol)" (presumably PVAL) in 2*M* NaCl, and Lambe and co-workers[162] that of PVAA with 12 mol % VAC in 2*M* $NaNO_3$; these cosolute effects are considered in more detail in Chapter 2 of Volume II.

3. Partition Equilibria

Studies on the partition behavior of polymers in two-phase, two-liquid systems are useful in giving an indication of the relative solvent powers of the two (mutually saturated) liquids towards the polymer. Dobry,[162a] in studies of this type with various polymers including PVAL, and with 17 such systems, found that in not one of these systems was PVAL appreciably soluble in *both* of the liquids involved. Water was involved as one component in four of these systems (the second components involved being amyl alcohol, acetic anhydride, diethyl ether, and phenol); this is a further example of the hydrophilic (and lipophobic) character of the polymer, since apparently unless a liquid is miscible with water it will not be a solvent for PVAL.

I. Chemical Behavior[30,163]

The thermal stability and decomposition reactions of PVAL and PVAA in the bulk have been discussed by Dunn and co-workers[21] and Tubbs and Wu.[29] For PVAL, on heating there is a dehydration reaction which sets in around or somewhat below 200°C; the decomposition temperature for PVAL produced by acid-catalyzed hydrolysis is lower than that from alkali-catalysis, which may result from traces of residual acid. With PVAA there is the possibility of the loss of acetic acid from the acetate groups; the ease of this reaction depends upon the distribution of these groups in the polymer chain. Each elimination reaction leads to unsaturation in the polymer chain, and correspondingly increased unsaturation makes the elimination reactions easier because of conjugation effects.

PVAL seems to be quite stable in aqueous solution at normal temperatures, even for example in the joint presence of sodium chloride, ferrous ions, and oxygen which induces the decomposition of several other water-soluble polymers.[77]

As already noted (Section II.C.4), poly(vinyl alcohol) normally contains between 1 and 2 mol % of 1,2-diol (vicinal hydroxyl) groups, so that the polymer is chemically sensitive to reagents that attack such groups; one such reagent is periodic acid, which may be therefore used to determine the content of such groups via either the consumption of reagent or the concurrent decrease in the molecular weight of the polymer (see Section II.C.4), while the degradation effect has also been used as a means of obtaining low molecular weight samples of PVAL.[148]

The effects of gamma radiation on PVAL in the solid state and in solution have been reviewed by Pritchard.[163] Thiourea has been shown by Sakurada and Ikada[164] to be an efficient agent for protecting PVAL from the effects of such radiation. The effects of X- and γ-rays on PVAL (and several other water-soluble synthetic polymers) in aqueous solution have also been studied by Alexander and Charlesby.[164a]

We now turn to consider those reactions which are used to produce chemically modified forms of PVAA or PVAL for various purposes; the polymers lend themselves to a wide variety of such modification reactions, which generally involve attack either on individual hydroxyl groups, or else on the pairs of these on adjacent monomer units on the polymer chain. Such reactions have been reviewed by Pritchard[163] and Finch.[30]

Many of the methods of chemical modification applied to PVAA and PVAL, such as reaction of pairs of adjacent hydroxyl groups with aldehydes to form acetals, have been developed to lead to a reduction in water solubility or sensitivity. Some, however, lead to this property being retained or even enhanced. For example, reaction of PVAL with an alkoxyacetaldehyde, RO–CH_2–CHO, leads to the formation of vinylalkoxyacetal units in the chain of the type:

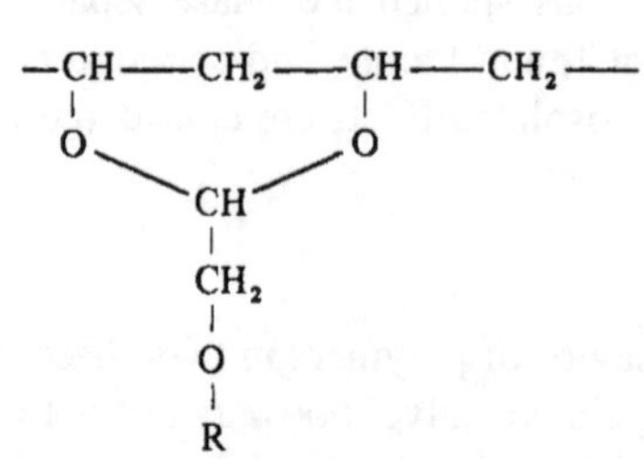

The products are readily water soluble at room temperature but show the inverse solubility-temperature behavior in aqueous solution which is characteristic of polyethers (see Chapter

2); the synthesis and solubility behavior of PVAL so modified, with a range of R groups, have been described by Beal and Thor,[165] while the water vapor uptake for R = Me with a range of degrees of substitution has been measured and discussed.[166,167]

Similarly, reaction of the individual hydroxyl groups in PVAL with urea gives vinyl urethane units:

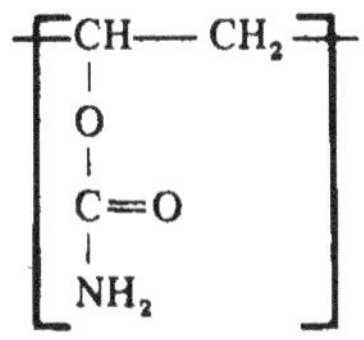

These are of particular interest in the present context because they represent an oxygenated form of the acrylamide unit (see Chapter 3). The synthesis and solution properties of PVAL so modified, with degrees of substitution up to 13 mol %, have been described by Sakurada and co-workers;[78] the MHS exponent a for aqueous solution increased with increase in the extent of urethane substitution, indicative of the expected increasing hydrophilic character of the polymer.

The water solubility of PVAL may also be enhanced by partial or complete conversion to poly(vinyl sulfuric acid) (see Chapter 1 of Volume II).

J. Biological Behavior

The microbial decomposition of PVAL has been the subject of several studies.[168-170] In general, however, the polymer is biologically much more stable than the natural or modified natural (semisynthetic) polymers which are used for the same purposes.[9] Other biological aspects of the polymer have been reviewed.[2]

K. Analysis[2,45,62,171]

For PVAL in aqueous solution at concentrations of 1% and upwards, the estimation of the polymer content by evaporation to dryness followed by heating to constant weight at 105°C seems to be a satisfactory procedure.[45]

For the assay of more dilute aqueous solutions of PVAL and PVAA, use is commonly made of the green complex (or more probably, complexes) formed by the polymers with mixtures of iodine, iodide ions, and boric acid. This system has been investigated and discussed by Zwick,[57,172] Hayashi and co-workers,[173] and Joshi and co-workers.[148] The formation of the complex has been developed by Finley[174] into an analytical method in which the final solution contains 3 mM molecular iodine, 9 mM potassium iodide, and 0.2 M boric acid together with 5 to 20 ppm of the polymer; the complex is then determined spectrophotometrically at the λ_{max} of 690 nm. This procedure has been further developed by Chene and co-workers[175] (see also Pritchard[62]), and it has been used, e.g., by Garvey and co-workers[90] and Suzuki.[170] A similar method was developed by Horacek,[176] and has been used, e.g., by Lyklema and co-workers.[55,177]

The Finley method has been examined in detail more recently by Joshi and co-workers,[148] who showed that the results obtained using the recommended concentrations of reagents were essentially insensitive to the VAC content over the range 0 to 25 mol %. Furthermore, if the concentration of molecular iodine is reduced to 0.5 mM, then the results become sensitive to the VAC content; thus measurements at each iodine concentration enable both the total amount of PVAA and/or PVAL, and its VAC content, to be determined.[148]

In the absence of boric acid, mixtures of iodine and iodide ions do not give a colored complex with PVAL, but they nevertheless give a red-violet color (λ_{max} = 490 nm) with PVAA, which can be used for its analysis.[178,179] However, this color seems to be sensitive to the distribution of the acetate groups, depending as it does upon the presence of blocks of these groups;[180] in fact, a similar color is produced when iodine is absorbed into PVAC.[181]

Other methods which are used for the analysis of PVAL have been listed and reviewed;[2] the procedures of the official Japanese Industrial Standard have been given by Finch.[45]

III. POLYVINYLPYRROLIDONE — PVP

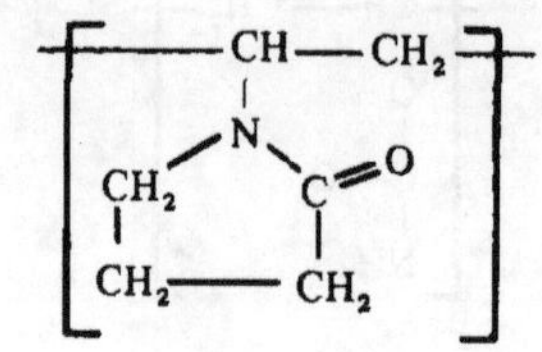

M_0 (C_6H_9NO) = 111.14

Other chemical names, abbreviations, and trade names are* 1-ethenyl-2-pyrrolidinone homopolymer *(Chemical Abstracts);* poly[1-(2-oxopyrrolidin-1-yl)-ethylene] (IUPAC); poly(vinylpyrrolidinone); poly(vinyl pyrrolidone); poly-vinylpyrrolidone; poly(vinyl-pyrrolidone); poly(1-vinyl-2-pyrrolidone); poly(N-vinylpyrrolidone); poly(N-vinyl-2-pyrrolidone); and other similar forms, as well as misspellings ("polyvinylpyrollidone", "polyvinylpyrrolidine", etc.); PVDD; P.V.P.; *Albigen; Isoplasma; Kollidon; Luviskol; Macrose; Periston; Plasdone; Polyclar;* Povidone (B.P., U.S.P.); Plasmosan; Peregal.

A. General Features

This polymer has been the subject of numerous reviews and general surveys.[182-191] Technical literature on the commercially available grades is available from the manufacturers,[192,193] while one of the major manufacturers, General Aniline & Film Corporation (GAF), has published annotated bibliographies[194-196] and literature surveys[197,198] on the properties and applications of the polymer.

As well as being readily soluble in water, PVP is soluble in a range of organic liquids. However, the solubility depends markedly upon the water content of the polymer sample so that media can be divided according to their solvent power into three classes:[187,191,192]

1. Media which are solvents both for commercial PVP (containing 5% water) and for desiccated PVP (<0.5% water): alcohols, diols, and glycerol; aliphatic acids (C_1—C_3); partially chlorinated alkanes (e.g., CH_2Cl_2, $CHCl_3$, CH_2ClCH_2Cl); nitroalkanes; amines, lactams, and lactones
2. Media which are solvents only for desiccated PVP (<0.5% water) and/or low molecular weight polymer: cyclic ethers (dioxane, tetrahydrofuran); aliphatic esters; ketones (only low molecular weight with acetone); fully chlorinated alkanes; aromatic hydrocarbons
3. Media which are nonsolvents for all forms of PVP: alkanes; aliphatic (noncyclic) ethers

The molecular weight characteristics of the main commercial grades of PVP are discussed in Section III.D.

From the scientific viewpoint, polyvinylpyrrolidone is interesting in several respects. Considering the structure of the monomer unit, it is seen to have an amphiphilic character since it contains a highly polar amide group with a dipole moment of close to 4 D (see Section III.B.1.a.) conferring hydrophilic and polar-attracting properties, and also apolar methylene and methine (CH) groups in the backbone and the ring conferring hydrophobic properties. Furthermore, the dipolar amide group has a different environment at each end,

* For a more comprehensive list, see Reference 182.

since as an examination of molecular models readily shows, the oxygen end is exposed while the nitrogen end is largely buried in the surrounding methylene and methine groups.

PVP is also interesting from the biological viewpoint, since it has structural features similar to those of proteins, which have led to its being proposed on several occasions as a simple protein model.[199-202] Indeed, even in its actual behavior it also shows similarities to proteins, i.e., its water solubility, its complexing ability towards many types of small-molecule substances, and its precipitation by many protein precipitants such as ammonium sulfate, trichloracetic acid, tannic acid, and other phenols (see Chapter 2 of Volume II). However, despite these parallels, PVP is physiologically inactive, which has led to its use as a plasma substitute and in other parenteral applications (although these applications are now very much restricted since its inertness means that the higher molecular weight fractions are retained in the body rather than being metabolized).[203-206] The analogy between PVP and the proteins, however, breaks down when we compare their conformations in solution, since PVP in aqueous solution has the loose, random-coil type of conformation usually possessed by flexible chain polymers, in contrast with the much more compact conformations of most proteins in their native states in solution.

B. Monomer Unit Analogs and Other Model Compounds for PVP

Because of the relatively complex structure of the monomer unit of PVP compared with that of most other common water-soluble synthetic polymers, it is useful to consider in this context the properties and behavior of a number of small-molecule compounds which are similar in structure to the monomer unit or to a short section of the polymer chain; these can provide information which is useful in interpreting the behavior of PVP itself, e.g., in indicating which are the features that would be expected for the polymer even if the component monomer units were not linked into a chain, and which are the features which have arisen or have been modified because of their being linked into the chain structure.

The two main types of such model compounds for PVP are the various N-alkyl derivatives of pyrrolidone, and the N,N′-disubstituted derivatives of diketopiperazine of which the most important is the dimethyl derivative, commonly referred to by the trivial name sarcosine anhydride. As well as outlining the comparative features of these model compounds in this section, the relevant studies in aqueous and nonaqueous (including mixed solvent) systems are included alongside those for PVP in Tables 3 and 4.

1. N-Substituted Derivatives of Pyrrolidone (γ-Butyrolactam)

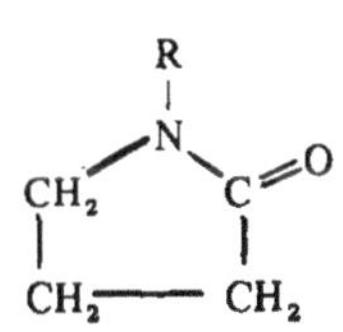

NMP : R = CH_3
NEP : R = CH_3CH_2
NVP : R = $CH_2{:}CH$

a. N-Methylpyrrolidone (NMP)

This is a compound whose properties have been quite widely studied, while its structure shows it to be a fairly close analog of the monomer unit of PVP. The main physical properties of NMP have been listed by Dyke and co-workers[207] and also by Rosenfarb and Baker,[208] while both of these groups have studied the behavior of solutions of electrolytes in the liquid. The infrared carbonyl stretching frequency has been reported to be 1680 cm^{-1} (Reference 209),* which is closely similar to that for PVP in the bulk solid state (see Section III.E).

* The value actually reported[209] was 1770 cm^{-1}, but from the context this is quite evidently an error for 1680 cm^{-1}.

The dipole moment in solution has been measured as 4.09 ± 0.04 D in benzene[210] and 4.06 ± 0.03 D in dioxane,[209] giving an average value of 4.07 ± 0.04 D. This last piece of information enables us to get an idea of the electronic distribution in the ring of NMP; assuming the dipole to result from equal fractional positive and negative charges centered on the nitrogen and oxygen atoms (nuclei) respectively (as in the structure below) resulting from the resonance between structures II and III:

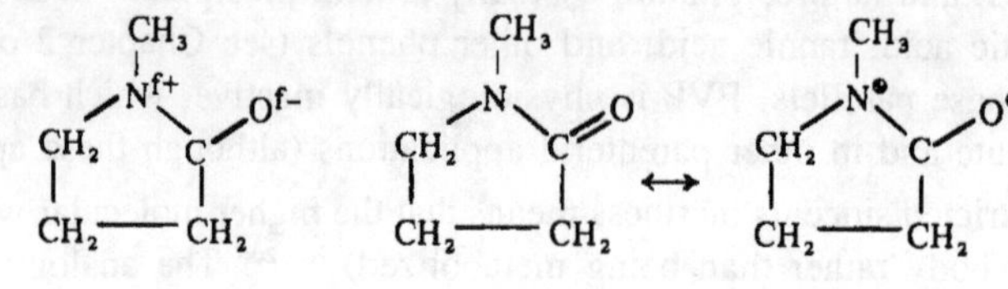

I II III

Taking the distance between the N and O atoms to be 2.28 Å (228 pm) as in sarcosine anhydride,[211] then this gives $f = 0.37$ as the fractional charge on each atom. This same charge distribution should therefore also hold for the pyrrolidone ring in PVP.

Examples of relevant studies on NMP are those of Takayama and co-workers[212] on complexes with aminobenzoic acids; of Randall and co-workers[213] on complexes with phenols; and of Hosono and co-workers[214] on complexes with the alkaloid ajmaline. Heats of solution[215,216] and NMR studies in solution[214] have also been reported (see Tables 3 and 4).

b. N-Ethylpyrrolidone (NEP)

From the viewpoint of its structure, this compound is an even closer monomer unit analog for PVP than NMP. Some of its physical properties have been listed by Frank,[217] who has also studied the equilibria for the hydrolytic ring opening in aqueous solution to N-ethyl, γ-aminobutyric acid.[217] Studies on its physical behavior in solution include those of Oster and Immergut[218] on UV and IR spectra; Saidel[219] on UV spectra; Rothschild[220] on viscosity, and IR and NMR spectra; and Kaatze[221] on specific volume and dielectric relaxation (see Tables 3 and 4). Takayama and co-workers[212] have studied complexes between NEP and aminobenzoic acids.

c. N-Vinylpyrrolidone (NVP)

This compound might, superficially, seem to be the closest possible monomer unit analog for PVP, and has accordingly been used with this intent; however, the inevitable electronic interaction of the unsaturated vinyl group with the adjacent amide group must modify the electronic distribution in the latter, hence leading to significant alterations in its behavior. This is demonstrated by the differences that are seen between the NMR chemical shifts of the ring protons in NVP and those in NMP and PVP,[214] and also by what appears to be a shift in the position of the carbonyl stretching frequency in the IR spectrum.[218] Studies which have been carried out on the behavior of NVP in solution include those of Breitenbach and Schmidt[222] on heats of solution; Frank[223] on heats of solution, and viscosity; Oster and Immergut[218] on UV spectra; Eley and co-workers[224] on NMR; and Hosono and co-workers[214] on NMR (see Tables 3 and 4). Hosono and co-workers[214] have also studied complex formation between NVP and the alkaloid ajmaline.

d. Other Substituted Pyrrolidones

Compounds of this type which have been studied from various aspects include N-isopropyl-5-methylpyrrolidone,[225,226] pyrrolidone-N-acetic acid,[225,226] N-cyclohexyl-pyrrolidone,[213] and hexamethylene-*bis* (2-pyrrolidone).[213]

*2. Sarcosine Anhydride (SA)**

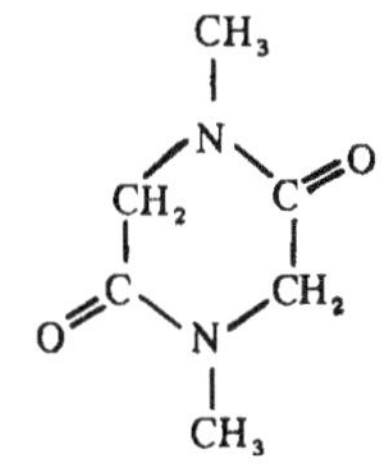

This compound may be viewed as a condensed version of two molecules of NMP, the removal of four of the methylene groups giving greater prominence to the amide groups, but the two methylene groups retained at the same time preventing the amide groups interacting directly with one another; it therefore simulates the greater density of these amide groups to be found close to the PVP chain in compact conformations. The crystal structure of SA has been determined using X-ray diffraction by Groth,[211] who has shown that the atoms in each $-N(CH_3)-CO-CH_2-$ group are (with the exception of the hydrogen atoms) coplanar, with the planes of the two groups parallel but displaced to give an extremely flattened chair form. The physical properties and the behavior of SA (and of several related compounds) in aqueous solution have been studied by Crescenzi and co-workers (see Table 3),[227] while Poole and Higuchi[228] have studied complex formation in the solid state and in aqueous solution between SA and a wide range of aromatic compounds.

C. Preparation and Purification

Vinylpyrrolidone may be polymerized to give PVP by free radical, anionic, and cationic mechanisms.[190] Considering the most commonly used route of free radical polymerization, the undiluted monomer was originally stated to polymerize spontaneously (i.e., in the absence of added initiator) at 140°C and above;[229] however, later work with more rigorously purified monomer gave no sign of spontaneous polymerization even at these high temperatures.[230] With azobisisobutyronitrile (AIBN) as initiator, the polymerization rate is also sensitive to monomer purity,[230] so that once again the earlier studies[229] may no longer be valid; the initiator efficiency for this system has been determined more recently.[231]

Studies have been carried out on the AIBN-initiated polymerization in benzene and in aqueous solutions;[232] the effect of variation in pH (over the range 6 to 11) in the latter case has also been studied.[233] Addition of various amounts of diallyl (with the undiluted monomer, and AIBN as initiator) gives products with correspondingly controllable degrees of cross-linking.[222] Addition of *p*-toluidine (with ethanol as diluent and the same initiator) leads to the incorporation of the additive, which enables the polymer to be radiolabeled subsequently with the radioisotope ^{131}I.[234]

Polymerization of NVP in aqueous solution may also be initiated with hydrogen peroxide, either alone[232] or with the addition of ammonia or an amine as an "activator";[183] this last type of system is apparently used currently for the commercial production of PVP.[235] It should be noted that NVP is unstable in aqueous *acidic* solution, being converted into acetaldehyde and other products.[230]

The polymerization of vinylpyrrolidone may also be induced photochemically in solution (hexane + acetone) using UV light.[236] However, Shizuka and co-workers[237] found that irradiation in ethanol solution using a 254-nm source gave only ring expansion to 4-azacyclohepten-1-one:

* Also variously named N,N'-dimethyldiketopiperazine; 1,4-dimethyl-2,5-diketopiperazine; N,N'-dimethyl-2,5-piperazinedione; N,N'-dimethylpiperazine-2,5-dione.

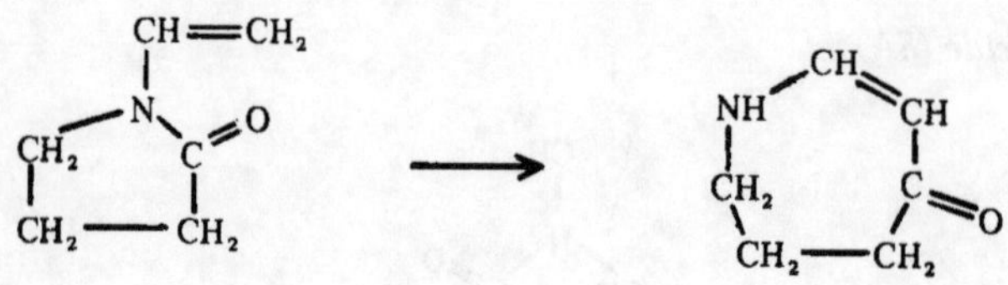

(5)

with apparently no production of polymer.

Free radical copolymerizations involving vinylpyrrolidone as one monomer have been reviewed by Tomalia.[190] Measurements of reactivity ratios give for NVP monomer the Alfrey-Price parameters $Q = 0.086 \pm 0.045$; $e = -1.08 \pm 0.18$.[238] Surface-active behavior of these copolymers with hydrophobic monomers has been studied by Breitenbach and Edelhauser (see also Chapter 4 of Volume II).[239,240] The preparation of VP copolymers for enzyme immobilization has been studied by Manecke and Korenzecker.[241]

For most ordinary applications, PVP is readily available commercially in a number of grades of different average molecular weights (see Section II.D) so that production of the polymer by the above discussed methods is only necessary when a material is required with special properties or whose provenance is better known than that of the commercial material.

The cross-linked polymer is also available commercially in bead form under such names as polyvinylpolypyrrolidone, crospovidone, Kollidon CL (BASF), and *Polyclar* (GAF).

PVP is generally purified (i.e., freed of low molecular weight contaminants) by precipitation using a variety of solvent/nonsolvent pairs, such as water/acetone,[217,242] ethanol/acetone,[243] ethanol/diethyl ether,[234] chloroform/diethyl ether,[244] and chloroform + ethanol (85:15)/diethyl ether.[245,246] Acetone as a precipitant has the drawback that it will not precipitate the lowest molecular weight fraction of the polymer sample.

Other purification methods that have been used include extraction of the solid with diethyl ether,[247-250] and treatment of its aqueous solution with mixed ion-exchange resins;[251-253] each of these methods has evident limitations in the types of impurities so removed. Exhaustive dialysis of aqueous solutions, used either alone or in combination with other methods, [244,252-254] is probably the most efficient method, albeit a rather laborious and cumbersome one, and it may be followed by freeze-drying to isolate the polymer in a convenient form.[244,253]

D. Fractionation and Characterization

A wide variety of different solvent/nonsolvent systems has been used for the fractionation of PVP by stepwise precipitation; the most commonly used system is water/acetone,[199,200,205,245,255-262] while others that have been used include water/acetone + 2-propanol (90:10),[256] water/sodium sulfate (25% aqueous solution),[261] chloroform/diethyl ether,[200] chloroform/petroleum ether,[261,263] ethanol/benzene,[200] ethanol/decalin,[261] ethanol/petroleum ether,[261] and methanol/diethyl ether.[205,259,264] Scholtan,[261] in an intercomparison of five systems, decided that water/acetone and ethanol/petroleum ether were the two best ones out of those studied. However, despite its evident popularity, the water/acetone system has the drawback that the low molecular weight material is soluble in the precipitant, so that there may be a low overall recovery (~85%) arising from the loss of this material.[199,200] Also, the system gives rather broad fractions, and to obtain narrow fractions it should be replaced by water/acetone + 2-propanol (90:10) at the low molecular weight end;[256] alternatively the fractions obtained with acetone/water may be refractionated with methanol/diethyl ether.[259]

The solubility of low molecular weight PVP in acetone may be used as a means of isolating this material from samples of the polymer by simple extraction of the solid polymer with the anhydrous liquid.[200,264]

The molecular weight distribution of PVP may be determined by preparative fractionation techniques as discussed above.[245,256,258] Turbidimetric titration methods using water/aqueous sodium sulfate have been developed for distribution studies by Scholtan,[261,265] while the simpler but more limited cloud-point titrations with water/ammonium sulfate (2*M* aqueous solution) have been used by Jirgensons[200] to characterize fractions. GPC[266-271] and sedimentation[264] have also been used to evaluate the distribution. Concurrent measurements of $\overline{M}_n$ and $\overline{M}_w$ have been used by Elias[272] to characterize various commercial samples.

For the evaluation of PVP for quality control purposes, e.g., for the clinical grades, it is generally considered sufficient to carry out an abbreviated fractionation procedure in which the top 10% and the bottom 10% of the distribution are isolated and their intrinsic viscosities determined; these values must then lie with certain limits if the distribution is to be satisfactorily narrow.[256]

Commercial samples of PVP are commonly designated in terms of the *K* value of Fikentscher,[273] which was originally proposed as a concentration-independent parameter, and which is derived from the relative viscosity η_{sp}/c_2 of the solution by the relation:

$$\log (\eta_{sp}/c_2) = [75\, k^2 c_2/(1 + 1.5kc_2)] + kc_2 \qquad (6)$$

where c_2 is the polymer concentration in g dl^{-1} (% w/v) and *k* is *K*/1000; this parameter has now become essentially obsolete in the polymer field except in the case of PVP. Levy and colleagues[256] have given a table relating the relative viscosity for a 1% polymer solution to the corresponding *K* value. The relation between intrinsic viscosity and *K* value was shown by Matthes[274] to be

$$[\eta] = 2.303\,(75\, k^2 + k) \quad \text{dl g}^{-1} \qquad (7)$$

$$[\eta] = 230.3\,(75\, k^2 + k) \quad \text{cm}^3\ \text{g}^{-1} \qquad (8)$$

where again *k* is *K*/1000. These relations for *K* values have also been quoted in numerous other later references.[187,191,192,255-257]

The four most common commercial grades of PVP, with their *K* values, the molecular weights generally quoted for them, and the intrinsic viscosities derived from the *K* value, are[191-193]

$$K\text{-}15 : M = 10{,}000 : [\eta] = 7.3\ \text{cm}^3\ \text{g}^{-1}$$

$$K\text{-}30 : M = 40{,}000 : [\eta] = 22.5\ \text{cm}^3\ \text{g}^{-1}$$

$$K\text{-}60 : M = 160{,}000 : [\eta] = 76\ \text{cm}^3\ \text{g}^{-1}$$

$$K\text{-}90 : M = 360{,}000 : [\eta] = 160\ \text{cm}^3\ \text{g}^{-1}$$

It appears that these molecular weights are if anything number-average values, since two independent estimates of $\overline{M}_w$ for different samples of PVP K-90 both give a much higher value, close to 1D6.[272,275] However, it is difficult to discover the origin of these commonly quoted values; at least in the case of the three highest molecular weight grades (K-30, K-60, and K-90) they seem to derive from the approximate values given (without further experimental details) by Jirgensons.[199] Furthermore, because of variations customarily allowed in commercial specifications, the actual *K* value and hence, molecular weight, of a particular sample supplied may be appreciably different even from the nominal values; e.g., one commercial specification for PVP K-90[192] allows the actual *K* value of the material supplied to lie between 80 and 100, corresponding to [η] values ranging from 130 to 200 $\text{cm}^3\ \text{g}^{-1}$.

Thus the molecular weights quoted for the various grades should be considered as largely nominal values (in the absence of any firm indication to the contrary) and in situations where

Table 2
SPECIFIC VOLUME, DENSITY, THERMAL EXPANSION COEFFICIENT, AND GLASS TRANSITION TEMPERATURE FOR PVP[a]

	$t < t_g$		$t = 25°C$		$t = t_g$		$t > t_g$	
Grade	v_o/cm³g⁻¹	$10^4\alpha$/deg⁻¹	v/cm³g⁻¹	ρ/g cm⁻³	t_g/°C	v/cm³g⁻¹	v_o/cm³g⁻¹	$10^4\alpha$/deg⁻¹
K-15	0.827	1.66	0.831	1.203[b]	85[c]	0.838	0.809	4.28
K-30	0.825	1.47	0.829	1.207[b]	86[c]	0.835	0.817	2.53
K-90	0.818	1.85	0.823	1.216[b]	87[c]	0.831	0.810	3.04

[a] Data of Sugiura and Fujii[276] except where otherwise indicated — see Section III.E (Equation 9).
[b] Other literature values: 1.20 ± 0.01 (Reference 262); 1.3 ± 0.1 (Reference 267); 1.13 (Reference 264); 1.25 (Reference 191).
[c] Other literature values: 54 (Reference 277); 66 (for K-90—much lower values are quoted for K-30 and K-15 grades).[267] See also postscript to this chapter.

the actual molecular weight value is important the samples should be characterized at the least by viscometry, so that the value of [η] so obtained can be used to give an estimate of the actual molecular weight.

The normal commercial grades of PVP have fairly broad molecular weight distributions; however, the distribution of the material produced for clinical purposes with a *K* value of about 30 is appreciably narrower.[256,258]

E. Bulk Polymer

The *specific volume, v* of PVP (and correspondingly its reciprocal, the *density* ρ) has been measured by Sugiura and Fujii[276] over the range 10 to 100°C for samples of the three grades K-15, K-30, and K-90; in each case plotting v against temperature gave two straight lines, with the intersection point taken to be the glass-transition temperature (see below). The data, presented only graphically in the original reference,[276] have been fitted to the standard form of equation:

$$v = v_0 (1 + \alpha t) \tag{9}$$

for each linear section, and the derived values of v_o and the thermal expansion coefficient α are given in Table 2. The footnote to the table compares corresponding values for the density at 25°C with other literature data; where differences occur, those of Sugiura and Fujii are to be preferred in view of the consistent pattern and evident precision of their data.[276]

The *glass-transition data* for PVP from the literature are also summarized in Table 2.* The results of Sugiura and Fujii[276] again show a consistent pattern, with $t_g = 86 \pm 1°C$ for the three grades; they also found that the value of t_g is markedly reduced by the presence of absorbed water, falling to 64°C with 10° water and to 20°C with 20% water. It is likely that the lower values of t_g reported by Jenckel[277] and Kaneniwa and Ikekawa[267] (see Table 2) resulted from the presence of absorbed water in the samples they used. It is also evident that drying PVP should be done above 85°C to give mobility to the absorbed water molecules throughout the drying process.

The refractive index of solid PVP has been reported as $n_D^{25} = 1.53$.[191]

The *X-ray diffraction pattern* of bulk PVP has been reported for values of the scattering angle 2θ from 10 to 26° by Simonelli and co-workers,[278] and from 10 to 40° by Sekikawa and co-workers.[246] These patterns indicate the polymer to be essentially fully amorphous,

* See also the postscript to this chapter.

Table 3
STUDIES ON PVP AND SMALL-MOLECULE ANALOGS IN AQUEOUS SOLUTION[a]

Entry	Year	Polymer samples: No.	Polymer samples: Range of $\overline{M}$	Techniques used	Temperature (t/°C)	Polymer concentration	Comments	Ref.
1	1951	3	1D4—9D4	SS	27	0.5%	$\rightarrow s^o$	245
2	1951	1	~5D4?	SV	20?	0.5, 1%	$\rightarrow \bar{v}_2^o = 0.775 \pm 0.005$ cm³g⁻¹	245
3	1951	3	1D4—9D4	TD	27	0.5%	$\rightarrow D^o$; $\overline{M}_{sD}$	245
4	1951	3	1D4—9D4	VI	20	0.5—2%	$\rightarrow[\eta]$; MHS (0.58; 6.45d2)	245
5	1952	7	2D4—1D6	LS		0.05—0.25%	436,546 nm: $\rightarrow \overline{M}_w$; A_2; $\hat{R}$	255
6	1952	32	1D4—2D5	OP	25		$\rightarrow \overline{M}_n$; A_2	255
7	1952	?	1D4—1D6?	RI			436, 546 nm: $\rightarrow \tilde{n} = 0.175$ cm³g⁻¹	255
8	1952	?	1D4—1D6?	SV	25?		$\rightarrow \bar{v}_2^o = 0.807$ cm³g⁻¹	255
8a	1952	6	2D4—2D5	VI	25?		$\rightarrow[\eta]$	255
9	1952	1	~3D4?	RI	20	≥6%	589 nm: $\rightarrow \tilde{n} = 0.171$ cm³g⁻¹	256
10	1952	1	~3D4	VI	17—37		$\rightarrow[\eta]$	256
11	1952	13	1D4—3D6	SS	20	0.04—1%	+ NaCl (0.2*M*): $\rightarrow s^o$	257
12	1952	1	2D4	SV	20, 30		$\rightarrow \bar{v}_2^o = 0.776, 0.784$ cm³g⁻¹	257
13	1952	1	5D5	SV	20, 30		$\rightarrow \bar{v}_2^o = 0.783, 0.801$ cm³g⁻¹	257
14	1952	13	1D4—3D6	TD	20	0.2—1.5%	$\rightarrow D^o$	257
15	1952	13	1D4—3D6	VI	30	0.06—2%	+ NaCl (0,1.7*M*): $\rightarrow[\eta]$; MHS (0.70; 1.4d2)[b]	257
16	1953	21	7D3—1D6[c]	VI	25	0.3—2%	$\rightarrow[\eta]$; MHS (0.68; 1.8d2)[c]	258
17	1953	9	1D4—1D6	SS	23.5	0.2—2%	$\rightarrow s^o$; MWD	264
18	1953	3	3D4—1D6	SV	25	0.1—5%	$\rightarrow \bar{v}_2^o = 0.802 \pm 0.008$ cm³g⁻¹	264
19	1953	7	1D4—4D4	TD	5—21	0.25—1%	$\rightarrow D^o$; M_{sD}	264
20	1953	7	1D4—1D6	VI	25—35	0.1—1.5%	$\rightarrow[\eta]$; MHS (25°C: 0.85; 4.1d3)	264
21	1954	?	Cross-linked	SV	20—70	$w_2 = 0.05—0.08$	$\rightarrow Q_w$	222, 230
22	1954	—	NVP	HS	20?	$x_2 = 0.01—0.144$	$\rightarrow \Delta H$	222
23	1954	—	NVP	HS	20	$x_2 = 0.15—0.70$	$\rightarrow \Delta H$ (<0; max. at $x_2 \approx 0.25$)	223
24	1954	—	NVP	VI	20	$x_2 = 0.15—0.70$	$\rightarrow \eta$ (max. at $x_2 \approx 0.35$)	223
25	1954	2	4D4	UV		0.001—0.02	210—320 nm; pH 0.8—12: $\rightarrow A$; ϵ[d]	218, 291
26	1954	—	NEP	UV			210—320 nm?; pH 0.8—12?	218, 291
27	1954	—	NVP	UV			210—330 nm; pH 0.8—12: $\rightarrow A$; ϵ	218
28	1955	7	7D3—2D5	LS			436, 546 nm: $\rightarrow \overline{M}_w$	259

Table 3 (continued)
STUDIES ON PVP AND SMALL-MOLECULE ANALOGS IN AQUEOUS SOLUTION[a]

		Polymer samples						
Entry	Year	No.	Range of $\overline{M}$	Techniques used	Temperature (t/°C)	Polymer concentration	Comments	Ref.
29	1955	7?	7D3—2D5?	RI			436, 546 nm: →$\tilde{n}$ = 0.185 cm³g⁻¹	259
30	1955	6	7D3—2D5	VI	25	<2%	→[η]; MHS (0.55; 6.76d2)	259
31	1955	3	4D4—3D5	UV		0.001—0.02	200—230 nm; pH 6—12: →A	219
32	1955	—	NEP	UV			200—260 nm	219
33	1956	3	8D3—6D5	TD		2—24%	→D (sucrose); critical overlap concentration c_2^*	292
34	1956	3	8D3—6D5	VI	25?		→[η]	292
35	1957	15	2D4—2D5	LS		0.1—1.8%	436, 546 nm: →M_w	242
36	1957	15?	2D4—2D5?	RI			436 nm?: →$\tilde{n}$ = 0.185 cm³g⁻¹	242
37	1957	15	2D4—4D5	VI	25		→[η]; MHS (0.68; 1.575d2)	242
38	1957	2	1D6, 3D6	SS	(20)	0.1—1.5%	→s^o (footnote e)	260
39	1957	2	1D6, 3D6	TD	20	0.4—1.5%	→ D^o; $\overline{M}_{sD}$	260
40	1957	2	1D6, 3D6	VI	25	0.02—1%	→[η]; k'; MHS (0.70; 1.4d2)[f]	260
41	1959		Review of MHS parameters — correlation between a and K					293
42	1960	6	8D4—1D6	OP	30	0.1—2%	→$\overline{M}_n$; A_2	262
43	1960	6	8D4—1D6	VI	30		→[η]; MHS (0.59; 3.92d2)	262
44	1960	3	1D4—3D5	SS	20	0.1—1%	Phosphate buffer (pH 7): → s^o; MWD	294
45	1961	1	4D4	LS	~19	4—16%	436, 546 nm: → footnote g	295
46	1961	1	4D4	VI	~19	4—18%	→ η_c (footnote h)	295
47	1961	1	1D4	RI	25	0.5—3%	546 nm: →$\tilde{n}$ = 0.135 cm³g⁻¹	272
48	1961	1	1D4	SV	25	<3%	→$\bar{v}_2^o$ = 0.820 cm³g⁻¹	272
49	1961		Compilation of MHS parameters					296
50	1961	1	1D6	LS	25		→$\overline{M}_w$; A_2; $\hat{R}$	275
51	1961	1	1D6	VI	30	0.1—0.5%	→[η]; k'	275
52	1963	1	3D4	SV	23	$w_2 < 0.03$	1—4500 atm: → $\bar{v}_2^o$ = 0.781 cm³g⁻¹ (1 atm)	139
53	1963	1?	?	EP	25	1.4—10%	0.2—30 MHz: →ε; ε′; ε″	297
54	1964	1	~2D4	OP	25	0.6—1.2%	→ $\overline{M}_n$?	298
55	1964	1	~2D4	VI	25	2—6%	→[η]; k'	298
56	1965	1	?	IR	30	20, 40%	1.4—1.5 μm: → Δυ (OH)	201
57	1966	1	5D4	SV	5—55		→ $\bar{v}_2^o$ = 0.783—0.821 cm³g⁻¹	299

58	1966	1	5D4	USV	5—55		$\rightarrow \kappa_2^o$	299
59	1966	1	5D4	VI	5—55		$\rightarrow [\eta]$	299
60	1966	1	4D4	IR	7—61	w_2 = 0.02, 0.3	5*M* H_2O in D_2O; 1.3—1.8 μm: $\rightarrow A$; Δυ (OH)	300
61	1967	1	6D4	DTA	−45 to +30	w_2 = 0.05—0.37	Calorimetry on freezing and melting	301
62	1967	1	6D4	HC	25	w_2 = 0.1—0.4	$\rightarrow C_p$	301
63	1967	1	6D4	SV	0 to −22	w_2 = 0.05—0.61	→ Freezing points	301
64	1967	1	9D5	VI	35	0.5—1.5%	$\rightarrow \eta$; $[\eta]$	302
65	1968	1	3D4	HC	20?	w_2 = 0.02—0.20	$\rightarrow C_p$	303
66	1968	1	3D4	HS	20?	0.5%	$\rightarrow \Delta H$	303
67	1968	1	3D4	SV	25	0.5—3%	$\rightarrow \bar{v}_2^o = 0.780$ cm^3g^{-1}	303
68	1968	1	3D4	VI	20—60	>0.1%	$\rightarrow [\eta]$; k'	303
69	1970	3	1D4—4D5	NMR	31	3—10%	D_2O: $\rightarrow T_1(^2H)$	304
70	1970	1	7D5	VI	27	5—14%	$\rightarrow \eta$ (Rheological studies)	305
71	1971	2	3D4, 1D5	OP	20—30	0.1—17%	$\rightarrow \overline{M}_n$; A_2; A_3; A_4 (= 0)	306
72	1971	2	3D4, 1D5	SV	25	0.1—17%	$\rightarrow \bar{v}_2^o = 0.785$ cm^3g^{-1}	306
73	1971	2	3D4, -D5	VI	25?		$\rightarrow [\eta]$	306
74	1972	1	1D4	NMR	0—55	w_2 = 0.08—0.4	D_2O; 4, 8 MHz: $\rightarrow T_1(^2H)$, $T_2(^2H)$	307, 308
75	1972	1	1D4	DTA	−74 to +3	w_2 = 0.50—0.75	→ Glass-transition behavior	308a
76	1972	1	4D4	DTA	−60 to 0	w_2 = 0.2—0.4	→ Freezing/melting of crystalline and amorphous ice	309
77	1972	1	1D4	IR	~30	w_2 = 0.05	D_2O: → υ(C=0) = 1648 cm^{-1}	220
78	1972	1	1D4	NMR/TD	30	w_2 = 0.24—0.65	$\rightarrow D(H_2O)$	220
79	1972	—	NEP	IR	~30	x_2 = 0.06—1	D_2O; 1350—1600 cm^{-1}; → υ(C-N)	220
80	1972	—	NEP	IR	~30	x_2 = 0.05—1	D_2O; 1570—1750 cm^{-1}: → υ(C=O)	220
81	1972	—	NEP	IR	~30	x_2 = 0.06—0.86	D_2O; 2000—2900 cm^{-1}: → υ(O-D)	220
82	1972	—	NEP	NMR/TD	30	x_2 = 0.04—0.5	$\rightarrow D(H_2O)$	220
83	1972	—	NEP	NMR/TD	30	x_2 = 0.05—1	D_2O: → D(NEP)	220
84	1972	—	NEP	VI	23	x_2 = 0.04—1	D_2O: $\rightarrow \eta$ (max. at $x_2 \approx 0.25$)	220
85	1973	—	SA	HD	25	0.4—3.4m	→ΔH	227
86	1973	—	SA	HS	25	0.1—4m	→ΔH (footnote i)	227
87	1973	—	SA	SV	25	0.27—7m	$\rightarrow \bar{v}_2^o$	227
88	1973	—	SA	VI	25	1.2—7.3m	$\rightarrow \eta$; $[\eta]$	227
89	1974	—	NMP	HS	25	x_2 = 0.002—0.98	$\rightarrow \Delta H$	215
90	1974	1	7D5	EB		2%	→ Kerr constant	309a
91	1975	2	1D4,2D4	EP	25	w_2 = 0.11—0.54	0.05—70 GHz: $\rightarrow \epsilon$; ϵ'; ϵ''	221
92	1975	2	1D4,2D4	SV	25	w_2 = 0.11—0.54	$\rightarrow \psi_2$	221
93	1975	—	NEP	EP	25	w_2 = 0.61	0.05—70 GHz: $\rightarrow \epsilon$; ϵ'; ϵ''	221
94	1975	—	NEP	SV	25	w_2 = 0.61	$\rightarrow \psi_2$	221

Table 3 (continued)
STUDIES ON PVP AND SMALL-MOLECULE ANALOGS IN AQUEOUS SOLUTION[a]

Entry	Year	Polymer samples No.	Polymer samples Range of $\overline{M}$	Techniques used	Temperature (t/°C)	Polymer concentration	Comments	Ref.
95	1975	—	NMP	HS	25	$x_2 \rightarrow 0$	$\rightarrow \Delta H$	216
96	1975	2	1D4,4D4	HD	25	$\phi_2 = 0.01 - 0.2$	$\rightarrow \Delta H$	310
97	1975	4	1D4—4D5	USA	10—40	~1%	5—130 MHz	311
98	1976	1	6D5	LS		0.1—0.4%	$\lambda = ?$: $\rightarrow \overline{M}_w$; A_2	312
99	1976	1	2D5	NMR	22—70	5—40%	D_2O; 5 MHz: $\rightarrow T_1$ (^2H)	224
100	1976	1	2D5	NMR	23	~20%	D_2O; 5—70 MHz: $\rightarrow T_1$(^1H)	224
101	1976	—	NVP	NMR	11—71	4—20%	D_2O; 5 MHz: $\rightarrow T_1$(^2H)	224
102	1976	—	NMP	HS	25	$x_2 = 0.002$—0.01	$\rightarrow \Delta H$	215
103	1976	16	1D5—1D7	LS	25	0.02—0.4%	436 nm: $\rightarrow \overline{M}_w$; A_2; $\hat{R}$	313
104	1976	16	1D5—1D7	VI	25		$\rightarrow [\eta]$; MHS[j]	313
105	1977	6	1D5—5D5	VI	25		$\rightarrow [\eta]$; k'	263
106	1977	6	1D5—5D5	VI	28		+ Na_2SO_4(0.55*M*) (Θ system): $\rightarrow [\eta]$; MHS (0.50; 5.8d2)	263
107	1979	2	1D4, 1D5	HD	25	$\phi_2 = 0.02$—0.11	$\rightarrow \Delta H$	314

[a] Small-molecule analogs as listed in column four: NEP, N-ethylpyrrolidone; NMP, N-methylpyrrolidone; NVP, N-vinylpyrrolidone; SA, sarcosine anhydride.
[b] MHS parameters derived from measurements in 1.7*M* NaCl, but these should also apply to pure H_2O, since it was stated that $[\eta](NaCl)/[\eta]\ (H_2O) = 1.02 \pm 0.02$.[257]
[c] See Reference 259 for revised $\overline{M}$ values.
[d] See Reference 219 for critical comments on these results.
[e] One sample also studied in 0.1*M* phosphate buffer (pH 7).
[f] MHS parameters obtained by combining these data with those from Reference 257 (see Entry 15).
[g] LS scattering at 90° angle — also absorption of "white light" and "red light"; breaks in the intensity or absorption plots at $c_2 \approx 14\%$.
[h] Break in viscosity plots at $c_2 \approx 16\%$.
[i] Heat of solution also obtainable from solubility values measured at 25 and 35°C.
[j] MHS parameters different for $\overline{M}$ = 1D5—1D6 (0.60; 5.75d2) and for $\overline{M}$ = 1D6—1D7 (0.19; 0.134).

Table 4
STUDIES ON PVP AND SMALL-MOLECULE ANALOGS IN NONAQUEOUS AND MIXED SOLVENTS[a]

Entry	Year	Polymer samples No.	Polymer samples Range of $\overline{M}$	Techniques used	Temperature (t/°C)	Polymer concentration	Comments	Ref.
						$H_2O + CH_4O$: Water + Methanol		
1	1974		NMP	HS	25	x_2 = 0.001—0.018	$x(H_2O)$ = 0.10—0.93: →ΔH	215
2	1975	2	1D4,4D4	HD	25	ϕ_2 = 0.01—0.2	$x(H_2O)$ = 0.1—0.9: →ΔH	310
						$H_2O + C_2H_6O$: Water + Ethanol		
3	1975	—	NMP	HS	25	($x_2 \to 0$)	$x(H_2O)$ = 0.1—0.9: →ΔH	216
						$H_2O + C_3H_6O$: Water + Acetone		
4	1961	3	1D4—1D6	VI	25		$\phi(H_2O)$ = 0.67 (Θ system): →[η]; k'; MHS (0.50; 7.5d2)	138, 272, 296
5	1977	6	1D5—5D5	VI	25		$\phi(H_2O)$ = 0.67 (Θ system): → [η]; k'; MHS (0.50; 9(±2)d2)	263
6	1977	6	1D5—5D5	VI	25		$\phi(H_2O)$ = 0.80: →[η]; k'; MHS (see text)	263
						$H_2O + C_3H_8O$: Water + 1-Propanol		
7	1951	1	4D4	VI	28	0.5%	$\phi(H_2O)$ = 0.1—0.9: →η	199
8	1967	1	9D5	VI	35	0.5—1.5%	$\phi(H_2O)$ = 0.5, 0.8: →η	302
9	1975	—	NMP	HS	25	($x_2 \to 0$)	$\phi(H_2O)$ = 0.1—0.9: →ΔH	216
						$H_2O + C_3H_8O$: Water + 2-Propanol		
10	1975	—	NMP	HS	25	($x_2 \to 0$)	$\phi(H_2O)$ = 0.1—0.9: → ΔH	216

Table 4 (continued)
STUDIES ON PVP AND SMALL-MOLECULE ANALOGS IN NONAQUEOUS AND MIXED SOLVENTS[a]

Entry	Year	Polymer samples: No.	Polymer samples: Range of $\overline{M}$	Techniques used	Temperature (t/°C)	Polymer concentration	Comments	Ref.
					H_2O + $C_4H_8O_2$: Water + Dioxane			
11	1954	1?	Cross-linked	SV	20	w_2 = 0.05—0.5	$w(H_2O)$ = 0.06—0.5: → Q_w	222
12	1979	1	7D5	GPC		0.2—1%	$\phi(H_2O)$ = 0.28—0.90: → PS	315
13	1979	1?	7D5?	ED			$\phi(H_2O)$ = 0.1—0.9: → PS	315
					H_2O + $C_4H_{10}O$: Water + n-Butanol			
14	1975	—	NMP	HS	25	($c_2 \to 0$)	$\phi(H_2O)$ = 0.1—0.5, 0.98, 0.99: →ΔH	216
					H_2O + $C_4H_{10}O$: Water + t-Butanol			
15	1975	—	NMP	HS	26.5	($c_2 \to 0$)	$\phi(H_2O)$ = 0.015—0.92: →ΔH	216
					CCl_4 : Carbon Tetrachloride			
16	1954	1?	Cross-linked	SV	20	$w_2 \approx 0.98$	→Q_w	222, 230
					CCl_4 + $CHCl_3$: Carbon Tetrachloride + Chloroform			
17	1974	1	1D4	NMR	21		$CDCl_3$; $\phi(CCl_4)$ = 0.33—0.75: →$T_1(^2H)$	316
					$CHCl_3$: Chloroform (See Also CCl_4 + $CHCl_3$)			
18	1954	1?	Cross-linked	SV	20, 50	w_2 = 0.04—0.06	→Q_w	222, 230
19	1954	—	NVP	HS	20?	x_2 = 0.02—0.09	→ΔH	222
20	1955	7	7D3—2D5	LS			436, 546 nm: →$\overline{M}_w$	259
21	1955	7?	7D3—2D5?	RI			436, 546 nm?: →$\tilde{n}$ = 0.108 cm^3g^{-1}	259
22	1955	6	7D3—2D5	VI	25		→[η]; MHS (0.64; 1.94d2)	259

23	1968	1	4D4	HS	20?	0.5%	$\to \Delta H$	303
24	1971	1	4D4	IR		0.1—5%	650—1800 cm^{-1}: $\to \upsilon(C=O)$ = 1680 cm^{-1}	317
25	1972	1	1D4	NMR	−30 to +40	w_2 = 0.025—0.15	$CDCl_3$; 4,8 MHz: $\to T_1(^2H)$; line width (^{35}Cl)	307, 308
26	1973	1	2D4	IR		0.05—0.2%	1600—2000 cm^{-1}: $\to \upsilon(C=O)$ = 1660 cm^{-1}	318
27	1974	1	1D4	NMR	21		$CDCl_3$: $\to T_1$ (^2H); line width (^{35}Cl)	316
28	1974	1	1D4	NMR/TD	21		Spin echo: $\to D(CHCl_3)$	316
29	1974	2	1D4, 7D5	NMR	−57 to +50		$CDCl_3$; 8—60 MHz: $\to T_1$ (^1H)	316
30	1975	1	7D5	LS	25		546 nm: $\to \overline{M}_w$	319, 320
31	1975	1	7D5	RI	25		546 nm: $\to \tilde{n}(=?)$	319
32	1975	1	6D4	IR		1%	1600—1725 cm^{-1}: $\to \upsilon(C=O)$ = 1675 cm^{-1}	320, 321
33	1977	6	1D5—5D5	VI	25		$\to [\eta]$; k'; MHS (see text)	263
34	1979	4	900—8D4	NMR			$CDCl_3$: $\to \delta$	214
35	1979	—	NMP	NMR			$CDCl_3$: $\to \delta$	214
36	1979	—	NVP	NMR			$CDCl_3$: $\to \delta$	214

$CHCl_3$ + $C_2HF_3O_2$: Chloroform + Trifluoroacetic Acid

37	1966	1	4D4	EC	25	1—9%	$\phi(CHCl_3)$ = 0—0.5, ~ 1.0: $\to$ protonation of PVP	322

$CHCl_3$ + C_2H_6O : Chloroform + Ethanol

38	1975	1	6D5	LS	25		546 nm; $\phi(CHCl_3)$ = 0.25—0.85: $\to \overline{M}_w$ (app); PS	319, 320
39	1975	1	6D5	RI	25		546 nm: $\phi(CHCl_3)$ = ?: $\to \tilde{n}$ (see text)	319
40	1975	1	6D4	IR			1600—1725 cm^{-1}; $\phi(CHCl_3)$ = 0.10—0.95: $\to \upsilon(C=O)$	320, 321

$CHCl_3$ + C_3H_6O : Chloroform + Acetone

41	1977	6	1D5—5D5	VI	25		$\phi(CHCl_3)$ = 0.4: $\to [\eta]$; MHS (see text)	263
42	1977	1	5D5	VI	25		$\phi(CHCl_3)$ = 0.1—0.9: $\to [\eta]$	263

$CHCl_3$ + C_6H_6 : Chloroform + Benzene

43	1979	4	900—8D4	NMR			$CDCl_3$ + C_6D_6; $x(CDCl_3)$ = 0.57, 0.79: $\to \delta$	214
44	1979	—	NMP	NMR			$CDCl_3$ + C_6D_6; $x(CDCl_3)$ = 0.02—0.77: $\to \delta$	214
45	1979	—	NVP	NMR			$CDCl_3$ + C_6D_6; $x(CDCl_3)$ = 0.02—0.77: $\to \delta$	214

Table 4 (continued)
STUDIES ON PVP AND SMALL-MOLECULE ANALOGS IN NONAQUEOUS AND MIXED SOLVENTS[a]

Entry	Year	Polymer samples: No.	Polymer samples: Range of $\overline{M}$	Techniques used	Temperature (t/°C)	Polymer concentration	Comments	Ref.
					$CHCl_3 + C_6H_{12}$: Chloroform + Cyclohexane			
46	1974	1	1D4	NMR	21		$CDCl_3$; $\phi(CDCl_3) = 0.50$—0.67: $\rightarrow T_1(^2H)$	316
47	1974	1	1D4	NMR	21	$w_2 = 0.07$—0.27	$CDCl_3 + C_6H_{12}$, $CHCl_3 + C_6D_{12}$; $\phi(CHCl_3, CDCl_3) = 0.67$: $\rightarrow T_1(^2H)$	316
48	1974	1	1D4	NMR	21		$\phi(CHCl_3) = 0.67$: $\rightarrow$ line width (^{35}Cl)	316
49	1974	1	1D4	NMR/TD	21		$CDCl_3 + C_6H_{12}$, $CHCl_3 + C_6D_{12}$: $\rightarrow D(C_6H_{12})$; $D(CHCl_3)$	316
					$CHCl_3 + C_7H_{16}$: Chloroform + Heptane			
50	1974	1	1D4	NMR	21		$CDCl_3$: $\phi(CDCl_3) = 0.75$: $\rightarrow T_1(^2H)$	316
					CH_2Cl_2 : Dichloromethane (Methylene Chloride)			
51	1972	1	1D4	IR	~30		~1650 cm^{-1}: $\rightarrow \nu(C{=}O)$	220
52	1972	1	1D4	NMR/TD	30	$w_2 = 0.2$—0.5	$\rightarrow D(CH_2Cl_2)$	220
					CH_4O : Methanol (See Also $H_2O + CH_4O$)			
53	1953	21	7D3—1D6[b]	LS			$\lambda = ?$: $\rightarrow \overline{M}_w$; [b]$A_2$	258
54	1953	6	7D4—8D4	OP			$\rightarrow \overline{M}_n$	258
55	1953	21	7D3—1D6	VI	30	0.3—2%	$\rightarrow [\eta]$; MHS (0.68; 1.8d2)[c]	258
56	1954	1?	Cross-linked	SV	20	$w_2 \approx 0.05$	$\rightarrow Q_w$	222, 230
57	1955		Reevaluation of VI data of Frank and Levy (entry 55):[258]				$\rightarrow$MHS(0.65; 2.3d2)	259
58	1957	8	2D4—2D5	LS			436 nm: $\rightarrow \overline{M}_w$	242
59	1957	8?	2D4—2D5?	RI			436 nm: $\rightarrow \tilde{n} = 0.197\ cm^3g^{-1}$	242
60	1961	1	1D4	LS	25		546 nm: $\rightarrow \overline{M}_w$	138, 272
61	1961	1?	1D4?	RI	25		546 nm: $\rightarrow \tilde{n} = 0.1765\ cm^3g^{-1}$	272

62	1961	2	2D4, 1D6	SS	25		$\rightarrow s^{o}$	272
63	1961	2?	2D4, 1D6?	SV	25		$\rightarrow \bar{v}_2^o = 0.730\ cm^3g^{-1}$	272
64	1961	2	2D4, 1D6	TD	25		$\rightarrow D^o;\ \overline{M}_{sD}$	272
65	1972	1	1D4	NMR	30	$w_2 = 0.4$—0.6	$\rightarrow D^o\ (CH_4O)$	220
66	1974	—	NMP	HS	25	$x_2 = 0.004$—0.99	$\rightarrow \Delta H$	125
67	1975	2	1D4, 4D4	HD	25	$\phi_2 = 0.01$—0.2	$\rightarrow \Delta H$	310
				$C_2HF_3O_2$: Trifluoroacetic Acid (See Also $CHCl_3$ + $C_2HF_3O_2$)				
68	1966	1	4D4	EC	25	0.1—9%	→ Protonation of PVP	322
				$C_2H_4Cl_2$: 1,2-Dichloroethane				
69	1975	2	5D5, 6D5	LS	25		546 nm: $\rightarrow \overline{M}_w$	319, 320
70	1975	1	6D4	IR		1%	1600—1725 cm^{-1}: $\rightarrow \upsilon(C{=}O) = 1683\ cm^{-1}$	320, 321
71	1976	1	6D5	NMR	30	5%	$\rightarrow T(^{13}C)$	323
				$C_2H_4Cl_2$ + $C_2H_4O_2$: 1,2-Dichloroethane + Acetic Acid				
72	1975	1	6D5	LS	25		546 nm: $\phi(C_2H_4Cl_2) = 0.20$—0.97: $\rightarrow \overline{M}_w$ (app); PS	319, 320
				$C_2H_4Cl_2$ + C_2H_6O : 1,2-Dichloroethane + Ethanol				
73	1975	1	5D5	LS	25		546 nm; $\phi(C_2H_4Cl_2) = 0.10$—0.98: $\rightarrow \overline{M}_w$ (app); PS	319, 320
74	1975	1	5D5	RI	25		546 nm: $\rightarrow \bar{n}$ (see text)	319
75	1975	1	6D4	IR		1%	1600—1725 cm^{-1}; $\phi(C_2H_4Cl_2) = 0.25$—0.95: $\rightarrow \upsilon(C{=}O)$	320, 321
76	1976	1	6D5	NMR	30	5%	$\phi(C_2H_4Cl_2) = 0.15$—0.95: $\rightarrow T_1(^{13}C)$	323
77	1979	1	7D5	GPC		0.4—1%	$\phi(C_2H_4Cl_2) = 0.10$—0.85: → PS	315
				$C_2H_4O_2$: Acetic Acid (See Also $C_2H_4Cl_2$ + $C_2H_4O_2$)				
78	1972	1	1D4	NMR/TD	30	$w_2 = 0.16$—0.43	$\rightarrow D(C_2H_4O_2)$	220
79	1975	1	6D5	LS	25		546 nm: $\rightarrow \overline{M}_w$	319
80	1975	1?	6D5?	RI	25		546 nm: $\rightarrow \bar{n}$ (see text)	319

Table 4 (continued)
STUDIES ON PVP AND SMALL-MOLECULE ANALOGS IN NONAQUEOUS AND MIXED SOLVENTS[a]

Entry	Year	Polymer samples: No.	Polymer samples: Range of $\overline{M}$	Techniques used	Temperature (t/°C)	Polymer concentration	Comments	Ref.
				C_2H_6O : Ethanol (See Also H_2O + C_2H_6O; $CHCl_3$ + C_2H_6O; $C_2H_4Cl_2$ + C_2H_6O)				
81	1954	1?	Cross-linked	SV	20,50	w_2 = 0.04—0.06	$\rightarrow Q_w$; ΔH (swelling)	222, 230
82	1975	—	NMP	HS	25	$c_2 \rightarrow 0$	$\rightarrow \Delta H$	216
83	1975	2	5D5, 6D5	LS	25		546 nm: $\rightarrow \overline{M}_w$	319, 320
84	1975	2?	5D5,6D5?	RI	25		546 nm: $\rightarrow \tilde{n}$ (see text)	319
85	1975	1	6D4	IR		1%	1600—1725 cm^{-1}: $\rightarrow \upsilon(C{=}O)$ = 1665 cm^{-1}	321
86	1976	1	7D5	NMR	30	5%	$\rightarrow T_1(^{13}C)$	323
				$C_2H_8N_2$: Ethylene Diamine				
87	1954	1?	Cross-linked	SV	20	w_2 = 0.06—0.07	$\rightarrow Q_w$	222, 230
				C_3H_6O : Acetone (See Also H_2O + C_3H_6O; $CHCl_3$ + C_3H_6O)				
88	1954	1?	Cross-linked	SV	20	ϕ_2 = 0.5—0.6	$\rightarrow Q_w$	222, 230
				C_3H_7NO : Dimethylformamide				
89	1975	1	4D5	USA	10—35	1%	4—100 MHz	311
				C_3H_8O : 1-Propanol (See Also H_2O + C_3H_8O)				
90	1954	1?	Cross-linked	SV	20	w_2 = 0.04—0.06	$\rightarrow Q_w$	222, 230
91	1967	1	9D5	VI	35	0.5—1.5%	$\rightarrow \eta$	302
92	1975	—	NMP	HS	25	$c_2 \rightarrow 0$	$\rightarrow \Delta H$	216
				C_3H_8O : 2-Propanol (See Also H_2O + C_3H_8O)				
93	1975	—	NMP	HS	25	$c_2 \rightarrow 0$	$\rightarrow \Delta H$	216

94	1977	7	1D5—5D5	OP	25	0.2—1.6%	→$\overline{M}_n$; A_2	263
95	1977	7	1D5—5D5	VI	25		→[η]; k'; MHS (see text)	263
				C_4H_8O : Butanone (Methyl Ethyl Ketone)				
96	1954	1?	Cross-linked	SV	20, 50	w_2 = 0.35—0.6	→Q_w	222, 230
				$C_4H_8O_2$: Dioxane (See Also H_2O + $C_4H_8O_2$)				
97	1954	1?	Cross-linked	SV	20—80	w_2 = 0.3—0.7	→Q_w	222, 230
98	1954	—	NVP	HS	20	x_2 = 0.25—0.65	→ΔH	222
99	1961	—	NMP	EP	30	w_2 = 0.002—0.003	→μ = 4.07 ± 0.03 D	209
				$C_4H_{10}O$: n-Butanol (See Also H_2O + $C_4H_{10}O$)				
100	1975	—	NMP	HS	25	$c_2 \to 0$	→ΔH	216
				$C_4H_{10}O$: t-Butanol (See Also H_2O + $C_4H_{10}O$)				
101	1975	—	NMP	HS	26.5	$c_2 \to 0$	→ΔH	216
				$C_5H_{12}O$: Isoamyl Alcohol (3-Methyl-1-Butanol)				
102	1954	1?	Cross-linked	SV	20	w_2 = 0.40—0.55	→Q_w	222, 230
				C_6H_6 : Benzene				
103	1954	1?	Cross-linked	SV	20	w_2 = 0.98	→Q_w	222, 230
104	1955	—	NMP	EP	30	w_2 = 0.0003—0.02	→μ = 4.09 ± 0.04 D	210
				C_6H_9NO : N-Vinylpyrrolidone (NVP)				
105	1954	1?	Cross-linked	SV	20	w_2 = 0.07	→Q_w	222
				$C_6H_{10}O$: Cyclohexanone				
106	1952	4	2D4—4D4	OP	25	0.25—1%	→$\overline{M}_n$; A_2	255

Table 4 (continued)
STUDIES ON PVP AND SMALL-MOLECULE ANALOGS IN NONAQUEOUS AND MIXED SOLVENTS[a]

Entry	Year	Polymer samples No.	Polymer samples Range of $\overline{M}$	Techniques used	Temperature (t/°C)	Polymer concentration	Comments	Ref.
107	1954	1?	Cross-linked	SV	20	$w_2 = 0.57$	$\rightarrow Q_w$	222, 230
						C_6H_{12} : Cyclohexane (See Also $CHCl_3$ + C_6H_{12})		
108	1972	—	NEP	IR	~30	$x_2 = 0.01$—0.8	1600—1800 cm^{-1}: $\rightarrow \nu(C{=}O)$	220
						$C_6H_{14}O$: Isopropyl Ether		
109	1954	1?	Cross-linked	SV	20	$w_2 = 1.0$	$\rightarrow Q_w$ (No swelling)	222, 230
						$C_6H_{15}N$: Triethylamine		
110	1954	1?	Cross-linked	SV	20	$w_2 = 0.93$	$\rightarrow Q_w$	222, 230
						C_7H_8O : Benzyl Alcohol		
111	1954	1?	Cross-linked	SV	20	$w_2 = 0.05$	$\rightarrow Q_w$	222, 230
						C_7H_9N : Benzylamine		
112	1954	1?	Cross-linked	SV	20	$w_2 \approx 0.06$	$\rightarrow Q_w$	222, 230
						C_7H_{16} : Heptane (See $CHCl_3$ + C_7H_{16})		
						$C_8H_{10}O$: 2-Phenylethanol		
113	1954	1?	Cross-linked	SV	20	$w_2 \approx 0.06$	$\rightarrow Q_w$	222, 230

$C_8H_{18}O$: n-Octanol

114	1954	1?	Cross-linked	SV	20	w_2 = 0.05—0.07	$\rightarrow Q_w$	222, 230

[a] Small-molecule analogs as designated in column four; NEP, N-ethylpyrrolidone; NMP, N-methylpyrrolidone; NVP, N-Vinylpyrrolidone.
[b] See Reference 259 for revised $\overline{M}$ values.
[c] See Entry 57 (Reference 259) and also text for revised MHS parameters.

and this has been taken to be the case throughout the present discussion; however, Tadokoro and co-workers[279] quoted an estimated fractional crystallinity of 0.18 for PVP films from their X-ray diffraction patterns, although they gave no further specific details of their patterns or of the intermediate data they obtained.

The *infrared absorption spectrum* of solid PVP has been published by several groups: Behen and co-workers (650 to 4000 cm^{-1});[280] Boyer-Kawenoki (1540 to 4000 cm^{-1});[281] Kahela and Forsén (625 to 4000 cm^{-1}, with a list of the 22 principal bands);[282] Nyquist (625 to 4000 cm^{-1});[101a] and Oster and Immergut (1100 to 5000 cm^{-1}).[218] The KBr disc technique was used in the first three of these studies, while the last two involved films cast from aqueous solution onto AgCl plates. More limited regions have been studied by Rothschild (~1680 and ~1400 cm^{-1})[220] and Biedermann and Graf (~1670 cm^{-1}).[283] The center of the C=O stretching band has been located variously at 1661 cm^{-1} (Reference 282), 1666 cm^{-1} (Reference 283), 1672 cm^{-1} (Reference 281), 1678 cm^{-1} (Reference 218), and 1680 cm^{-1} (Reference 220); Rothschild[220] has pointed out that the value of 35 cm^{-1} for its bandwidth is extraordinarily large, indicative of strong interactions between the amide groups, and similar behavior is seen with the monomer unit analog N-ethylpyrrolidone.[220] Behen and co-workers[280] have used the "amide III" band at 7.75 μm (1290 cm^{-1}), attributed to a C–N stretching mode, for the determination of PVP. The band at 3500 cm^{-1} (2.8 μm), which appears in all the published spectra,[101a,218,280-282] has been assigned to bound (absorbed) water.[218,252,282] This assignment is confirmed by the fact that this same band which occurs in the spectrum of PVP dissolved in NMP is used for the determination of the water content of PVP-iodine complexes.[284] Molyneux and Frank[252] have described the procedure necessary to obtain films of anhydrous PVP (i.e., with this band absent) for IR spectral studies.

F. Amorphous Solid Solutions and Other Solid-State Complexes

Polyvinylpyrrolidone is distinguished by the readiness with which it forms solid-state complexes with a wide variety of substances, including iodine[284,285] and β-carotene,[286] as well as many pharmaceutical compounds such as tolbutamide and phenytoin,[287] ajmaline, griseofulvin, and reserpine,[249] and various sulfa drugs.[246,278,288-290] Where X-ray diffraction studies have been carried out they indicate that the complexes are essentially amorphous,[246,278,287,288,290] showing that the small-molecule component is molecularly dispersed in an amorphous matrix of the polymer; these complexes may therefore be referred to as amorphous solid-state complexes or as *amorphous solid solutions* (the simple term *solid solution* having, of course, already been appropriated for mixed crystals). The ease of formation of these "complexes" with PVP is clearly the result of the high solvent power of the amorphous polymer, with the monomer units containing amide groups (having a highly polar character and able to accept hydrogen bonds) as well as nonpolar groups, and with each type of group thus able to "solvate" corresponding groups on the small-molecule component.

In the case of iodine the situation is made more complex by the possibility of reaction between the additive and reducing components in the polymer, possibly mediated by the absorbed water, leading to iodide ions which can also complex with the iodine to form triiodide ions.[284,285]

These amorphous solid-state complexes have found various practical applications, particularly in the pharmaceutical field (where they are referred to rather misleadingly as "high energy" complexes[278,290]), both because the complexed small-molecule substance has a lowered thermodynamic activity (e.g., in the case of iodine, a lowered volatility), and since the substance is molecularly dispersed the complexed form dissolves much more readily than in the crystalline state.

PVP also forms remarkable solid complexes with the group of metals Mn, Fe, Co, and Ni, which are obtained by interaction of the polymer with the corresponding metal carbonyl; e.g., the complex formed with Fe has the molecular formula $Fe(C_6H_9NO)_5$ which represents the replacement of the coordinating CO groups in $Fe(CO)_5$ by the carbonyl groups of the pyrrolidone rings.[283]

G. Solution Behavior

As with the other polymers, it is convenient to consider the behavior of PVP in aqueous solution separately from that in nonaqueous and mixed solvents. Although in the present context the major interest in these polymers lies in their water solubility and behavior in aqueous solution, that in other media is also significant both in giving a more complete picture of the solution behavior of the polymer, and also in showing other media which are suitable for the handling, characterization and analysis of the polymer. These features apply particularly to PVP, which is distinguished among the water-soluble synthetic polymers for the range of its nonaqueous solvents (see Section III.A).

1. Aqueous Solution

The results of various studies on the behavior of PVP in aqueous solution are summarized in Table 3; the table also contains reference to certain parallel studies (entries 11 and 15) in aqueous sodium chloride solutions (0.2*M* and 2*M*),[257] which are included since the addition of uni-univalent inorganic salts seems to make little difference to such conformational properties as the intrinsic viscosity.[242,257,259,275] The studies by Meza and Gargallo (entries 105 and 106)[263] in 0.55 *M* sodium sulfate at 28°C are also included since this solution had been shown by concurrent precipitation temperature measurements to be a Θ system for PVP (see also Section III.H.4 and Chapter 2 of Volume II).

The partial *specific volume* of PVP in dilute aqueous solution ($c_2 < 5\%$) has been studied by nine groups: Dialer and Vogler (entry 2);[245] Hengstenberg and Schuch (entry 8);[255] Scholtan (entries 12 and 13);[257] Miller and Hamm (entry 18);[264] Elias (entry 48);[272] Andersson (entry 52);[139] Nomura and Miyahara (entry 57);[299] Goldfarb and Rodriguez (entry 67);[303] and Vink (entry 72).[306] The variation of $\bar{v}_2^\circ$ with temperature was shown by Nomura and Miyahara[299] to be essentially linear over the temperature range 5 to 55°C, with the temperature coefficient as 9.5d4 deg^{-1} (the scatter about their correlation line is $\pm$ 0.004 $cm^3\ g^{-1}$, giving an indication of the experimental uncertainty in the individual values); this coefficient enables values determined at 20 or 30°C to be adjusted to 25°C. For this last temperature, the individual values (after any such adjustment) seem to form two groups clustered around 0.78 and 0.80 $cm^3\ g^{-1}$, with the value of 0.82 $cm^3\ g^{-1}$ obtained by Elias[272] as an "outlier"; omitting this last value, the overall average is 0.791 $\pm$ 0.010 $cm^3\ g^{-1}$. The reciprocal of this value, i.e., 1.264 $\pm$ 0.016 g cm^{-3}, thus represents the "effective density" of the polymer in aqueous solution; this is appreciably greater than the actual density of about 1.21 $cm^3\ g^{-1}$ (see Table 2), which is an indication of the compacting effect of the polymer on the solvent.

The specific volume (density) behavior of more concentrated solutions (up to w_2 = 0.54) has been studied by Kaatze,[221] while some parallel, less precise data are also available from the technical literature.[192] The data are plotted in Figure 5 for the full composition range, including for the limit w_2 = 1 the values of ρ_2 for the solid polymer (see Table 2) and the reciprocal of $\bar{v}_2^\circ$ corresponding to the extrapolation from the extreme dilution region. The plotted data show a break at $w_2 \approx 0.3$, which seems to correspond with the point of loss of "free" water deduced from dielectric relaxation measurements;[221] the linear section above the break point extrapolates above the level of the solid polymer at w_2 = 1, and hence must ultimately curve over to approach this level.

The *specific refractive index increment* for PVP in aqueous solution has been determined by four groups: Hengstenberg and Schuch (entry 7);[255] Levy and co-workers (entries 9 and 29);[256,259] Graham (entry 36);[242] and Elias (entry 47).[272] The values of $\bar{n}$ at 436 and 546 nm are essentially the same[259] so that at 589 nm[256] they should be comparable. The overall

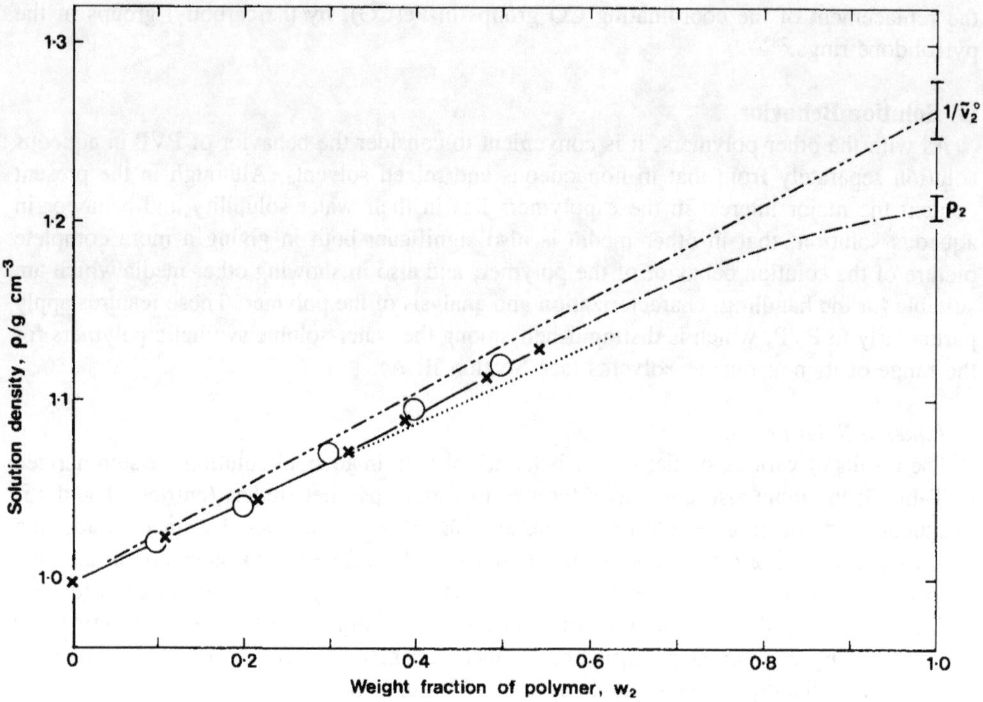

FIGURE 5. Density, ρ, vs. polymer weight fraction, w_2, for concentrated aqueous solutions of polyvinylpyrrolidone at 25°C. (○) GAF Technical Bulletin;[192] (×) Kaatze.[221] The bars at w_2 = 1 represent the ranges of the experimental values of ρ_2, the density of the bulk polymer, and $1/\bar{v}_2^0$, the reciprocal of the partial specific volume for extreme dilution, as labeled. The shape of the chain-dotted section of curve between w_2 = 0.54 and 1 is conjectural.

average, omitting the "outlier" value of 0.135 cm^3 g^{-1}, reported by Elias,[272] is 0.179 ± 0.007 cm^3 g^{-1} for the three wavelengths quoted and the temperature range 20 to 25°C.

Considering the *intrinsic viscosity* values for PVP in aqueous solution, numerous and diverse sets of parameters for the MHS relation have been quoted (Table 3 — entries 4, 15, 16, 20, 30, 37, 40, 43, and 104) with, e.g., the *a* values ranging from 0.55[259] to 0.85,[264] and the *K* values ranging correspondingly. Before considering these in any detail, it is necessary to look first at the effect of temperature upon [η], since the data quoted relate variously to 20, 25, and 30°C. The results of Goldfarb and Rodriguez[303] for the range 20 to 60°C give the temperature coefficient for [η] as −0.013 deg^{-1}; the results of Levy and co-workers[256] and Nomura and Miyahara[299] are in broad agreement with this, except for a greater scatter in the [η] values. Thus over the temperature range 20 to 30°C the value of [η] will fall by only 13%, which is a change sufficiently small to be neglected for the purposes of obtaining the MHS parameters. It may also be noted that the negative sign of this temperature coefficient is in agreement with the fall in the solvent power of water for PVP with rise in temperature which is seen from other measurements.

The main literature data for intrinsic viscosity and molecular weight of PVP samples in aqueous solution, for the temperature range 20 to 30°C, are plotted in Figure 6.*

* In this compilation, the point for one sample (their fraction E) from the work of Meza and Gargallo[263] has been omitted from the plot since it lies well below the trend of the other points; the same applies to the data for all three samples from the work of Nishijima and Oster.[292] Furthermore, it has only been possible to plot data for the three samples from the more extensive work of Graham[242] for which the actual numerical values of both [η] and M were published; similarly only the data for one sample from the work of Nakagaki et al.[313] could be plotted.

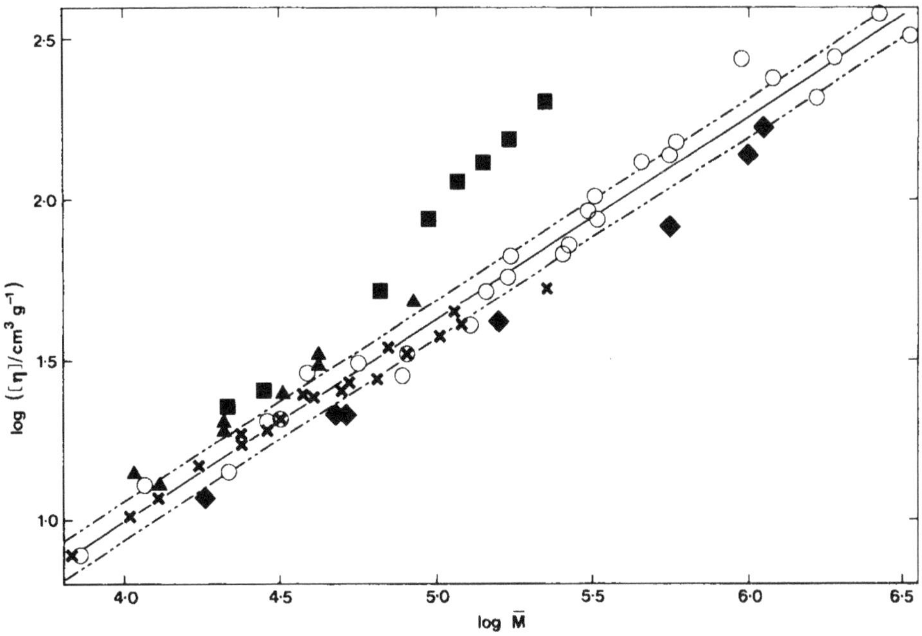

FIGURE 6. Double logarithmic plot of intrinsic viscosity vs. molecular weight for polyvinylpyrrolidone in aqueous solution at 25(±5)°C. (×) Sharp fractions;[258,259] (○) medium fractions;[242,257,259,260,262,263,313] (■, ▲, ♦) broad fractions and whole polymers, with molecular weight values respectively, $\overline{M}_n$,[255,306] $\overline{M}_{sD}$ (i.e., from sedimentation plus diffusion),[245,264] and $\overline{M}_w$.[258,259,275]

In view of the suggestion by Frank and Levy[258,259] that the correlation is unusually sensitive to the breadth of the molecular weight distribution, three categories of polymer sample have been distinguished in the Figure 6: (1) unfractionated (whole) polymer and broad fractions (filled symbols); (2) medium fractions (open circles); and (3) sharp fractions (crosses). The data in the first category have been further distinguished according to the type of molecular weight, i.e., whether obtained by light scattering (giving $\overline{M}_w$), sedimentation plus diffusion ($\overline{M}_{sD}$, approximating to $\overline{M}_n$), and osmotic pressure ($\overline{M}_n$). These distinctions having been made, the plot shows that within the limits of experimental uncertainty there is little difference between the correlations for sharp fractions and for medium fractions; the best fitting line through the 47 data points for the two combined categories gives the MHS parameters as $a = 0.63 \pm 0.06$; $K = 3.0\ (\pm 0.4)$d2 cm^3 g^{-1}.

Nakagaki et al.[313] reported that whereas up to $\overline{M}$ = 1D6, their MHS plot had a fairly normal slope (in their case, giving $a = 0.60$), at around this molecular weight the plot changed fairly abruptly to one of much lower slope, giving $a = 0.19$. It is likely, however, that this behavior is the result of the presence of branching in the samples studied; the solvent/nonsolvent precipitation method used to obtain the fractions leads to the branched material being concentrated in the earliest, i.e., highest molecular weight fractions obtained, and it is the data from these fractions which give the line of anomalously low slope in the MHS plot.

For the ideal system, 0.55 *M* aqueous Na_2SO_4 at 28°C established by Meza and Gargallo[263] by precipitation measurements, their parallel viscosity data give the MHS parameters $a = 0.50$; $K = 5.8\ (\pm 0.1)$d2 cm^3 g^{-1}.

2. *Nonaqueous and Mixed Solvents*

The results of various studies on PVP and its small-molecule analogs in nonaqueous and mixed solvent systems are summarized in Table 4.

The table also includes the studies by Breitenbach and Schmidt[222,230] on the swelling of cross-linked samples of PVP; although strictly speaking these are not solution studies, and indeed in several cases the media chosen are not even solvents for PVP, nevertheless the results obtained give valuable intercomparisons of the relative solvent properties of the liquids studied.

The studies on mixed solvent systems in Table 4 include several by Chaufer and co-workers[315,319-321,323] on preferential solvation, leading to the preferential solvation parameter (symbolized λ in the original references). It is notable that with mixtures of liquids both of which are solvents (e.g., chloroform + ethanol; 1,2-dichloroethane + ethanol), there is found to be a particular mixture which shows no preferable solvation and which therefore seems to have the optimum solvating power for the different types of groups on the polymer; if the composition is made richer in one component then that component is excluded from the polymer domain so as to maintain as far as possible the optimum composition within the domain.

The only published value of the *partial specific volume*, $\bar{v}_2^\circ$, for PVP in a nonaqueous medium seems to be that in methanol, reported as 0.730 cm^3 g^{-1} by Elias.[272]

For the *specific refractive index increment*, $\tilde{n}$, for PVP in nonaqueous media, that in methanol has been reported variously as 0.1765 cm^3 g^{-1} (Reference 272) and 0.197 cm^3 g^{-1} (Reference 242), and that in chloroform as 0.108 cm^3 g^{-1} (Reference 259). Chaufer and co-workers[319] state that for mixtures of chloroform + ethanol and 1,2-dichloroethane + ethanol there is a common linear relation between $\tilde{n}$ and the refractive index n_1 of the solvent, i.e., the Gladstone-Dale law is obeyed; the equation as obtained from their graphically presented results (25°C; 546 nm) is

$$\tilde{n}/\text{cm}^3\ \text{g}^{-1} = 1.3304 - 0.864n_1 \tag{10}$$

Comparing this with the values obtained for other systems, for water (n_1 = 1.333) this equation predicts $\tilde{n}$ = 0.179 cm^3 g^{-1} in agreement with the experimental value (see Section III.G.1); for methanol (n_1 = 1.327) the predicted value is $\tilde{n}$ = 0.184 cm^3 g^{-1}, which lies between the two values quoted above; and for chloroform (n_1 = 1.443) the predicted value is 0.084 cm^3 g^{-1}, in disagreement with the single experimental value quoted above. Extrapolation to $\tilde{n}$ = 0 (polymer and solvent isorefractive) gives n_1 = 1.540, which is thus the effective value of the refractive index of PVP in these systems and which may be compared with the value n_D^{25} = 1.53 reported for the solid polymer.[191] However, there is an anomaly in the graph of $\tilde{n}$ vs. n_1 published by Chaufer and co-workers[319] in that the span of the values of n_1 ranges from 1.32 to 1.42 on the graph, whereas it would be expected to range from 1.36 (for ethanol) to 1.44 (for chloroform and 1,2-dichloroethane).

Considering the *intrinsic viscosity-molecular weight* relations for PVP in nonaqueous media, the best established system is that for methanol where there are extensive data for 30°C from the work of Frank and Levy[258,259] together with data for one unfractionated sample from Molyneux and Frank;[275] as will be seen from the double logarithmic plot in Figure 7, the correlation is much less sensitive to the breadth of the molecular weight distribution than with water as solvent (cf. Figure 6), and the best fitting line gives the MHS parameters a = 0.645; K = 2.4 (±0.2)d2 cm^3 g^{-1}. This solvent therefore seems on the whole to be a more suitable one than water for molecular weight determinations.

For chloroform, data for 25°C are available from both Levy and Frank[259] and Meza and Gargallo;[263] as Figure 7 shows, the scatter is much greater than for methanol, the line drawn giving the MHS parameters a = 0.70; K = 9.8 (±1.3)d3 cm^3 g^{-1}. Some data are also available for 2-propanol,[263] but they are not sufficiently extensive to make any MHS parameters that might be deduced trustworthy.

For mixed solvent media, the system water (ϕ = 0.332) + acetone at 25°C has been established by Elias[272] to be an ideal system, i.e., with a = 0.50 and hence with K = 7.5

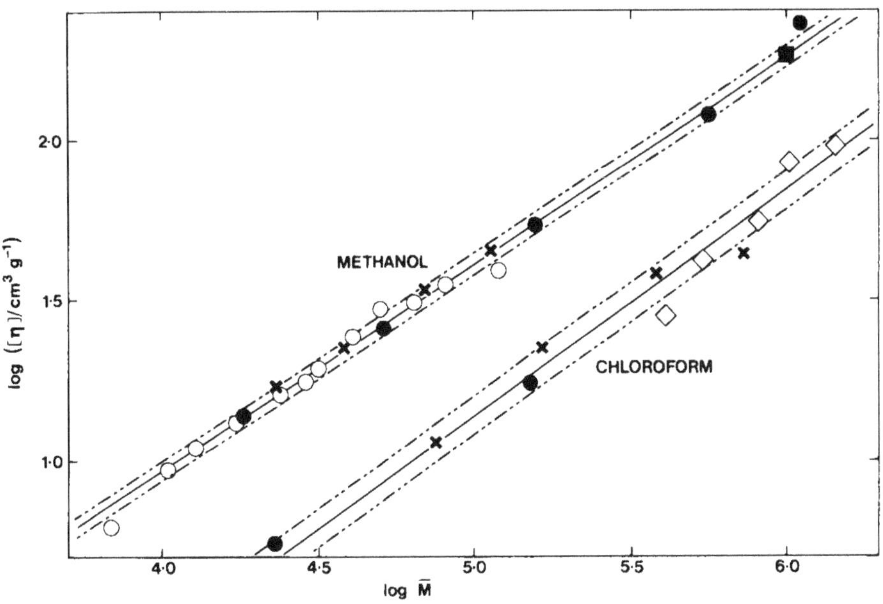

FIGURE 7. Double logarithmic plots of intrinsic viscosity vs. molecular weight for polyvinylpyrrolidone in methanol at 30°C and in chloroform at 25°C, as labeled (for clarity, on the chloroform plot the log $\overline{M}$ values have been increased by 0.5). (x, ○, ●) Frank and Levy (sharp fractions, medium fractions, and whole polymer, respectively);[258,259] (◇) Meza and Gargallo (medium fractions);[263] (■) Molyneux and Frank (whole polymer).[275]

(±0.2)d2 $cm^3\ g^{-1}$; the corresponding data obtained subsequently for this system by Meza and Gargallo[263] are substantially in agreement, but with greater experimental scatter. Data are also available from this pair of workers[263] for water (ϕ = 0.80) + acetone and for chloroform (ϕ = 0.40) + acetone, but in each case the data are not such as to allow trustworthy MHS parameters to be obtained.

H. Phase Equilibria

1. Water Vapor Uptake

Data for the equilibrium uptake of water vapor by PVP at various water activities have been reported by six groups, comprising the four listed in the caption to Figure 8 together with Jellinek and co-workers[324] who studied the uptake at various temperatures between −20 and +7.5°C, and MacKenzie and Rasmussen[308a] whose studies covered the range −40 to +22°C; in Figure 8 the values of the uptake, U_1, for 20 to 25°C are plotted on a logarithmic scale both because of the wide span of the values and because this gives an almost linear plot over the middle section. There is seen to be fairly good agreement between the data shown in the figure obtained from the different sources. Judging from the results of Sugiura and Fujii[276] there is little systematic dependence of U_1 upon the molecular weight of the polymer; the marked deviations between the data of this last group and those of the other groups at a_1 = 0.15, and to a lesser extent at a_1 = 0.47, may be due to systematic errors in the a_1 values assigned to the hygrostatting solutions used by this group.

The measurements of Tadokoro and co-workers,[279] which were carried out at 30, 40, and 50°C, showed that the uptake increases with rise in temperature by about 1.2%/degree for $a_1 < 0.5$ and by progressively lesser extents at higher values of a_1; on the other hand, Jellinek and co-workers[324] state that they obtained higher uptakes working at temperatures between −20 and +7.5°C than those reported by Dole and Faller[166] for 25°C.

The kinetic aspects of the uptake have also been reported by Tadokoro and co-workers,[279] together with studies of the effect of the uptake on the creep properties of the polymer.

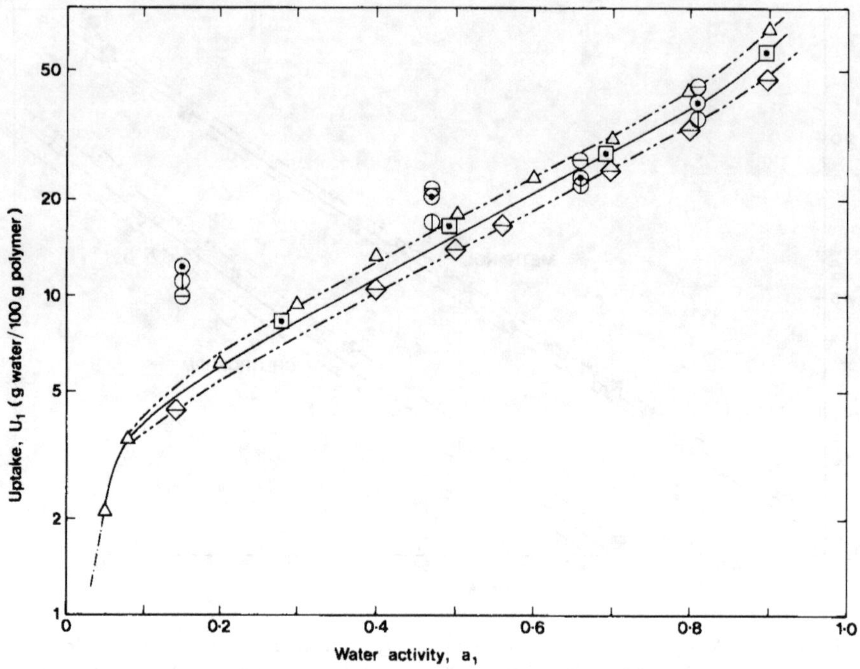

FIGURE 8. Equilibrium uptake of water, U_1 (logarithmic scale) by polyvinylpyrrolidone vs. water activity, a_1. Key (with temperature and grade of polymer): (⊡) Azorlosa et al. (temperature not stated; grade K-30);[187,191] (△) Dole and Faller (25°C; grade not stated);[166] (⦶, ⊙, ⊖) Sugiura and Fujii (20°C; grades K-15, K-30, and K-90, respectively);[276] (⦵) Tadokoro et al. (extrapolated to 25°C; grade K-90).[279]

Sugiura and Fujii[276] also studied the effect of the uptake on the glass-transition temperature (see Section III.E).

2. Nonaqueous Vapor Uptake

Miyagi and Tanaka[243] have measured the vapor sorption isotherms for PVP with methanol at 25 and 30°C and with ethanol at 25, 30, and 35°C.

3. Partition Behavior

Fikentscher and Herrle,[183] in their early work on the development of PVP, showed that with the system water/methylene chloride the polymer passes essentially entirely into the aqueous phase; they also obtained phase diagrams for PVP/water/acetone for the three grades K-15, K-33, and K-71, which are useful in connection with the fractionation of the polymer with this system.

Dobry[162a] has studied the partition of PVP in four two-phase two-liquid systems. With water/amyl alcohol (pentanol) and water/acetic anhydride the polymer passed entirely into the aqueous phase; with water/phenol, where the two phases are more exactly described as water-rich and phenol-rich, the partition was in favor of the phenol-rich phase with a partition coefficient of 10. The fourth system studied was glycerol/isoamyl alcohol (3-methyl-1-butanol) where the polymer passed entirely into the glycerol phase.

4. Precipitation and Solubility Equilibria

PVP in aqueous solution does not, in the absence of an additive (cosolute), show any sign of precipitation over the readily accessible temperature range 0 to 100°C. However, the fact that the solvent power decreases with rise in temperature, as shown by the negative signs of the temperature coefficients of intrinsic viscosity for PVP in solution (see Section

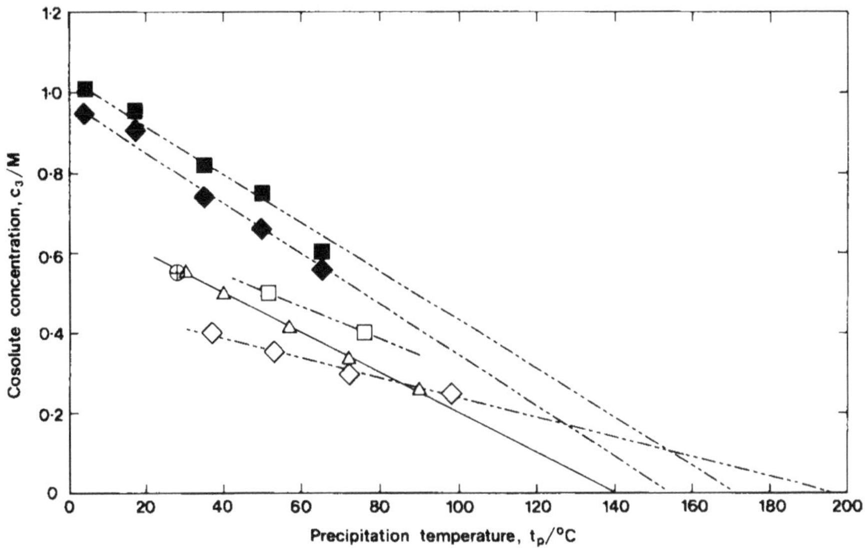

FIGURE 9. Precipitation temperature, t_p/°C, vs. cosolute concentration, c_3, for polyvinylpyrrolidone in aqueous solution with the cosolutes ammonium sulfate (filled symbols) and sodium sulfate (open symbols). Key (with polymer molecular weight and concentration): (□, ◇) Goldfarb and Rodriguez (3D4; 3 and 10%);[303] (■, ◆) Jirgensons (6D4; 0.4 and 2%);[200] (⊕) Meza and Gargallo (ideal system, i.e., $M_2 \rightarrow \infty$; $c_2 \rightarrow 0$);[263] (△) Nakagaki and Shimabayashi (1.5D6; 0.5%).[326]

III.G.1) and degree of swelling for cross-linked polymers,[222,230] suggests that any ideal temperature for the system should be above 100°C. Moreover, the polymer is precipitated out by the addition of various salts, particularly readily by those such as sodium sulfate,[250,263,265,303,325,326] ammonium sulfate,[200,250] and others which contain polyvalent anions;[250,325] similar precipitation effects are seen with glycine as cosolute.[325] (Although, in general, the specific discussion of cosolute effects is dealt with in Chapter 2 of Volume II, in the present instance it is useful to consider such effects with PVP at this stage because, by extension, they give information about the behavior of the polymer alone in aqueous solution.) All these systems show inverse temperature dependence of the precipitation point, i.e., cooling leads to the precipitate dissolving, and heating to it reappearing, while lower concentrations of cosolute are required for precipitation at higher temperatures. The results of the main studies with ammonium sulfate and sodium sulfate are plotted in Figure 9, showing that in each case there is an essentially linear dependence of the precipitation temperature t_{p+} on the cosolute concentration c_3; similar dependence is also shown in the graphically presented data for these and other salts by Sekikawa and co-workers.[250] The most significant of the correlations in Figure 9 is that drawn as a full line, which represents the data (triangles) for sodium sulfate obtained by Nakagaki and Shimabayashi[326] with a high molecular weight sample (K-90: $\overline{M} \approx$ 1D6) at relatively low concentration (~0.5%); this line passes very close to the point (crossed circle) representing the corresponding ideal system, 0.55*M* sodium sulfate at 28°C, established by Meza and Gargallo.[263] These features suggest that this same line represents a sequence of ideal systems, so that extrapolation to $c_3 = 0$ gives the ideal temperature for PVP in water as $\Theta_+ = 140 \pm 5$°C. In agreement with this, all the other lines in Figure 9 (which relate to lower polymer molecular weights and/or higher polymer concentrations) extrapolate as expected to higher limiting t_{p+} values.

I. Chemical Behavior

Polyvinylpyrrolidone seems to be quite stable in the solid state under normal conditions of storage. Ball milling leads to degradation of the polymer, a process which has been studied extensively by Kaneniwa and co-workers.[266-270]

In aqueous solution, osmotic pressure measurements indicate that the polymer is stable on storage at normal temperatures and even upon autoclaving (30 min at 10 lb of steam pressure, i.e., at 115°C).[204] However, longer term viscosity measurements on somewhat higher molecular weight samples reveal a slow degradation at normal temperatures leading to a fall in intrinsic viscosity;[275,325] this seems to involve the hydrolysis of "weak links" in the polymer chain, since it can be brought to completion by treatment with acid.[325]

The effect of radiation-produced hydroxyl radicals on PVP in aqueous solution has been studied by Behzadi and co-workers.[327] The effects of radiation on PVP (and several other water-soluble polymers) in aqueous solution has also been studied by Charlesby and co-workers,[164a,327a,328] who have shown that it is protected from radiation damage by the addition of thiourea.[328a]

In the presence of acid or alkali in aqueous solution at 100°C, some of the pyrrolidone rings are opened to give N-vinyl, γ-aminobutyric acid units, e.g., in alkaline solution giving mainly the anionic form:

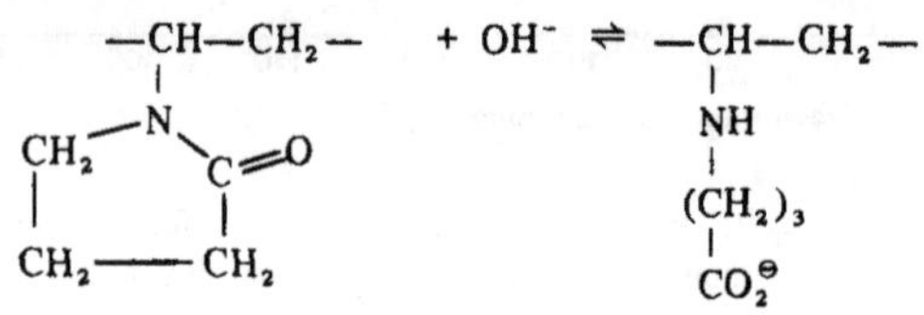

(11)

together with contributions from the zwitterion and from the protonated (cationic) forms varying in extent with pH. This equilibrium has been studied by Frank[217] over the pH range from −0.5 (25% HCl) up to 12.9 (0.1*M* NaOH), together with parallel studies at pH 1.25 to 12.70 with NEP; the equilibrium was not studied at higher pH with PVP because the polymer becomes insoluble (apparently, a reversible salting-out effect) under these conditions.

The kinetics of the alkaline hydrolysis reaction in 1*M* NaOH (i.e., pH ≈ 14) have been studied at 100 to 130°C by Conix and co-workers,[225,226] together with some parallel studies carried out by them on related polymers and small-molecule analogs; the precipitation of the PVP that occurs under these highly alkaline conditions did not seem to affect the kinetic behavior. The reaction followed pseudo first-order dependence (the equilibrium under these conditions was stated to be shifted almost completely in favor of the amino acid form, although such an extreme shift would not have been expected from the pH dependence observed by Frank[217]); the Arrhenius parameters for the stated temperature range are given as A = 8d9 $M^{-1}s^{-1}$; E = 25 ± 1 kcal mol^{-1} (105 ± 4 kJ mol^{-1}).[225,226]

This ring opening reaction has been used by Klotz and co-workers to covalently conjugate dyes[329,330] and azo mercurials[330] to the polymer chain; the pK_a values of the groups are found to be markedly changed on conjugation, which has been attributed to modifications in the water structure within the polymer coil,[329,330] although some contribution might also be anticipated from the expected reversible intramolecular noncovalent binding of the conjugated aromatic group with other sections of the same polymer chain.

Molecular iodine reacts with PVP in aqueous solution, the reaction leading to its incorporation in the polymer;[205] this apparently involves the addition of I_2 molecules to unsaturated end groups derived from the monomer, a reaction which does not occur with simple C=C bonds, but which is confirmed as a possibility for PVP by the fact that I_2 reacts similarly with NVP monomer.[256] This reaction enables the polymer to be radiolabeled by using the radioisotope ^{131}I.[205] The polymer reacts similarly (as would be expected) with iodine monochloride.[205]

J. Biological Behavior

The biological behavior of PVP, and more specifically its physiological and pharmacological aspects, are important from two viewpoints: first, because of the use of the polymer pharmaceutically and clinically in a variety of applications including that as a blood-plasma substitute,[203-206,256] and second, because of its use in hair sprays and other aerosol products, droplets of which may be inhaled into the lungs.[331] In general, the polymer does not seem to be metabolized appreciably in the body, so that when given orally it is excreted unchanged, while when given intravenously the fraction with molecular weight below about 4D4 is excreted within a few days and the remaining higher molecular weight material stored essentially indefinitely in the tissues; this last aspect has to be borne in mind in any projected routine or long-term intravenous use of PVP or derived preparations.[203-205] Further aspects of the biological behavior of the polymer are dealt with in papers listed in the bibliographies already cited.[182,194,197,198]

K. Analysis

1. General Methods

For the assay of moderately concentrated aqueous solutions of PVP, evaporation followed by heating at 105°C, or at a lower temperature in a vacuum oven, seems to be a satisfactory procedure;[256] any heating in vacuum should be above the glass-transition temperature (see Section III.E) to ensure mobility of the absorbed molecules.

Refractometry has also been used for the assay of aqueous PVP solutions of concentrations between about 1 and 6%.[256,278]

The Kjeldahl method has also been used for the determination of the PVP content of PVP itself,[332-334] VP copolymers,[238-240] and PVP-iodine complexes,[284] via the nitrogen content of the sample (12.61% for pure PVP) which is given by this method. Such an approach has the advantage that it may be scaled down to the microlevel, and in this form it is applied in some of the more specific and sensitive methods discussed later. However, the Kjeldahl method has been widely reported to give low values for PVP and its copolymers, necessitating the use of empirical "correction factors";[238-240,332,333] these low values presumably arise from the circumstance that conversion of the nitrogen in the polymer into a form that is volatile in the later distillation step of the Kjeldahl procedure requires the breaking of all *three* C–N bonds in each monomer unit in the preliminary digestion step. Breaking of only *two* of these either leaves the nitrogen atom still attached to the polymer chain, or released as γ-aminobutyric acid which is still involatile even in alkaline solution. From this viewpoint the Dumas method would be preferred, although it does not seem to have been applied at all commonly for this purpose.[264]

From the more general analytical viewpoint, Müller[335] has listed a variety of identification tests which may be used for PVP, most of them involving inorganic reagents, while Levy and co-workers[256] have described and discussed the procedures for the technical evaluation and quality control of PVP, particularly that of grades to be used for clinical purposes.

2. Specific and Sensitive Methods

As a result of the many and various applications and uses of PVP, it is frequently necessary to determine the polymer specifically and at low concentration; typical cases are where the polymer occurs as residues after its introduction into biological systems, either in body fluids (blood, sweat, tears, urine, etc.) or in tissue samples, while in other cases the analysis may involve relatively small amounts of the polymer in tablets or other pharmaceutical products. A wide variety of methods have consequently been developed which are more specific and/or more sensitive than the methods (solids content, refractometry, and nitrogen content) previously described. These methods are listed and discussed briefly below from the viewpoint of the physical chemistry of the procedures used in relation to the properties and behavior of PVP; the original papers should be consulted for the more strictly analytical aspects, and for details of the techniques and their applicability to particular samples and systems.

Before considering the individual methods, some general points require to be discussed. First, there do not seem to have been any previous detailed intercomparisons of the various methods even from the analytical viewpoint, although some limited critical comments have been made on various other methods used in a number of the publications,[331,335-342] while in some cases two or more methods have been recommended as alternatives for particular applications.[280,331,337,340,343,344] Second, in most of the methods the results which are obtained show a sensitivity to molecular weight, so that they require calibrating with standards having the molecular weight of the particular sample being assayed; this is clearly an unsatisfactory situation, and it seems in many cases to be an unresolved problem. Third, in many biological applications it is necessary to separate the PVP from proteins which are also present and which can give similar reactions; several more or less elaborate procedures have been developed for this purpose using inorganic reagents which precipitate the proteins,[333,339,343,345-347] but these procedures have a number of attendant disadvantages including adsorption of PVP by the precipitate,[343,346] and in fact the simplest method is precipitation with methanol or ethanol since PVP is readily soluble in these alcohols and their mixtures with water, while there is little if any adsorption onto the precipitate formed.[280,343,348]

Considering now the various methods that have been proposed for the specific or sensitive analysis of PVP in different samples, these methods may be conveniently classified into ten main categories.

Zipf method — This method was introduced by Zipf[332] early in the history of PVP, and subsequently applied and developed by Ammon and co-workers[349,350] and Meijer.[346] In the method, the sample of solution (previously freed of protein if necessary) is treated with trichloroacetic acid to precipitate the PVP, which is then redissolved in water or methanol and determined by the micro-Kjeldahl method. The drawbacks of this method are that the precipitation is not complete with lower molecular weight samples,[350] while it also has the limitations of the Kjeldahl method as already discussed.

Chloroform extraction — Here the sample (assuming it to be an aqueous solution or suspension) is evaporated to dryness, extracted with diethyl ether (which removes certain possible contaminants but not the PVP), and then extracted with chloroform to remove the PVP which is determined by the micro-Kjeldahl method.[348] This type of chloroform extraction procedure has also been used as a preliminary to IR assay (see the infrared method below). It should be noted that the chloroform extraction cannot be carried out directly on an aqueous solution or a wet sample because the partition coefficient of PVP in chloroform/water is essentially zero;[348] correspondingly, the evaporated sample must be absolutely anhydrous if the extraction of the PVP is to be complete. The method is also not absolutely specific since other nitrogenous substances (e.g., urea) can be extracted by chloroform.[346,349]

Colorimetry with iodine — This method has been widely used and developed.[204,242,331,335,337-339,347,351,352] It depends upon the intensification of the color of an aqueous solution of molecular iodine and an alkali iodide in the presence of PVP; this intensification presumably results from the binding either of iodine molecules or tri-iodide ions to the dissolved polymer (see Chapter 2 of Volume II). In using the method, care has to be taken not to add too much reagent, otherwise the complex is precipitated (see the precipitation with iodine method below). The method has also been automated.[337] One potential drawback is the *chemical* reaction that occurs between PVP and iodine molecules, although this involves a consumption of iodine which only takes place essentially in a 1:1 molecular ratio (see Section III.I).

Colorimetry with dyes — This type of method, like that of the preceding method, depends upon the intensification of the color of a dye (often accompanied by a shift in λ_{max}) in the presence of the polymer, again presumably as a result of the reversible binding of the dye to the dissolved polymer. Examples of the dyes that have been used include Congo Red,[341] Vital Red,[340] and Brilliant Vital Red.[331,345]

Precipitation with iodine —This method was originally proposed apparently by Cannan,[342,343] and was subsequently developed by Dwyer and Lewandowski.[343] Like the colorimetry with iodine method, it involves the addition of a reagent containing molecular iodine and iodide ions, but in this case sufficient reagent is added to precipitate out the complex, which is isolated and titrated with thiosulfate for its content of molecular iodine; this is then converted into the content of PVP using empirical factors.

Turbidimetric and nephelometric methods — Here, as in the preceding method, the polymer is precipitated using a reagent, but by making the polymer concentration sufficiently low the precipitate will remain dispersed and suspended in the system; the amount of suspended precipitate can then be obtained either by turbidimetric (light-transmission) or nephelometric (light-scattering) techniques. Precipitating agents that have been used for this purpose include perchloric acid,[333,337,343] phenol,[342] and trichloroacetic acid (see also the Zipf method).[334,337,340] The method using perchloric acid has also been automated.[337] It should be noted that phenol, and probably several other precipitants for PVP, will also precipitate other water-soluble polymers so that this method is not absolutely specific; however, the concentration required for incipient precipitation depends upon the nature of the polymer, which may be used to distinguish between such polymers.[353] A further feature is that the precipitation point depends upon molecular weight, which has been applied to the estimation of molecular weight of PVP by Jirgensons[200] using ammonium sulfate as the precipitant, and to molecular weight distributions via turbidimetric titrations by Scholtan[265] using sodium sulfate.

Infrared methods — Certain specific bands in the IR absorption spectrum of PVP (see Section III.E) may be used for its analysis. The band at 1430 cm^{-1} (~7 μm), assigned to CH_2 scissor bending vibration, is not suitable for this purpose since it is too susceptible to interference from organic contaminants;[280,317,318] however, Ridgway and Rubinstein[317] and Minamikawa and Yamagishi[318] have successfully used the carbonyl stretching band at ~1670 cm^{-1} (6.0 μm) with PVP in chloroform solution for this purpose, whereas Behen and coworkers[280] preferred to use the 1290 cm^{-1} (7.75 μm) C–N stretching band ("amide III" band) with solid PVP in KBr discs because there was less interference from other organic compounds. In this latter case a preliminary extraction of the PVP with chloroform (see the chloroform extraction method) may be necessary to free it of nitrogen-containing contaminants (e.g., creatinine).[280] In general, it would seem to be good practice to use each of these two recommended bands as a mutual check, when calculating the PVP content from the recorded spectrum.

Fluorimetric methods — The fluorescence of 1-anilinonaphthalene-8-sulfonate (ANS; 1-phenylnaphthylamine-8-sulfonate) that appears in the presence of PVP has been used for the assay of the polymer.[287,354] The method has the advantage that the results are independent of the molecular weight of the polymer, but it suffers interference from detergents and proteins, although not from poly(vinyl alcohol), sodium polyacrylate, or potassium poly(vinyl sulfate) among the synthetic polymers.[354] The basis of the method is clearly the fluorogenic effect on ANS of its binding to the PVP (see Chapter 2 of Volume II).

Chromatographic methods — In one chromatographic method, the PVP is adsorbed on the top of a column of silica gel or *Micro-Cel* E, the column stained with a solution of the dye Vital Red or Brilliant Vital Red and eluted with sodium borate to remove protein and any excess dye, and then the dye that remains complexed onto the adsorbed PVP is eluted with NMP or dimethylformamide and estimated colorimetrically.[336,340,355] In another chromatographic method, the ability of aqueous PVP solutions to elute the dye Pyramine Orange 2R adsorbed on filter paper is used to give an estimate of the PVP concentration;[344] this effect is evidently the result of reversible binding of the dissolved dye molecules by the polymer (see Chapter 2 of Volume II).

Miscellaneous methods — These include a polarographic method;[356] a precipitation method using silicotungstic acid;[357] a spectrophotometric method at 212 nm;[334] and a colorimetric method using a reaction with *p*-dimethylaminobenzaldehyde.[334] Insoluble (cross-linked) PVP has been determined by its absorption of tannic acid from solution.[358]

3. Other Analytical Aspects

Colorimetric reagents of the types listed in the above methods, i.e., iodine (tri-iodide) and dyes have also been used in histology as stains for the demonstration of the presence of PVP in tissue preparations and sections from organisms that have been exposed to the polymer.[331,359-363]

Some of the assay methods listed here for PVP have also been applied to the analysis of vinyl acetate/vinylpyrrolidone copolymers.[335,336]

IV. POLYVINYLOXAZOLIDONE (PVO) AND POLYVINYLMETHYLOXAZOLIDONE (PVMO)

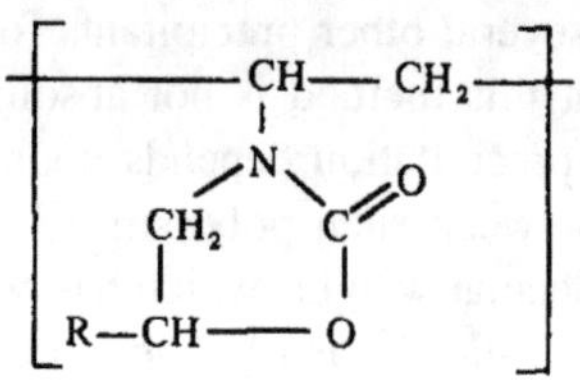

PVO : R = H : $M_0(C_5H_7NO_2)$ = 113.12
PVMO : R = CH_3 : $M_0(C_6H_9NO_2)$ = 127.14

Other chemical names, abbreviations, and trade names for PVMO are poly(5-methyl,3-vinyloxazolidine-2-one); polyvinylmethyloxazolidinone; poly-vinyl-5-methyl-oxazolidinone; poly(N-vinyl-5-methyl-2-oxazolidone); poly(vinyloxazolidonemethyl); PVO; PVOM; *Devlex* 130 (Dow).

Of this pair of related polymers, the methyl-substituted one, PVMO, was manufactured and marketed in the 1960s,[364] and various aspects of its properties and behavior studied at that time, but it is apparently no longer available commercially; the two polymers are interesting in the present context because of their evident structural similarities to PVP. The general features of PVO[365] and PVMO[364,365] have been described.

In each case, the polymer may be produced by free radical polymerization of the parent monomer.[365] The copolymerization of VO with a variety of other monomers has been studied;[238,365] these studies lead to the following average values of the Alfrey-Price parameters $Q = 0.05 \pm 0.02$; $e = -0.7 \pm 0.2$.[238,365]

The infrared spectrum of bulk PVMO over the range 1300 to 5000 cm^{-1} has been reported.[366]

PVMO is readily soluble in water at normal temperatures but precipitates out on heating to about 40°C, i.e., it shows inverse solubility-temperature behavior.[201,365] This may be compared with the case of PVP, where the corresponding Θ_+ value has already been deduced to be about 140°C (see Section III.H.4), i.e., about 100° higher. The NMR and viscosity behavior of PVMO in aqueous solution has been compared with that of PVP, PMAA, and several biopolymers by Glasel.[304]

PVMO, like PVP, forms complexes in the solid state and in solution,[364-367] and its precipitation temperature is affected by the presence of various cosolutes;[201] these cosolute effects in solution are considered more specifically in Chapter 2 of Volume II.

POSTSCRIPT

II.B. Review of the properties and applications of PVAA and PVAL[369]

II.G.1. Vapor pressure (24% VAC);[370] viscosity and density (0.6 to 4.5% VAC);[371] electron-spin resonance (spin-labeled gels)[372]

II.H.2.	Phase diagrams (20 to 26% VAC)[370]
II.I.	Partial conversion to polyvinylcarboxyformal by treatment with glyoxylic acid;[373] wet oxidation[374]
II.K.	Effects of potentially interfering substances on colorimetric analysis with boric acid + iodine + iodide mixture[375]
III.A.	Reviews of the properties and applications of PVP[376,377]
III.B.1.a.	Heats of mixing, viscosity, density, and phase diagrams for water/N-methylpyrrolidone[378-380]
III.B.1.d.	Viscosity and density for water/N-butylpyrrolidone[378,379]
III.B.2.	Crystal structures, and NMR and CD in aqueous solution, of C-substituted N,N′-dimethyldiketopiperazines[381]
III.C.	Polymerization of NVP induced by X-rays[382,383]
III.E.	There are doubts about the value of the glass-transition temperature, $t_g = 86 \pm 1°C$ for PVP obtained from specific volume measurements by Sugiura and Fujii (see also Table 2).[276] More recently, Tan and Challa[384] have obtained the much higher value $t_g = 175 \pm 1°C$ by differential scanning calorimetry — they find that this value is markedly reduced by absorbed water ($\Delta t_g/\Delta w_1 \approx 950°C$) and they suggest that lower values quoted in the literature are the consequence of absorbed water in the samples studied; although they quote other high literature values, ranging from 145 to 175°C, in support of their own, it is unfortunate that none of the references they cite gives details of the experimental techniques used. Their data would require that the samples of Sugiura and Fujii[276] each contained precisely 9.4 ($\pm$0.1)% water to give the consistency in t_g values ($\pm$ 1°C) observed by the latter; also the corresponding dependence of t_g upon water content ($\Delta t_g/\Delta w_1 \approx 222°C$) is much less than that reported by Tan and Challa.[384] It is evident that further studies are required to clarify this situation.
III.I.	Effects of X-rays[383,383]

REFERENCES

1. **Anon.,** *IARC Monographs on the Evaluation of the Carcinogenic Risk of Chemicals to Humans,* Vol. 19, World Health Organization/International Agency for Research on Cancer, Lyon, France, 1979.
2. **Anon.,** Vinyl acetate, polyvinyl acetate and polyvinyl alcohol, in *IARC Monographs on the Evaluation of the Carcinogenic Risk of Chemicals to Humans,* Vol. 19, World Health Organization/International Agency for Research on Cancer, Lyon, France, 1979, 341.
3. **Finch, C. A., Ed.,** *Polyvinyl Alcohol: Properties and Applications,* John Wiley & Sons, London, 1973.
4. **Toyoshima, K.,** General properties of polyvinyl alcohol in relation to its applications, in *Polyvinyl Alcohol: Properties and Applications,* Finch, C. A., Ed., John Wiley & Sons, London, 1973, chap. 2.
5. **Anon.,** Recommendations for abbreviations of terms relating to plastics and elastomers, *Pure Appl. Chem.,* 18, 581, 1969.
6. **Anon.,** List of standard abbreviations (symbols) for synthetic polymers and polymer materials — 1974, *Pure Appl. Chem.,* 40, 475, 1974.
7. **Leeds, M.,** Poly(vinyl alcohol), in *Encyclopedia of Chemical Technology,* Vol. 21, 2nd ed., Kirk, R. E. and Othmer, D. F., Eds., John Wiley & Sons, New York, 1970, 353.
8. **Lindemann, M. K.,** Vinyl alcohol polymers: poly(vinyl alcohol), in *Encyclopedia of Polymer Science and Technology: Plastics, Resins, Rubbers, Fibers,* Vol. 14, Mark, H. M., Gaylord, N. G., and Bikales, N. M., Eds., Interscience, New York, 1971, 149.
9. **Argana, C. P. and Czerwin, E. P.,** Polyvinyl alcohol, in *Water-Soluble Resins,* 2nd ed., Davidson, R. L. and Sittig, M., Eds., Van Nostrand Reinhold, New York, 1968, chap. 6.
10. **Pritchard, J. G.,** *Poly(Vinyl Alcohol): Basic Properties and Uses, (Polymer Monographs,* Vol. 4), Gordon & Breach, New York, 1970.

11. **Finch, C. A., Ed.,** *Properties and Applications of Polyvinyl Alcohol*, S.C.I. Monograph No. 30, Society of Chemical Industry, London, 1968.
12. **Hackel, E.,** Industrial methods for the preparation of polyvinyl alcohol, in *Properties and Applications of Polyvinyl Alcohol*, S.C.I. Monograph No. 30, Finch, C. A., Ed., Society of Chemical Industry, London, 1968, 1.
13. **Shohata, H.,** Continous polymerisation of vinyl acetate for polyvinyl alcohol production, in *Properties and Applications of Polyvinyl Alcohol*, S.C.I. Monograph No. 30, Finch, C. A., Ed., Society of Chemical Industry, London, 1968, 18.
14. **Warson, H.,** Polyvinyl alcohols from copolymers, in *Properties and Applications of Polyvinyl Alcohol*, S.C.I. Monograph No. 30, Finch, C. A., Ed., Society of Chemical Industry, London, 1968, 46.
15. **Moore, W. R. A. D. and O'Dowd, M.,** Factors affecting aqueous solubility of polyvinyl alcohol and partially hydrolysed polyvinyl acetate, in *Properties and Applications of Polyvinyl Alcohol*, S.C.I. Monograph No. 30, Finch, C. A., Ed., Society of Chemical Industry, London, 1968, 77.
16. **Tubbs, R. K., Inskip, H. K., and Subramanian, P. M.,** Relationships between structure and properties of polyvinyl alcohol and vinyl alcohol copolymers, in *Properties and Applications of Polyvinyl Alcohol*, S.C.I. Monograph No. 30, Finch, C. A., Ed., Society of Chemical Industry, London, 1968, 88.
17. **Tsunemitsu, K. and Shohata, H.,** Useful properties and industrial uses of polyvinyl alcohol as a water-soluble polymer, in *Properties and Applications of Polyvinyl Alcohol*, S.C.I. Monograph No. 30, Finch, C. A., Ed., Society of Chemical Industry, London, 1968, 104.
18. **Reynolds, G. E. J. and Gulbekian, E. V.,** The role of polyvinyl alcohol in emulsion polymerisation, in *Properties and Applications of Polyvinyl Alcohol*, S.C.I. Monograph No. 30, Finch, C. A., Ed., Society of Chemical Industry, London, 1968, 131.
19. **Toyoshima, K.,** Characteristics of the aqueous solutions and solid properties of polyvinyl alcohol and their applications, in *Properties and Applications of Polyvinyl Alcohol*, S.C.I. Monograph No. 30, Finch, C. A., Ed., Society of Chemical Industry, London, 1968, 154.
20. **Zwick, M. M., Duiser, J. A., and Van Bochove, C.,** Phase separation spinning of polyvinyl alcohol fibres, in *Properties and Applications of Polyvinyl Alcohol*, S.C.I. Monograph No. 30, Finch, C. A., Ed., Society of Chemical Industry, London, 1968, 188.
21. **Dunn, A. S., Coley, R. L., and Duncalf, B.,** Thermal decomposition of polyvinyl alcohol, in *Properties and Applications of Polyvinyl Alcohol*, S.C.I. Monograph No. 30, Finch, C. A., Ed., Society of Chemical Industry, London, 1968, 208.
22. **Finch, C. A.,** Introduction, in *Polyvinyl Alcohol: Properties and Applications*, Finch, C. A., Ed., John Wiley & Sons, London, 1973, xiii.
23. **Winkler, H.,** Historical development of polyvinyl alcohol, in *Polyvinyl Alcohol: Properties and Applications*, Finch, C. A., Ed., John Wiley & Sons, London, 1973, chap. 1.
24. **Noro, K.,** Manufacture of polyvinyl acetate for polyvinyl alcohol, in *Polyvinyl Alcohol: Properties and Applications*, Finch, C. A., Ed., John Wiley & Sons, London, 1973, chap. 3.
25. **Noro, K.,** Hydrolysis of polyvinyl acetate to polyvinyl alcohol, in *Polyvinyl Alcohol: Properties and Applications*, Finch, C. A., Ed., John Wiley & Sons, London, 1973, chap. 4.
26. **Noro, K.,** Manufacture and engineering aspects of the commercial production of polyvinyl alcohol, in *Polyvinyl Alcohol: Properties and Applications*, Finch, C. A., Ed., John Wiley & Sons, London, 1973, chap. 5.
27. **Noro, K.,** Preparation of polyvinyl alcohol from monomers other than vinyl acetate, in *Polyvinyl Alcohol: Properties and Applications*, Finch, C. A., Ed., John Wiley & Sons, London, 1973, chap. 6.
28. **Noro, K.,** Preparation of modified polyvinyl alcohol from copolymers, in *Polyvinyl Alcohol: Properties and Applications*, Finch, C. A., Ed., John Wiley & Sons, London, 1973, chap. 7.
29. **Tubbs, R. K. and Wu, T. K.,** Thermal properties of polyvinyl alcohol, in *Polyvinyl Alcohol: Properties and Applications*, Finch, C. A., Ed., John Wiley & Sons, London, 1973, chap. 8.
30. **Finch, C. A.,**Chemical properties of polyvinyl alcohol, in *Polyvinyl Alcohol: Properties and Applications*, Finch, C. A., Ed., John Wiley & Sons, London, 1973, chap. 9.
31. **Finch, C. A.,** Stereochemistry of polyvinyl alcohol, in *Polyvinyl Alcohol: Properties and Applications*, Finch, C. A., Ed., John Wiley & Sons, London, 1973, chap. 10.
32. **Tsunemitsu, K., Kishimoto, H., and Toyoshima, K.,** Use of polyvinyl alcohol in warp sizing and processing of textile fibres, in *Polyvinyl Alcohol: Properties and Applications*, Finch, C. A., Ed., John Wiley & Sons, London, 1973, chap. 11.
33. **Tsunemitsu, K., Murakami, Y., and Toyoshima, K.,** Use of polyvinyl alcohol in paper manufacture, in *Polyvinyl Alcohol: Properties and Applications*, Finch, C. A., Ed., John Wiley & Sons, London, 1973, chap. 12.
34. **Toyoshima, K.,** Reactions of polyvinyl alcohol with clay, in *Polyvinyl Alcohol: Properties and Applications*, Finch, C. A., Ed., John Wiley & Sons, London, 1973, chap. 13.
35. **Toyoshima, K.,** Properties of polyvinyl alcohol films, in *Polyvinyl Alcohol: Properties and Applications*, Finch, C. A., Ed., John Wiley & Sons, London, 1973, chap. 14.

36. **Toyoshima, K.,** Acetalization of polyvinyl alcohol, in *Polyvinyl Alcohol: Properties and Applications,* Finch, C. A., Ed., John Wiley & Sons, London, 1973, chap. 15.
37. **Toyoshima, K.,** Applications of polyvinyl alcohol in adhesives, in *Polyvinyl Alcohol: Properties and Applications,* Finch, C. A., Ed., John Wiley & Sons, London, 1973, chap. 16.
38. **Gulbekian, E. V. and Reynolds, G. E. J.,** Polyvinyl alcohol in emulsion polymerization, in *Polyvinyl Alcohol: Properties and Applications,* Finch, C. A., Ed., John Wiley & Sons, London, 1973, chap. 17.
39. **Duncalf, B. and Dunn, A. S.,** Photosensitized reactions of polyvinyl alcohol used in printing technology and other applications, in *Polyvinyl Alcohol: Properties and Applications,* Finch, C. A., Ed., John Wiley & Sons, London, 1973, chap. 18.
40. **Haas, H. C.,** Polyvinyl alcohol in optical films, in *Polyvinyl Alcohol: Properties and Applications,* Finch, C. A., Ed., John Wiley & Sons, London, 1973, chap. 19.
41. **Toyoshima, K.,** Moulded products from polyvinyl alcohol, in *Polyvinyl Alcohol: Properties and Applications,* Finch, C. A., Ed., John Wiley & Sons, London, 1973, chap. 20.
42. **Toyoshima, K.,** Miscellaneous applications of polyvinyl alcohol, in *Polyvinyl Alcohol: Properties and Applications,* Finch, C. A., Ed., John Wiley & Sons, London, 1973, chap. 21.
43. **Toyoshima, K.,** Compatibility of polyvinyl alcohol with other water-soluble high polymers, in *Polyvinyl Alcohol: Properties and Applications,* Finch, C. A., Ed., John Wiley & Sons, London, 1973, append. 1.
44. **Toyoshima, K.,** Preparation of polyvinyl alcohol solutions, in *Polyvinyl Alcohol: Properties and Applications,* Finch, C. A., Ed., John Wiley & Sons, London, 1973, append. 2.
45. **Finch, C. A.,** Analytical methods for polyvinyl alcohol, in *Polyvinyl Alcohol: Properties and Applications,* Finch, C. A., Ed., John Wiley & Sons, London, 1973, append. 3.
46. **Fujiwara, Y., Fujiwara, S., and Fujii, K.,** NMR Study of model compounds of vinyl polymers, *J. Polym. Sci. Part A-1,* 4, 257, 1966.
47. **Friedlander, H. N., Harris, H. E., and Pritchard, J. G.,** Structure-property relationships of poly(vinyl alcohol). I. Influence of polymerization solvents and temperature on the structure and properties of poly(vinyl alcohol) derived from poly(vinyl acetate), *J. Polym. Sci. Part A-1,* 4, 649, 1966.
48. **Harris, H. E., Kenney, J. F., Willcockson, G. W., Chiang, R., and Friedlander, H. N.,** Structure-property relationships of poly(vinyl alcohol). II. The influence of molecular regularity on the crystallization-dissolution temperature relationships of poly(vinyl alcohol), *J. Polym. Sci. Part A-1,* 4, 665, 1966.
49. **Kenney, J. F. and Willcockson, G. W.,** Structure-property relationships of poly(vinyl alcohol). III. Relationsips between stereoregularity, crystallinity, and water resistance in poly(vinyl alcohol), *J. Polym. Sci. Part A-1,* 4, 679, 1966.
50. **Kenney, J. F. and Holland, V. F.,** Crystallization and dissolution temperatures of poly(vinyl alcohol) crystal lamellae, *J. Polym. Sci. Part A-1,* 4, 699, 1966.
51. **Pritchard, J. G., Vollmer, R. L., Lawrence, W. C., and Black, W. B.,** Fluorine NMR spectra of poly(vinyl trifluoroacetate), *J. Polym. Sci. Part A-1,* 4, 707, 1966.
52. **Anon.,** *Gelvatol Polyvinyl Alcohol: Properties and Uses,* Tech. Bull. No. 6082D, Monsanto Polymers & Petrochemicals Co., St. Louis, 1976.
53. **Anon.,** *Elvanol Polyvinyl Alcohol* 71-30: *Properties and Uses,* Du Pont Co., Wilmington, Del., 1979.
54. **Anon.,** *Mowiol-Polyvinyl Alcohol,* Technical Brochure, Hoechst AG, Frankfurt, 1975.
55. **Lankveld, J. M. G. and Lyklema, J.,** Adsorption of polyvinyl alcohol on the paraffin-water interface. I. Interfacial tension as a function of time and concentration, *J. Colloid Interface Sci.,* 41, 454, 1972.
56. **Zwick, M. M. and van Bochove, C.,** Wet spinning of high-strength, low-elongation polyvinyl alcohol fibers, *Textile Res. J.,* 34, 417, 1964.
57. **Zwick, M. M.,** Poly(vinyl alcohol)-iodine complexes, *J. Appl. Polym. Sci.,* 9, 2393, 1965.
58. **Bunn, C. W.,** Crystal structure of polyvinyl alcohol, *Nature (London),* 161, 929, 1948.
59. **Pritchard, J. G.,** Properties in the solid state, in *Poly(Vinyl Alcohol): Basic Properties and Uses (Polymer Monographs,* Vol. 4), Gordon & Breach, New York, 1970, chap. 3.
60. **Bravar, M., Rolich, J., Ban, N., and Grijatovic, V.,** Studies of alcoholysis of poly(vinyl acetate) to poly(vinyl alcohol), *J. Polym. Sci. Polym. Symp.,* 47, 329, 1974.
61. **Melville, H. W. and Sewell, P. R.,** Intramolecular branching in polyvinyl acetate, *Makromol. Chem.,* 32, 139, 1959.
62. **Pritchard, J. G.,** Introduction, in *Poly(Vinyl Alcohol): Basic Properties and Uses (Polymer Monographs,* Vol. 4), Gordon & Breach, New York, 1970, chap. 1.
63. **Tubbs, R. K.,** Sequence distribution of partially hydrolyzed poly(vinyl acetate), *J. Polym. Sci. Part A-1,* 4, 623, 1966.
64. **Flory, P. J. and Leutner, F. S.,** Occurrence of head-to-head arrangements of structural units in polyvinyl alcohol, *J. Polym. Sci.,* 3, 880, 1948; 5, 267, 1950 (erratum).
65. **Harris, H. E. and Pritchard, J. G.,** Determination of 1,2-glycol units in polyvinyl alcohol, *J. Polym. Sci. Part A,* 2, 3673, 1964.
66. **Pritchard, J. G. and Lan Chun Fung, Y. L.,** Determination of vicinal hydroxyl groups in poly(vinyl alcohol) (PVA), *Talanta,* 23, 237, 1976.

67. **Beresniewicz, A.,** The relation between intrinsic viscosity and molecular weight for partially alcoholized polyvinyl acetates, *J. Polym. Sci.*, 39, 63, 1959.
68. **Dieu, H. A.,** Études des solutions d'alcool polyvinylique, *J. Polym. Sci.*, 12, 417, 1954.
69. **Cooper, W., Johnston, F. R., and Vaughan, G.,** Stereoregularity in polyvinyl alcohols and esters, *J. Polym. Sci. Part A*, 1, 1509, 1963.
70. **Nozakura, S., Sumi, M., Uoi, M., Okamoto, T., and Murahashi, S.,** Stereoregulation in the radical polymerization of vinyl esters, *J. Polym. Sci. Polym. Chem. Ed.*, 11, 279, 1973.
71. **Matsuzawa, S., Yamaura, K., Umeda, Y., Inoue, Y., Kitamura, H., and Pak, S.-Y.,** Mechanical denaturation of high polymers in solutions. 34. Nucleation and growth of poly(vinyl alcohol) crystals from flowing solution, *Makromol. Chem.*, 180, 2009, 1979.
72. **Ogasawara, K., Yuasa, K., and Matsuzawa, S.,** Gelation kinetics in aqueous solutions of *syndiotactic*-rich poly(vinyl alcohol), *Makromol. Chem.*, 177, 3403, 1976.
73. **Nozakura, S.-I. and Kida, S.,** Polymerization of vinyloxyaluminium and formation of syndiotactic poly(vinyl alcohol), *J. Polym. Sci. Polym. Chem. Ed.*, 12, 2337, 1974.
74. **Kawai, T.,** Freezing point depression of polymer solutions and gels, *J. Polym. Sci.*, 32, 425, 1958.
75. **Long, F. A. and Thompson, L. J.,** Diffusion of water vapor in polymers, *J. Polym. Sci.*, 15, 413, 1955.
76. **Tager, A. A., Anikeyeva, A. A., Adamova, L. V., Andreyeva, V. M., Kuz'mina, T. A., and Tsilipotkina, M. V.,** The effect of temperature on the water solubility of polyvinyl alcohol, *Vysokomol. Soedin.*, A13, 659, 1971; *Polym. Sci. (U.S.S.R.)*, 13, 751, 1971.
77. **Vink, H.,** Degradation of some polymers in aqueous solution, *Makromol. Chem.*, 67, 105, 1963.
78. **Sakurada, I., Nakajima, A., and Shibatani, K.,** Dilute solution properties of partly urethanized polyvinyl alcohol, *J. Polym. Sci. Part A*, 2, 3545, 1964.
79. **Al-Madfai, S. and Frisch, H. L.,** Surface tension of synthetic high polymer solutions. II, *J. Am. Chem. Soc.*, 80, 5613, 1958.
80. **Saito, S. and Mizuta, Y.,** Solubilization of polymers in aqueous cationic surfactant solution, *J. Colloid Interface Sci.*, 23, 604, 1967.
81. **Beresniewicz, A.,** Extraction fractionation of polyvinyl acetate and of polyvinyl alcohol, *J. Polym. Sci.*, 35, 321, 1959.
82. **Levi, D. W., Scherer, P. C., Hunter, W. L., and Washimi, K.,** Solvation of polyvinyl alcohol in solvent-nonsolvent mixtures, *J. Polym. Sci.*, 28, 481, 1958.
83. **Matsuo, T. and Inagaki, H.,** Über den Lösungzustand des Polyvinylalkohols in Wasser. 2. Mitt. Molekulargewichtsbestimmung durch Streulichtmessungen, *Makromol. Chem.*, 55, 150, 1962.
84. **Kolnibolotchuk, N. K., Klenin, V. J., and Frenkel, S. Ya.,** Formation of supermolecular order in aqueous solutions of poly(vinyl alcohol) in a turbulent flow, *J. Polym. Sci. Polym. Symp.*, 44, 119, 1974.
85. **Dialer, K., Vogler, K., and Patat, F.,** Zur Charakterisierung fraktionerter Polyvinylalkohole, *Helv. Chim. Acta*, 35, 869, 1952.
86. **Matsuo, T. and Inagaki, H.,** Über den Lösungzustand des Polyvinylalkohols in Wasser. I. Mitt.: Metastabiler Zustand der Lösung, *Makromol. Chem.*, 53, 130, 1962.
87. **Pritchard, J. G.,** Molecular structure of PVA, in *Poly(Vinyl Alcohol): Basic Properties and Uses (Polymer Monographs*, Vol. 4), Gordon & Breach, New York, 1970, chap. 2.
88. **Van den Boomgard, Th., King, T. A., Tadros, Th. F., Tang, H., and Vincent, B.,** The influence of temperature on the adsorption and adsorbed layer thickness of various molecular weight fractions of poly(vinyl alcohol) on polystyrene latex particles, *J. Colloid Interface Sci.*, 66, 68, 1978.
89. **Nord, F. F., Bier, M., and Timasheff, S. N.,** Investigations on proteins and polymers. IV. Critical phenomena in polyvinyl alcohol-acetate copolymer solutions, *J. Am. Chem. Soc.*, 73, 289, 1951.
90. **Garvey, M. J., Tadros, Th. F., and Vincent, B.,** A comparison of the volume occupied by macromolecules in the adsorbed state and in bulk solution. Adsorption of narrow molecular weight fractions of poly(vinyl alcohol) at the polystyrene/water interface, *J. Colloid Interface Sci.*, 49, 57, 1974.
91. **Fuchs, O.,** Die Fraktionierung von Hochpolymeren nach dem Löseverfahren, *Makromol. Chem.*, 7, 259, 1951.
92. **Matsumoto, M. and Imai, K.,** Viscosity of dilute solutions of polyvinyl alcohol, *J. Polym. Sci.*, 24, 125, 1957.
93. **Bovey, F. A. and Tiers, G. V. D.,** Polymer NSR spectroscopy. II. The high resolution spectra of methyl methacrylate polymers prepared from free radical and anionic initiators, *J. Polym. Sci.*, 44, 173, 1960.
94. **Bovey, F. A.,** The high resolution NMR spectroscopy of polymers, in *Progress in Polymer Science*, Vol. 3, Jenkins, A. D., Ed., Pergamon Press, Oxford, 1971, chap. 1.
95. **Sakurada, I., Nukushina, Y., and Sone, Y.,** Relation between crystallinity and swelling of polyvinyl alcohol, *Ric. Sci. Suppl. Simp. Int. Chim. Macromol. Milan/Turin (1954)*, 25, 715, 1955.
96. **Packter, A. and Nerurkar, M. S.,** Crystallization in films of polar vinyl polymers. I. Crystallinity of polyvinyl alcohol films prepared by evaporation of aqueous solutions, *Eur. Polym. J.*, 4, 685, 1968.
97. **Hauser, P. M. and McLaren, A. D.,** Permeation through and sorption of water vapor by high polymers, *Ind. Eng. Chem.*, 40, 112, 1948.

98. **Nakajima, A. and Furatate, K.**, Determination of molecular weight of polyvinylalcohol by osmotic method, *Kobunshi Kagaku*, 6, 460, 1949; *Chem. Abstr.*, 46, 1294b, 1952.
99. **Sakurada, I., Nukushina, Y., and Sone, Y.**, Swelling of poly(vinyl alcohol). I. Relation between crystallinity and swelling of poly(vinyl alcohol) with various degrees of polymerization, *Kobunshi Kagaku*, 12, 506, 1955; *Chem. Abstr.*, 51, 3174i, 1957.
100. **Sone, Y., Hirabayashi, K., and Sakurada, I.**, Specific gravity of polyvinyl alcohol and its derivatives, *Kobunshi Kagaku*, 10, 277, 1953; *Chem. Abstr.*, 49, 5075, 1955.
101. **Thompson Hughes, L. J. and Fordyce, D. B.**, Sorption of water vapor by water-soluble polymers: kinetic, equilibrium, and glass transition data, *J. Polym. Sci.*, 22, 509, 1956.
101a. **Nyquist, R. A.**, *Infrared Spectra of Plastics (Polymers) and Resins*, 2nd ed., The Dow Chemical Co., Midland, Mich., 1961.
102. **Tadokoro, H., Seki, S., and Nitta, I.**, Infrared absorption spectrum of deuterated polyvinyl alcohol film, *J. Chem. Phys.*, 23, 1351, 1955.
103. **Tadokoro, H., Seki, S., and Nitta, I.**, The crystallinity of solid high polymers. I. The crystallinity of polyvinyl alcohol film, *Bull. Chem. Soc. Jpn.*, 28, 559, 1955.
104. **Peppas, N. A. and Merrill, E. W.**, Development of semicrystalline poly(vinyl alcohol) hydrogels for biomedical applications, *J. Biomed. Mater. Res.*, 11, 423, 1977.
105. **Peppas, N. A.**, Turbidimetric studies of aqueous poly(vinyl alcohol) solutions, *Makromol. Chem.*, 176, 3433, 1975.
106. **Lloyd, D. G.**, Ultraviolet spectroscopic studies on aqueous polyvinyl alcohol solutions. I. Characterization of chromophores, *J. Appl. Polym. Sci.*, 1, 70, 1959.
107. **Colvin, B. G.**, Crystal structure of polyvinyl alcohol, *Nature (London)*, 248, 756, 1974.
108. **Sakurada, I., Nukushina, Y., and Mori, N.**, Estimation of crystallinity of poly(vinyl alcohol) fibers by an X-ray spectrograph with a Geiger counter. I. Preparation of specimens, which have different crystallinity but the same orientation of crystalline regions, *Kobunshi Kagaku*, 12, 302, 1955; *Chem. Abstr.*, 51, 756a, 1957.
109. **Sakurada, I., Nukushina, Y., and Mori, N.**, Estimation of crystallinity of poly(vinyl alcohol) fibers by an X-ray spectrograph with a Geiger counter. II. Calculation of crystallinity from intensity curves, *Kobunshi Kagaku*, 12, 307, 1955; *Chem. Abstr.*, 51, 756c, 1957.
110. **Sakurada, I., Nukushina, Y., and Mori, N.**, Estimation of crystallinity of poly(vinyl alcohol) fibers by an X-ray spectrograph with a Geiger counter. III. Effect of formalization on the crystallinity of poly(vinyl alcohol) fibers, *Kobunshi Kagaku*, 12, 311, 1955; *Chem. Abstr.*, 51, 756d, 1957.
111. **Sakurada, I. and Nukushina, Y.**, X-ray estimation of crystallinity of poly(vinyl alcohol) films with a counter, *Kobunshi Kagaku*, 12, 483, 1955; *Chem. Abstr.*, 51, 3175g, 1957.
112. **Sakurada, I., Nukushina, Y., and Sone, Y.**, Swelling of poly(vinyl alcohol). II. Behavior of the crystalline region during swelling, *Kobunshi Kagaku*, 12, 510, 1955; *Chem. Abstr.*, 51, 3175b, 1957.
113. **Sakurada, I., Nukushina, Y., and Sone, Y.**, Swelling of poly(vinyl alcohol). III. Change of crystallinity during drying after swelling, *Kobunshi Kagaku*, 12, 514, 1955; *Chem. Abstr.*, 51, 3175b, 1957.
114. **Sakurada, I., Nukushina, Y., and Sone, Y.**, Swelling of poly(vinyl alcohol). IV. Influence of the conditions of film preparation and heat-treatment on the swelling and crystallinity of poly(vinyl alcohol) films, *Kobunshi Kagaku*, 12, 517, 1955; *Chem. Abstr.*, 51, 3175c, 1957.
115. **Sakurada, I., Nakajima, A., and Fujiwara, H.**, Vapor pressures of polymer solutions. II. Vapor pressure of the poly(vinyl alcohol)-water system, *J. Polym. Sci.*, 35, 497, 1959.
116. **Valentine, L.**, Studies on the sorption of moisture by polymers. I. Effect of crystallinity, *J. Polym. Sci.*, 27, 313, 1958.
117. **Staudinger, H. and Warth, H.**, Über die Konstitution von hochpolymeren Kunststoffen, *J. Prakt. Chem.*, 155, 261, 1940.
118. **Timasheff, S. N., Bier, M., and Nord, F. F.**, Aggregation phenomena in polyvinyl alcohol-acetate solutions, *Proc. Natl. Acad. Sci. U.S.A.*, 35, 364, 1949.
119. **Hosono, M. and Sakurada, I.**, Diffusion in high polymers. III. Diffusion experiment with polyvinyl alcohol, *Kobunshi Kagaku*, 9, 64, 1952; *Chem. Abstr.*, 48, 25a, 1954.
120. **Kuroiwa, T.**, The viscosity of dilute aqueous solutions of polyvinyl alcohol, *Kobunshi Kagaku*, 9, 253, 1952; *Chem. Abstr.*, 48, 3107i, 1954.
121. **Tager, A. A. and Kargin, V. A.**, Thermodynamic studies of systems of polymers and hydrogenated monomers, *Kolloidn. Zh.*, 14, 367, 1952; *Chem. Abstr.*, 47, 1465b, 1953.
122. **Naito, R. and Kominami, T.**, The state of poly(vinyl alcohol) solution. I. The viscosity behavior of poly(vinyl alcohol) in mixed solvent, *Kobunshi Kagaku*, 11, 444, 1954; *Chem. Abstr.*, 50, 6145d, 1956.
123. **Oya, S.**, Heat of solution of synthetic high polymers. III. Heat of solution of partially acetylated poly(vinyl alcohol), *Kobunshi Kagaku*, 12, 122, 1955; *Chem. Abstr.*, 51, 1719f, 1957.
124. **Oya, S.**, Heats of solution of synthetic polymers. IV. Heats of swelling of heat-treated poly(vinyl alcohol), *Kobunshi Kagaku*, 12, 410, 1955; *Chem. Abstr.*, 51, 63b, 1957.

125. **Sakurada, I., Nakajima, A., and Takita, H.,** Thermodynamic studies of partially acetylated poly(vinyl alcohol). I. Osmotic pressure and viscosity of partially acetylated poly(vinyl alcohol) in aqueous solution, *Kobunshi Kagaku,* 12, 15, 1955; *Chem. Abstr.*, 51, 1694c, 1957.
126. **Stacey, K. A. and Alexander, P.,** Aggregation and hydrogen-bonding in solutions of polyvinyl alcohol, *Ric. Sci. Suppl., Simp Int. Chim. Macromol. Milan/Turin (1954),* 25, 889, 1955.
127. **Amaya, K. and Fujishiro, R.,** Heats of dilution of polyvinylalcohol solution. I, *Bull. Chem. Soc. Jpn.*, 29, 361, 1956.
128. **Amaya, K. and Fujishiro, R.,** Heats of dilution of polyvinylalcohol solutions. II, *Bull. Chem. Soc. Jpn.*, 29, 830, 1956.
129. **Meerson, S. I. and Lipatov, S. M.,** Dependence of the heat of solution of polymers on their physical state, *Kolloidn. Zh.*, 18, 447, 1956; *Colloid J. (U.S.S.R.),* 43, 437, 1956.
130. **Shiio, H., Ogawa, T., and Yoshihashi, H.,** Measurement of the amount of bound water by ultrasonic interferometer, *J. Am. Chem. Soc.*, 77, 4980, 1955.
131. **Shiio, H. and Yoshihashi, H.,** Measurement of the amount of bound water by ultrasonic interferometer. II. Polyvinyl alcohol and its partially substituted acetates, *J. Phys. Chem.*, 60, 1049, 1956.
132. **Freund, L. and Daune, M.,** Détermination de la fonction de distribution d'un polymère en solution. Étude et comparaison de deux méthodes, *J. Polym. Sci.*, 29, 161, 1958.
133. **Martin, W. G., Cook. W. H., and Winkler, C. A.,** The determination of partial specific volumes by differential sedimentation, *Can. J. Chem.*, 34, 809, 1956.
134. **Matsumoto, M. and Ohyanagi, Y.,** Refractive index of dilute aqueous solution of polyvinyl alcohol, *J. Polym. Sci.*, 31, 225, 1958.
135. **Levi, D. W., Scherer, P. C., Washimi, K., and Hunter, W. L.,** Viscosity of polyvinyl alcohol, *J. Appl. Polym. Sci.*, 1, 127, 1959.
136. **Matsumoto, M. and Ohyanagi, Y.,** Viscosity-molecular weight relationships of poly-(vinyl acetate) and poly-(vinyl alcohol). V. Viscosity-molecular weight relationship of poly-(vinyl alcohol), *Kobunshi Kagaku,* 17, 191, 1960; *Chem. Abstr.*, 55, 19319a, 1961.
137. **Nakanishi, K and Kurata, M.,** Density measurement in dilute aqueous solution of polyvinyl alcohol, *Bull. Chem. Soc. Jpn.*, 33, 152, 1960.
138. **Elias, H.-G.,** Konstitution und Lösungseigenschaften von Makromolekülen. III. Zur Bestimmung des Molekulargewichtes aus Staudinger-Index und Konstitution, *Makromol. Chem.*, 54, 78, 1962.
139. **Andersson, G. R.,** A study of the pressure dependence of the partial specific volume of macromolecules in solution by compression measurements in the range 1—8000 atm, *Arkiv Kemi,* 20, 513, 1963.
140. **Porter, R. S. and Johnson, J. F.,** The entanglement concept in polymer systems, *Chem. Rev.*, 66, 1, 1966.
141. **Pritchard, J. G.,** Properties in association with water, in *Poly(Vinyl Alcohol): Basic Properties and Uses (Polymer Monographs,* Vol. 4), Gordon & Breach, New York, 1970, chap. 4.
142. **Gruber, E., Soehendra, B., and Schurz, J.,** Ordering in solutions of poly(vinyl alcohol) in water as detected by light scattering, *J. Polym. Sci. Polym. Symp.*, 44, 105, 1974.
143. **Klenin, V. J., Klenina, O. V., Shvartsburd, B. I., and Frenkel, S. Ya.,** Thermodynamic nature of supermolecular order in the aqueous solutions of poly(vinyl alcohol), *J. Polym. Sci. Polym. Symp.*, 44, 131, 1974.
144. **Yamaura, K., Karaki, T., and Matsuzawa, S.,** Mechanical denaturation of high polymers in solutions. 32. Effect of flow pattern on shearing stress-induced crystallization of poly-(vinyl alcohol) from its aqueous solution, *Makromol. Chem.*, 175, 247, 1974.
145. **Inzelt, Gy. and Gróf, P.,** Investigation of hydration of macromolecules. I. Measurement of self-diffusion of water in solutions of polyvinyl alcohol, *Acta Chim. Acad. Sci. Hung.*, 88, 13, 1976.
146. **Masszi, G., Inzelt, G., and Gróf, P.,** Investigation of hydration of macromolecules. II. Studies on poly-(vinyl-alcohol) solutions by microwave dielectric method, *Acta Biochem. Biophys. Acad. Sci. Hung.*, 11, 45, 1976.
147. **Yamaura, K., Nakada, K., and Matsuzawa, S.,** Mechanical denaturation of high polymers in solutions. 33. Effect of antifoaming reagent on flow-induced crystallization of poly(vinyl alcohol) from its aqueous solution, *Makromol. Chem.*, 177, 927, 1976.
148. **Joshi, D. P., Lan-Chun-Fung, Y. L., and Pritchard, J. G.,** Determination of poly(vinyl alcohol) via its complex with boric acid and iodine, *Anal. Chim. Acta,* 104, 153, 1979.
149. **Naito, R.,** Studies on the state of PVA solution. IV. Viscosity measurements on polyvinyl alcohol solution in dimethylsulfoxide, *Kobunshi Kagaku,* 15, 597, 1958; *Chem. Abstr.*, 54, 16113i, 1960.

149a. **Naito, R.,** Studies on the state of PVA solution. V. Viscosity measurements on partially acetylated polyl(vinyl alcohol), polyl(vinyl acetate), and partially saponified poly(vinyl acetate) in dimethylsulfoxide, *Kobunshi Kagaku,* 15, 605, 1958; *Chem. Abstr.*, 54, 16114b, 1960.

150. **McLaren, A. D. and Rowan, J. W.,** Sorption of water vapor by proteins and polymers: a review, *J. Polym. Sci.*, 7, 289, 1951.

151. **Dole, M. and McLaren, A. D.**, The free energy, heat and entropy of sorption of water vapour by proteins and high polymers, *J. Am. Chem. Soc.*, 69, 651, 1947.
152. **Myers, A. W., Meyer, J. A., Rogers, C. E., Stannett, V., and Szwarc, M.**, Studies in gas and vapor permeability of plastic films and coated papers. VI. The permeation of water vapor, *Tappi*, 44, 58, 1961.
153. **Yamamura, H. and Hamano, K.**, Adsorption of water vapour by polyvinyl alcohol-acetate copolymers, *J. Sci. Hiroshima Univ. Ser. A*, 18, 113, 1954.
154. **Tadokoro, H., Seki, S., and Nitta, I.**, The densities of polyvinyl alcohol film at moist states, *Bull. Chem. Soc. Jpn.*, 27, 451, 1954; *Chem. Abstr.*, 49, 7930f, 1955.
154a. **Spencer, H. G., Beyerlein, A., and Honeycutt, S. C.**, Structure changes in glassy hydrophilic polymers under the stress of water sorption, *J. Polym. Sci. Polym. Symp.*, 44, 141, 1974.
155. **Saito, S.**, Solubilisation von Polymeren durch Nicht-Ionische Tenside, *Kolloid-Z. Z. Polym.*, 226, 10, 1968.
156. **Sakurada, I., Sakaguchi, Y., and Ito, Y.**, Phase equilibrium of partially acetylated polyvinyl alcohols, *Kobunshi Kagaku*, 14, 41, 1957; *Chem. Abstr.*, 52, 1676, 1958.
157. **Saito, S. and Otsuka, T.**, Dissolution of some polymers in aqueous solutions of urea, of its related compounds and of tetraalkylammonium salts, *J. Colloid Interface Sci.*, 25, 531, 1967.
158. **Saito, S. and Yukawa, M.**, Interactions of polymers and cationic surfactants with thiocyanate as counterions, *J. Colloid Interface Sci.*, 30, 211, 1969.
159. **Saito, S.**, Salt effect on polymer solutions, *J. Polym. Sci. Part A-1*, 7, 1789, 1969.
160. **Saito, S.**, Precipitation of some nonionic polymers by octylammonium thiocyante, *J. Polym. Sci. Part A-1*, 8, 263, 1970.
161. **Napper, D. H.**, A reversible transformation from enthalpic to entropic stabilization, *Kolloid-Z. Z. Polym.*, 234, 1149, 1969.
162. **Lambe, R., Tadros, Th. F., and Vincent, B.**, The effect of temperature, particle volume fraction, and polymer concentration on the stability of aqueous polystyrene latex dispersions in the presence of poly(vinyl alcohol), *J. Colloid Interface Sci.*, 66, 77, 1978.
162a. **Dobry, A.**, Partage des substances macromoléculaires entre deux solvants, *Makromol. Chem.*, 18/19, 317, 1956.
163. **Pritchard, J. G.**, Chemical properties, in *Poly(Vinyl Alcohol): Basic Properties and Uses (Polymer Monographs*, Vol. 4), Gordon & Breach, New York, 1970, chap. 5.
164. **Sakurada, I. and Ikada, Y.**, Effects of gamma radiation on polymer in solution. VI. Radiation protection and promotion in aqueous solutions of poly(vinyl alcohol), *Bull. Inst. Chem. Res. Kyoto Univ.*, 40, 123, 1963.
164a. **Alexander, P. and Charlesby, A.**, Effect of X-rays and γ-rays on synthetic polymers in aqueous solution, *J. Polym. Sci.*, 23, 355, 1957.
165. **Beal, K. F. and Thor, C. J. B.**, Water-soluble alkoxyacetals of polyvinyl alcohol, *J. Polym. Sci.*, 1, 540, 1946.
166. **Dole, M. and Faller, I. L.**, Water sorption by synthetic high polymers, *J. Am. Chem. Soc.*, 72, 414, 1950.
167. **Molyneux, P.**, Synthetic polymers, in *Water: A Comprehensive Treatise*, Vol. 4, Franks, F., Ed., Plenum Press, New York, 1975, chap. 7.
168. **Morita, M. and Watanabe, Y.**, A secondary alcohol oxidase: a component of a polyvinyl alcohol degrading enzyme preparation, *Agric. Biol. Chem.*, 41, 1535, 1977.
169. **Hosoya, H., Miyazaki, N., Sugisaki, Y., Takanashi, E., Tsurufuji, M., Yamasaki, M., and Tamura, G.**, Bacterial degradation of synthetic polymers and oligomers with the special reference to the case of polyethylene glycol, *Agric. Biol. Chem.*, 42, 1545, 1978.
170. **Suzuki, T.**, Oxidation of secondary alcohols by polyvinyl alcohol-degrading enzyme produced by *Pseudomonas* 0-3, *Agric. Biol. Chem.*, 42, 1187, 1978.
171. **Urbański, J., Czerwiński, W., Janicka, K., Majewska, F., and Zowall, H.**, *Handbook of Analysis of Synthetic Polymers and Plastics*, Ellis Horwood, Chichester, England, 1977.
172. **Zwick, M. M.**, The blue complexes of iodine with poly(vinyl alcohol) and amylose, *J. Polym. Sci. Part A-1*, 4, 1642, 1966.
173. **Hayashi, S., Tanabe, Y., and Hojo, N.**, The color reaction between partially saponified poly(vinyl acetate) and iodine-iodide in the presence of boric acid, *Makromol. Chem.*, 178, 1679, 1977.
174. **Finley, J. H.**, Spectrophotometric determination of polyvinyl alcohol in paper coatings, *Anal. Chem.*, 33, 1925, 1961.
175. **Chene, M., Martin-Borret, O., and Clery, M.**, Dosage spectrophotométrique de l'alcool polyvinylique, *Paperterie*, 88, 273, 1966.
176. **Horacek, J.**, Kolorimetrické stanovení polyvinylalkoholu, *Chem. Prum.*, 12, 385, 1962; *Chem. Abstr.*, 58, 10305f, 1963.
177. **Fleer, G. J., Koopal, L. K., and Lyklema, J.**, Polymer adsorption and its effect on the stability of hydrophobic colloids. I. Characterization of polyvinyl alcohol adsorption on silver iodide, *Kolloid Z. Z. Polym.*, 250, 689, 1972.

178. **Hayashi, S., Takama, M., Kaneko, I., and Hojo, N.,** The formation of a coloured complex of partially saponified poly(vinyl acetate) with iodine in the presence of potassium iodide, *Makromol. Chem.*, 176, 3221, 1975.
179. **Hayashi, S., Takama, M., and Hojo, N.,** The color reaction between partially saponified poly(vinyl acetate) and iodine-iodide in the presence of amylose, *Makromol. Chem.*, 177, 859, 1976.
180. **Hayashi, S., Nakano, C., and Motoyama, T.,** The reaction of iodine with partially saponified poly(vinyl acetate), *Kobunshi Kagaku*, 20, 303, 1963; *Chem Abstr.*, 61, 5802b, 1964.
181. **Hayashi, S., Kaneko, I., and Hojo, N.,** Reaction of iodine into poly(vinyl acetate) particles suspended in aqueous solution in the presence of potassium iodide, *Polym. J.*, 6, 33, 1974.
182. **Anon.,** *N*-Vinyl-2-pyrrolidone and polyvinyl pyrrolidone, in *IARC Monographs on the Evaluation of the Carcinogenic Risk of Chemicals to Humans*, Vol. 19, World Health Organization/International Agency for Research on Cancer, Lyon, France, 1979, 461.
183. **Fikentscher, H. and Herrle, K.,** Polyvinylpyrrolidone, *Mod. Plast.*, 23, 157, 1945.
184. **Anon.,** Polyvinylpyrrolidone and its precursors, *Ind. Chem.*, 29, 122, 1953.
185. **Reppe, W.,** *Polyvinylpyrrolidon, Monographien zu Angewandte Chemie und Chemie-Ingenieur-Technik*, No. 66, Verlag Chemie, Weinheim, Germany, 1954.
186. **Greenfield, I.,** The manufacture and uses of polyvinylpyrrolidone, *Ind. Chem.*, 32, 11, 1956.
187. **Chiddix, M. E. and Azorlosa, J. L.,** Polyvinylpyrrolidone, in *Kirk-Othmer Encyclopedia of Chemical Technology*, 2nd Suppl. Vol., Standen, A., Ed., Interscience, New York, 1960, 673.
188. **Wood, A. S.,** Vinyl polymers (pyrrolidone), in *Encyclopedia of Chemical Technology*, Vol. 21, 2nd ed., Kirk, R. E. and Othmer, D. F., Eds., John Wiley & Sons, New York, 1970, 427.
189. **Lorenz, D. H.,** *N*-Vinyl amide polymers, in *Encyclopedia of Polymer Science and Technology: Plastics, Resins, Rubbers, Fibers*, Vol. 14, Mark, H. M., Gaylord, N. G., and Bikales, N. M., Ed., Interscience, New York, 1971, 239.
190. **Tomalia, D. A.,** N-Vinylpyrrolidone, in *Functional Monomers: Their Preparation, Polymerization, and Application*, Vol. 2, Yocum, R. H. and Nyquist, E. B., Eds., Marcel Dekker, New York, 1974, 132.
191. **Azorlosa, J. L. and Martinelli, A. J.,** Polyvinylpyrrolidone, in *Water-Soluble Resins*, 2nd ed., Davidson, R. L. and Sittig, M., Eds., Van Nostrand Reinhold, New York, 1968, chap. 7.
192. **Anon.,** *PVP: Polyvinylpyrrolidone*, Tech. Bull. FO-13-2 (1961); 7543-113 (USA) and 7583-033 (International) (1964); 2550-006 (undated); 9-642-070 (undated), GAF Corp., New York.
193. **Anon.,** *Polyvinylpyrrolidone (PVP); Kollidon; Luviskol*, Tech. Bull., BASF, Ludwigshafen, Germany, 1979.
194. **Anon.,** *PVP: An Annotated Bibliography, 1951-1966*, Vol. 1, *Chemical, Physical, and Physiological Properties*, GAF Corp., New York, 1967.
195. **Anon.,** *PVP: An Annotated Bibliography, 1951-1966*, Vol. 2, *Food, Drug, and Cosmetic Uses*, GAF Corp., New York, 1967.
196. **Anon.,** *PVP: An Annotated Bibliography, 1951-1966*, Vol. 3, *Industrial Uses*, GAF Corp., New York, 1967.
197. **Anon.,** *Polyvinylpyrrolidone: Its Physiological Properties and Applications in Medicine*, Literature Survey LS-69-3, GAF Corp., New York, 1969.
198. **Anon.,** *Polyvinylpyrrolidone: Its Physiological Properties and Applications in Medicine and Biology*, Linden Literature Survey LS-70-1, GAF Corp., New York, 1970; *Update*, Literature Survey LLS 71-2, 1971; *Update II*, Literature Survey LLS-71-7, 1971; *Update III*, Literature Survey LLS-72-1, 1972; *Update IV*, Literature Survey WLS-72-12, 1972; *Update V*, Literature Survey WLS-73-29, 1973; *Update VI*, 1974.
199. **Jirgensons, B.,** Vikosität, Fällbarkeit und Schutzwirkung der Polyvinylpyrrolidone, *Makromol. Chem.*, 6, 30, 1951.
200. **Jirgensons, B.,** Solubility and fractionation of polyvinylpyrrolidone, *J. Polym. Sci.*, 8, 519, 1952.
201. **Klotz, I. M.,** Role of water structure in macromolecules, *Fed. Proc., Fed. Am. Soc. Exp. Biol.*, 24 (Suppl. 15), S-24, 1965.
202. **Molyneux, P.,** Studies on polyvinylpyrrolidone in relation to protein interactions in solution, in *The Chemistry and Rheology of Water-Soluble Gums and Colloids*, Stainsby, G., Ed., S.C.I. Monograph No. 24, Society of Chemical Industry, London, 1966, 91.
203. **Hecht, G. and Weese, H.,** Periston, ein neuer Blutflüssigkeitsersatz, *Muench. Med. Wochenschr.*, 90, 11, 1943; *Chem. Abstr.*, 38, 1607, 1944.
204. **Thrower, W. R. and Campbell, H.,** Plasmosan: a synthetic substitute for plasma, *Lancet*, 260, 1096, 1951.
205. **Ravin, H. A., Seligman, A. M., and Fine, J.,** Polyvinyl pyrrolidone as a plasma expander: studies on its excretion, distribution and metabolism, *N. Engl. J. Med.*, 247, 921, 1952.
206. **Reynolds, J. E. F. and Prasad, A. B., Eds.,** *Martindale: The Extra Pharmacopoeia*, 28th ed., The Pharmaceutical Press, London, 1982.
207. **Dyke, M. D., Sears, P. G., and Popov, A. I.,** Conductances of some uni-univalent electrolytes in N-methyl-2-pyrrolidone at 25°, *J. Phys. Chem.*, 71, 4140, 1967.

208. **Rosenfarb, J. and Baker, R. J.**, Investigation of solute-solvent interaction in *N*-methyl-2-pyrrolidone using proton magnetic resonance, *J. Phys. Chem.*, 82, 329, 1978.
209. **Lee, C. M. and Kumler, W. D.**, The dipole moment and structure of five- and six-membered lactams, *J. Am. Chem. Soc.*, 83, 4593, 1961.
210. **Fischer, E.**, The electric moment of 1-methyl-pyrrolid-2-one, *J. Chem. Soc.*, 1382, 1955.
211. **Groth, P.**, Crystal structure of N,N'-dimethyldiketopiperazine, *Acta Chem. Scand.*, 23, 3155, 1969.
212. **Takayama, K., Nambu, N., and Nagai, T.**, Interaction of N-methyl-2-pyrrolidone with aminobenzoic acids in solution and in solid state, *Chem. Pharm. Bull.*, 25, 879, 1977.
213. **Randall, D. I., Smolin, E. M., and Copes, J. P.**, New complexes of phenols and lactams, *Nature (London)*, 244, 369, 1973.
214. **Hosono, T., Tsuchiya, S., and Matsumaru, H.**, Molecular complex formation between N-vinyl-2-pyrrolidone and ajmaline, *Chem. Pharm. Bull.*, 27, 58, 1979.
215. **Murakami, S., Tanaka, R., and Fujishiro, R.**, Thermodynamics of aqueous solutions of nonelectrolytes. I. Enthalpies of transfer of 1-methyl-2-pyrrolidinone from water to aqueous methanol, *J. Solution Chem.*, 3, 71, 1974.
216. **Kimura, F., Murakami, S., and Fujishiro, R.**, Thermodynamics of aqueous solutions of nonelectrolytes. II. Enthalpies of transfer of 1-methyl-2-pyrrolidone from water to many aqueous alcohols, *J. Solution Chem.*, 4, 241, 1975.
217. **Frank, H. P.**, The lactam-amino acid equilibria for ethylpyrrolidone and polyvinylpyrrolidone, *J. Polym. Sci.*, 12, 565, 1954.
218. **Oster, G. and Immergut, E. H.**, Ultraviolet and infrared spectral studies of polyvinylpyrrolidone, *J. Am. Chem. Soc.*, 76, 1393, 1954.
219. **Saidel, L. J.**, Ultraviolet absorption spectra of peptides. III. N,N-Dialkylamides including polyvinylpyrrolidone, *J. Am. Chem. Soc.*, 77, 3892, 1955.
220. **Rothschild, W. G.**, Binding of hydrogen donors by peptide groups of lactams. Identity of the interaction sites, *J. Am. Chem. Soc.*, 94, 8676, 1972.
221. **Kaatze, U.**, Dielectric relaxation in aqueous solutions of polyvinylpyrrolidone, *Adv. Mol. Relaxation Process.*, 7, 71, 1975.
222. **Breitenbach, J. W. and Schmidt, A.**, Quellungsgleichgewichte vernetzter Polyvinylpyrrolidone, *Monatsh. Chem.*, 85, 52, 1954.
223. **Frank, H. P.**, Studies on the binary system N-vinylpyrrolidone-water, *J. Polym. Sci.*, 13, 187, 1954.
224. **Eley, D. D., Hey, M. J., and Winteringham, B. L.**, Nuclear magnetic relaxation in aqueous solutions of vinylpyrrolidone and polyvinylpyrrolidone, *J. Solution Chem.*, 5, 787, 1976.
225. **Conix, A., Smets, G., and Moens, J.**, Alkaline hydrolysis of amide group containing polymers, *Ric. Sci. Suppl. Simposio Int. Chim. Macromol. Milan/Turin (1954)*, 25, 200, 1955; *Chem. Abstr.*, 54, 11545d, 1960.
226. **Conix, A. and Smets, G.**, Ring opening in lactam polymers, *J. Polym. Sci.*, 15, 221, 1955.
227. **Crescenzi, V., Cesàro, A., and Russo, E.**, On some physicochemical properties of diketopiperazines in water, *Int. J. Peptide Protein Res.*, 5, 427, 1973.
228. **Poole, J. W. and Higuchi, T.**, Complexes formed in aqueous solutions by sarcosine anhydride; interactions with organic acids, phenols, and aromatic alcohols, *J. Am. Pharm. Assoc. Sci. Ed.*, 48, 592, 1959.
229. **Breitenbach, J. W. and Schmidt, A.**, Die Polymerisation des N-Vinylpyrrolidons, *Monatsh. Chem.*, 83, 833, 1952.
230. **Breitenbach, J. W.**, Polymerization and polymers of N-vinylpyrrolidone, *J. Polym. Sci.*, 23, 949, 1957.
231. **Braun, D. and Quella, F.**, Bestimmung der Geschwindigkeitskonstanten und der Effektivitäten beim Zerfall von 2,2'-Azoisobutyronitril in Styrol, N-Vinyl-2-pyrrolidon und Methylmethacrylat, *Makromol. Chem.*, 179, 387, 1978.
232. **Breitenbach, J. W. and Schmidt, A.**, Die Polymerisation des N-Vinylpyrrolidons in Lösung, *Monatsh. Chem.*, 83, 1288, 1952.
233. **Kwei, K.-P. S.**, The influence of pH on the polymerization of VP, *J. Polym. Sci. Part B (Polym. Lett.)*, 1, 379, 1963.
234. **Gordon, R. S.**, The preparation of radioactive polyvinylpyrrolidone for medical use, *J. Polym. Sci.*, 31, 191, 1958.
235. **Elias, H.-G.**, *Macromolecules*, Vol. 2, Plenum Press, New York, 1977.
236. **Sébille, B. and Néel, J.**, Interactions entre la polyvinylpyrrolidone et les composés acides: quelques considérations sur la coacervation, *J. Chim. Phys.*, 60, 475, 1963.
237. **Shizuka, H., Ogiwara, T., and Morita, T.**, Mechanism of the photo-chemical ring expansion of *N*-vinyl-2-pyrrolidone. Estimation of the recombination-time of biradical, *Bull. Chem. Soc. Jpn.*, 50, 2067, 1977.
238. **Bork, J. F. and Coleman, L. E.**, Nitrogen-containing monomers. II. Reactivity ratios of N-vinyloxazolidone and N-vinylpyrrolidone with vinyl monomers, *J. Polym. Sci.*, 43, 413, 1960.
239. **Breitenbach, J. W.**, Makromolekulare, grenzflächenaktive Stoffe, *Z. Elektrochem.*, 59, 309, 1955.

240. **Breitenbach, J. W. and Edelhauser, H.,** Copolymere des N-Vinylpyrrolidons und ihre Grenzflächenaktivität, *Ric. Sci. Suppl. Simposio Int. Chim. Macromol. Milan/Turin (1954)*, 25, 242, 1955; *Chem. Abstr.*, 54, 11554h, 1960.
241. **Manecke, G. and Korenzecker, R.,** Reaktive Copolymer des N-vinyl-2-pyrrolidons zur Immobilisierung von Enzymen, *Makromol. Chem.*, 178, 1729, 1977.
242. **Graham, W. D.,** The weight average molecular weight of polyvinylpyrrolidone preparations as determined by light scattering, *J. Pharm. Pharmacol.*, 9, 230, 1957.
243. **Miyagi, Z. and Tanaka, K.,** Sorption of organic vapours by polyvinyl pyrrolidone, *Kolloid-Z. Z. Polym.*, 251, 739, 1973.
244. **Fishman, M. L. and Eirich, F. R.,** Interactions of aqueous poly(N-vinylpyrrolidone) with sodium dodecyl sulfate. I. Equilibrium dialysis measurements, *J. Phys. Chem.*, 7, 3135, 1971.
245. **Dialer, K. and Vogler, K.,** Molekulargewichtsverteilung von Polyvinylpyrrolidon, *Makromol. Chem.*, 6, 191, 1951.
246. **Sekikawa, H., Nakano, M., and Arita, T.,** Dissolution behaviors and gastrointestinal absorption of sulfisoxazole in sulfisoxazole-polyvinylpyrrolidone coprecipitate, *Yakugaku Zasshi*, 98, 62, 1978.
247. **Higuchi, T. and Kuramoto, R.,** Study of possible complex formation between macromolecules and certain pharmaceuticals. I. Polyvinylpyrrolidone with sulfathiazole, procaine hydrochloride, sodium salicylate, benzyl penicillin, chloramphenicol, mandelic acid, caffeine, theophylline, and cortisone, *J. Am. Pharm. Assoc. Sci. Ed.*, 43, 393, 1954.
248. **Cho, M. J., Mitchell, A. G., and Pernarowski, M.,** The interaction of bishydroxycoumarin with polyvinylpyrrolidone, *J. Pharm. Sci.*, 60, 720, 1971.
249. **Matsumaru, H., Tsuchiya, S., and Hosono, T.,** Interaction and dissolution characteristics of ajmaline-PVP coprecipitate, *Chem. Pharm. Bull.*, 22, 2504, 1977.
250. **Sekikawa, H., Hori, R., Arita, T., Ito, K., and Nakano, M.,** Application of the cloud point method to the study of the interaction of polyvinyl pyrrolidone with some organic compounds in aqueous solution, *Chem. Pharm. Bull.*, 26, 2489, 1978.
251. **Frank, H. P., Barkin, S., and Eirich, F. R.,** The interaction of polyvinylpyrrolidone with some azo dyes, *J. Phys. Chem.*, 61, 1375, 1957.
252. **Molyneux, P. and Frank, H. P.,** The interaction of polyvinylpyrrolidone with aromatic compounds in aqueous solution. Part I. Thermodynamics of the binding equilibria and interaction forces, *J. Am. Chem. Soc.*, 83, 3169, 1961.
253. **Hsiao, C. H., Rhodes, H. J., and Blake, M. I.,** Fluorescent probe study of sulfonamide binding to povidone, *J. Pharm. Sci.*, 66, 1157, 1977.
254. **Molyneux, P. and Ahmed, G. S.,** The interaction of water-soluble α,ω-disubstituted alkanes (bolaform compounds) with polyvinylpyrrolidone in aqueous solution, *Kolloid-Z. Z. Polym.* 251, 310, 1973.
255. **Hengstenberg, J. and Schuch, E.,** Molekulargewichtsbestimmung von Polyvinylpyrrolidonen (PVP) mittels des osmotischen Drucks und der Lichtstreuung iher Lösungen, *Makromol. Chem.*, 7, 105, 1951; 7, 236, 1952.
256. **Levy, G. B., Caldas, I., and Fergus, D.,** Evaluation of polyvinylpyrrolidone preparations, *Anal. Chem.*, 24, 1799, 1952.
257. **Scholtan, W.,** Moleculargewichtsbestimmung von Polyvinylpyrrolidon mittels der Ultrazentrifuge, *Makromol. Chem.*, 7, 4, 1951; 7, 209, 1952.
258. **Frank, H. P. and Levy, G. B.,** Determination of molecular weight of polyvinylpyrrolidone, *J. Polym. Sci.*, 10, 371, 1953.
259. **Levy, G. B. and Frank, H. P.,** Determination of molecular weight of polyvinylpyrrolidone. II, *J. Polym. Sci.*, 17, 247, 1955.
260. **Scholtan, W.,** Konzentrationsabhängigkeit der Sedimentationskonstanten von Polyvinylpyrrolidon, *Makromol. Chem.*, 23, 128, 1957.
261. **Scholtan, W.,** Bestimmung der Molekulargewichtsverteilung von Polyvinylpyrrolidon durch fraktionierte Fällung und experimentelle Prufung der theoretischen Grundlagen der Fällungsmethode, *Makromol. Chem.*, 24, 83, 1957.
262. **Cerny, L. C., Helminiak, T. E., and Meier, J. F.,** Osmotic pressures of aqueous polyvinylpyrrolidone solutions, *J. Polym. Sci.*, 44, 539, 1960.
263. **Meza, R. and Gargallo, L.,** Unperturbed dimensions of polyvinylpyrrolidone in pure solvents and binary mixtures, *Eur. Polym. J.*, 13, 235, 1977.
264. **Miller, L. E. and Hamm, F. A.,** Macromolecular properties of polyvinylpyrrolidone: molecular weight distribution, *J. Phys. Chem.*, 57, 110, 1953.
265. **Scholtan, W.,** Bestimmung der Molekulargewichtsverteilung von Polyvinylpyrrolidon durch Trübungstitration und Vergleich der verschiedenen Auswertungsmethoden, *Makromol. Chem.*, 24, 104, 1957.
266. **Kaneniwa, N. and Ikekawa, A.,** Degradation of poly-vinylpyrrolidone by ball-milling, *J. Soc. Mater. Sci. Jpn.*, 20, 720, 1971.

267. **Kaneniwa, N. and Ikekawa, A.,** Degradation of polyvinylpyrrolidone by ball-milling, *Yakuzaigaku*, 31, 201, 1971.
268. **Kaneniwa, N. and Ikekawa, A.,** Influence of ball-milling atmosphere on decrease of molecular weight of polyvinylpyrrolidone powders, *Chem. Pharm. Bull.*, 20, 1536, 1972.
269. **Kaneniwa, N. and Ikekawa, A.,** Influence of addition of organic and inorganic powders on degradation of polyvinylpyrrolidone by ball-milling in air, *Chem. Pharm. Bull.*, 21, 1539, 1973.
270. **Kaneniwa, N. and Ikekawa, A.,** Induction period for a decrease in molecular weight of polyvinylpyrrolidone by ball-milling in various kinds of atmosphere in the presence of organic and inorganic powders, *Chem. Pharm. Bull.*, 25, 1534, 1977.
271. **Kaneniwa, N. and Ikekawa, A.,** Interaction between iodine and polyvinylpyrrolidone or α-pyrrolidone, *Chem. Pharm. Bull.*, 22, 2990, 1974.
272. **Elias, H.-G.,** Konstitution und Lösungseigenschaften von Makromolekulen. I. Ermittlung von Θ-Lösungen, *Makromol. Chem.*, 50, 1, 1961.
273. **Fikentscher, H.,** Systematik der Cellulosen auf Grund ihrer Vislosität in Lösung, *Cellulosechemie*, 13, 58, 1932.
274. **Matthes, A.,** Zur viscosimetrischen Charakterisierung der Teilchengrösse von Hochpolymeren, *Angew. Chem.*, 54, 517, 1941.
275. **Molyneux, P. and Frank, H. P.,** The interaction of polyvinylpyrrolidone with aromatic compounds in aqueous solution. II. The effect of the interaction on the molecular size of the polymer, *J. Am. Chem. Soc.*, 83, 3175, 1961.
276. **Sugiura, M. and Fujii, E.,** Sorption of water and glass transition temperature in poly(vinylpyrrolidinone), *Kogyo Kagaku Zasshi*, 66, 1228, 1963; *Chem. Abstr.*, 60, 9369g, 1964.
277. **Jenckel, E.,** Die Einfriertemperatur hochmolekularer Gläser und ihr chemischer Aufbau, *Kolloid-Z.*, 100, 163, 1942.
278. **Simonelli, A. P., Mehta, S. C., and Higuchi, W. I.,** Dissolution rates of high energy polyvinylpyrrolidone (PVP)-sulfathiazole coprecipitates, *J. Pharm. Sci.*, 58, 538, 1969.
279. **Tadokoro, Y., Okazaki, S., and Yamamura, H.,** Diffusion of water vapour into polyvinyl pyrrolidone, *J. Sci. Hiroshima Univ. Ser. A-II*, 29, 89, 1965.
280. **Behen, J. J., Dwyer, R. F., and Bierl, B. A.,** Determination of polyvinylpyrrolidone by an infrared absorption method, *Anal. Biochem.*, 9, 127, 1964.
281. **Boyer-Kawenoki, F.,** Contribution à l'étude spectroscopique infrarouge du composé d'addition polyvinylpyrrolidone-acide polyacrylique, *C. R. Acad. Sci. Ser. C.*, 263, 278, 1966.
282. **Kahela, P. and Forsén, K.,** Nuclear magnetic resonance and infrared spectroscopic study of interaction of polyvinylpyrrolidone (PVP) with benzoic acid and salicylic acid, *Farm. Aikak.*, 80, 220, 1971.
283. **Biedermann, H.-G. and Graf, W.,** Metal complexes with polymeric ligands. XI. Reactions of metal carbonyls with poly-1-vinyl-2-pyrrolidinone, *Z. Naturforsch.*, 30b, 226, 1975.
284. **Siggia, S.,** The chemistry of polyvinylpyrrolidone-iodine, *J. Am. Pharm. Assoc. Sci. Ed.*, 46, 201, 1957.
285. **Cournoyer, R. F. and Siggia, S.,** Interaction of polyvinylpyrrolidone and iodine, *J. Polym. Sci. Polym. Chem. Ed.*, 12, 603, 1974.
286. **Tachibana, T. and Nakamura, A.,** A method for preparing an aqueous colloidal dispersion of organic materials by using water-soluble polymers: dispersion of β-carotene by polyvinylpyrrolidone, *Kolloid-Z. Z. Polym.*, 203, 130, 1965.
287. **Sekikawa, H., Naganuma, T., Fujiwara, J.-E., Nakano, M., and Arita, T.,** Dissolution behaviors and gastrointestinal absorption of tolbutamide in tolbutamide-polyvinylpyrrolidone coprecipitate, *Chem. Pharm. Bull.*, 27, 31, 1979.
288. **Sekikawa, H., Nakano, M., and Arita, T.,** Inhibitory effect of polyvinylpyrrolidone on the crystallization of drugs, *Chem. Pharm. Bull.*, 26, 118, 1978.
289. **Sekikawa, H., Nakano, M., and Arita, T.,** Dissolution mechanisms of drug-polyvinylpyrrolidone coprecipitates in aqueous solution, *Chem. Pharm. Bull.*, 26, 1223, 1979.
290. **Simonelli, A. P., Mehta, S. C., and Higuchi, W. I.,** Dissolution rates of high energy sulfathiazole-povidone coprecipitates. II. Characterization of form of drug controlling its dissolution rate *via* solubility studies, *J. Pharm. Sci.*, 65, 355, 1976.
291. **Oster, G.,** Spectral studies of polyvinylpyrrolidone (PVP), *J. Polym. Sci.*, 9, 553, 1952.
292. **Nishijima, Y. and Oster, G.,** Diffusion in concentrated polymer solutions. I.A. An interferometric micromethod for diffusion measurements. B. Diffusion of sucrose in solutions of polyvinylpyrrolidone, *J. Polym. Sci.*, 19, 337, 1956.
293. **Pavlova, S. A. and Rafikov, S. R.,** Relation between the molecular weight of polymers and the viscosity of their solutions, *Vysokomol. Soedin.*, 1, 623, 1959; *Chem. Abstr.*, 54, 14872b, 1960.
294. **Scholtan, W.,** Bestimmung der Molekulargewichtsverteilung von Polyvinylpyrrolidon mit kleinem und grossem Molekulargewicht und von Albumin in der Ultrazentrifuge, *Makromol. Chem.*, 36, 162, 1960.
295. **Dognon, A., Dutheil, R., and Guérin, H.,** Quelques particularités de solutions aqueuses concentrées de polyvinylpyrrolidone, *J. Chim. Phys.*, 58, 812, 1961; *Chem. Abstr.*, 56, 2948i, 1962.

296. **Elias, H.-G.,** Konstitution und Lösungseigenschaften von Makromolekulen. II. Die Beziehungen zwischen den Konstanten *K* and *a* der Staudingerschen Viskositätsgleichung, *Kunstst.-Plast. (Solothurn, Switz.),* 8, 441, 1961.

297. **Weyl, D. A. and Neale, S. M.,** A comparison of the dielectric behaviour of a charged and an uncharged polymer in aqueous solution, *Dielectrics,* 1, 22, 1963.

298. **Boyer-Kawenoki, F.,** Osmometrie avec une membrane non seléctive et évaluation de la mass moléculaire du soluté, *C. R. Acad. Sci.,* 258, 4274, 1964.

299. **Nomura, H. and Miyahara, Y.,** The partial specific compressibility of polyvinylpyrrolidone, *Bull. Chem. Soc. Jpn.,* 39, 1599, 1966.

300. **Worley, J. D. and Klotz, I. M.,** Near-infrared spectra of H_2O-D_2O solutions, *J. Chem. Phys.,* 45, 2868, 1966.

301. **Jellinek, H. H. G. and Fok, S. Y.,** Freezing of aqueous polyvinylpyrrolidone solutions, *Kolloid-Z. Z. Polym.,* 220, 122, 1967.

302. **Mineshita, T., Watanabe, T., and Ono, S.,** The flow properties of polyvinylpyrrolidone solutions, *Bull. Chem. Soc. Jpn.,* 40, 2217, 1967; *Chem. Abstr.,* 68, 30237z, 1968.

303. **Goldfarb, J. and Rodriguez, S.,** Aqueous solutions of polyvinylpyrrolidone, *Makromol. Chem.,* 116, 96, 1968.

304. **Glasel, J. A.,** Participation of water in conformational changes of biopolymers as studied by deuteron magnetic relaxation, *J. Am. Chem. Soc.,* 92, 375, 1970.

305. **Kassem, M. A. and Mattha, A. G.,** Rheological studies on dispersions of polyvinylpyrrolidone. I. General flow characteristics, *Pharm. Acta Helv.,* 45, 18, 1970.

306. **Vink, H.,** Precision measurements of osmotic pressure in concentrated polymer solutions, *Eur. Polym. J.,* 7, 1411, 1971.

307. **Sillescu, H. and Brüssau, R.,** NMR-Relaxation and solvation dynamics in macromolecular solutions, *Chem. Phys. Lett.,* 5, 525, 1970.

308. **Brüssau, R. G. and Sillescu, H.,** Solvation dynamics in macromolecular solutions as studied by NMR relaxation, *Ber. Bunsenges. Phys. Chem.,* 76, 31, 1972.

308a. **MacKenzie, A. P. and Rasmussen, D.,** Interactions in the water-polyvinylpyrrolidone system at low temperatures, in *Water Structure at the Water-Polymer Interface,* Jellinek, H. H. G., Ed., Plenum Press, New York, 1972, 146.

309. **Rasmussen, D. H. and MacKenzie, A. P.,** Effect of solute on ice-solution interfacial free energy; calculation from measured homogeneous nucleation temperatures, in *Water Structure at the Water-Polymer Interface,* Jellinek, H. H. G., Ed., Plenum Press, New York, 1972, 126.

309a. **Rudd, P. J. and Jennings, B. R.,** Electric birefringence for the study of polymer-surfactant interactions: the polyvinylpyrrolidone-sodium dodecyl sulphate system, *J. Colloid Interface Sci.,* 48, 302, 1974.

310. **Murakami, S., Kimura, F., and Fujishiro, R.,** Enthalpies of dilution of poly(1-vinyl-2-pyrrolidone) + aqueous methanol solution, *Makromol. Chem.,* 176, 3425, 1975.

311. **Nomura, H., Kato, S., and Miyahara, Y.,** Ultrasonic absorption in polymer solutions, *Mem. Fac. Eng. Nagoya Univ.,* 27, 72, 1975.

312. **Clark, A. T., Robb, I. D., and Smith, R.,** Influence of solvent on the conformaton of polymers adsorbed at the solid/liquid interface, *J. Chem. Soc. Faraday Trans. 1,* 72, 1489, 1976.

313. **Nakagaki, M., Shimabayashi, S., Yoshida, S., and Saito, M.,** Light scattering studies of aqueous solution of polyvinylpyrrolidone, *Yakugaku Zasshi,* 96, 757, 1976.

314. **Garvey, M. J. and Robb, I. D.,** Effect of electrolytes on solution behaviour of water soluble macromolecules, *J. Chem. Soc. Faraday Trans. 1,* 75, 993, 1979.

315. **Chaufer, B., Lesec, J., and Quivoron, C.,** Preferential solvation study of polyvinylpyrrolidone in organic or aqueous mixed solvents by GPC, *J. Liq. Chromatogr.,* 2, 633, 1979.

316. **Rockelmann, H. and Sillescu, H.,** NMR Investigation of molecular motion, preferential solvation and hydrogen bonding in polymer solutions, *Z. Phys. Chem. (Frankfurt am Main),* 92, 263, 1974.

317. **Ridgway, K. and Rubinstein, M. H.,** The quantitative analysis of polyvinylpyrrolidone by infrared spectrophotometry, *J. Pharm. Pharmacol.,* 23, 587, 1971.

318. **Minamikawa, T. and Yamagishi, N.,** Studies on the quality of pharmaceutical preparations. V. Determination of polyvinylpyrrolidone in drugs by infrared absorption spectroscopy, *Bunseki Kagaku,* 22, 109, 1973; *Chem. Abstr.,* 79, 97044, 1973.

319. **Chaufer, B., Sebille, B., and Quivoron, C.,** Solvatation preféréntielle dans les systèmes ternaires polyvinylpyrrolidone/solvant halogéné/solvant donneur de proton. I. Étude par diffusion de la lumière, *Eur. Polym. J.,* 11, 683, 1975.

320. **Chaufer, B., Sebille, B., and Quivoron, C.,** Solvatation preféréntielle dans les systèmes ternaires polyvinylpyrrolidone/solvant halogéné/solvant donneur de proton. III. Confrontation des resultats obtenus par diffusion de la lumière et par spectroscopie infra-rouge, *Eur. Polym. J.,* 11, 695, 1975.

321. **Lety-Sistel, C., Chaufer, B., Sebille, C., and Quivoron, C.,** Solvatation preféréntielle dans les systèmes ternaires polyvinylpyrrolidone/solvant halogéné/solvant donneur de proton. II. Étude, par spectroscopie infra-rouge, des interactions hydrogène, *Eur. Polym. J.*, 11, 689, 1975.
322. **Stake, M. A. and Klotz, I. M.,** Protonation of polypeptides in helix-breaking solvents. Electrical conductance, *Biochemistry*, 5, 1726, 1966.
323. **Lafuma, F., Chaufer, B., and Quivoron, C.,** Étude de la solvatation préférentielle de la polyvinylpyrrolidone dans des mélanges dichloro-1.2 éthane-éthanol par les temps de relaxation T_1 des solvants en RMN ^{13}C, *C. R. Acad. Sci. Ser. C.*, 283, 183, 1976.
324. **Jellinek, H. H. G., Luh, M. D., and Nagarajan, V.,** Sorbed water on polymers near 0°C, *Kolloid-Z. Z. Polym.*, 232, 758, 1969.
325. **Eliassaf, J., Eriksson, F., and Eirich, F. R.,** The interaction of poly(vinyl pyrrolidone) with cosolutes, *J. Polym. Sci.*, 47, 193, 1960.
326. **Nakagaki, M. and Shimabayashi, S.,** The effect of disodium sulfate on viscosity of concentrated solution of poly(vinylpyrrolidone) and poly(vinyl alcohol), *Nippon Kagaku Kaishi*, 2403, 1972.
327. **Behzadi, A., Borgwardt, U., Henglein, A., Schamberg, E., and Schnabel, W.,** Pulsradiolytische Untersuchung der Kinetik diffusionkontrollierter Reaktionen des OH-Radikals mit Polymeren und Oligomeren in wässriger Lösung, *Ber. Bunsenges. Phys. Chem.*, 74, 649, 1970.

327a. **Charlesby, A., Garratt, P. G., and Kopp, P. M.,** The use of sulphur as a protecting agent against ionizing radiations, *Int. J. Rad. Biol.*, 5, 439, 1962.

328. **Charlesby, A. and Kopp, P. M.,** Radiation protection in aqueous polymer solutions, *Int. J. Rad. Biol.*, 5, 521, 1962.

328a. **Kopp, P. M. and Charlesby, A.,** Radiation protection of aqueous polyvinylpyrrolidone solutions of thiourea, *Int. J. Rad. Biol.*, 7, 173, 1963.

329. **Klotz, I. M. and Stryker, V. H.,** "Denaturation" of a synthetic polymer and its relationship to protein denaturation, *J. Am. Chem. Soc.*, 82, 5169, 1960.
330. **Klotz, I. M., Stellwagen, E. C., and Stryker, V. H.,** Ionic equilibria in protein conjugates: comparison of proteins, *Biochim. Biophys. Acta*, 86, 122, 1964.
331. **Burnette, L. W.,** A review of the physiological properties of polyvinylpyrrolidone, *Proc. Toilet Goods Assoc. Sci. Sec.*, 34, 1, 1962.
332. **Zipf, K.,** Quantitative Kollidonbestimmung im Harn und Blut (Peristonnachweis), *Klin. Wochenschr.*, 23, 340, 1944.
333. **Ishikawa, S., Furuichi, Y., and Nakamura, S.,** Determination of polyvinylpyrrolidone, *Nippon Nogei Kagaku Kaishi*, 46, 267, 1972; *Chem. Abstr.*, 77, 137024v, 1972.
334. **Basu, K. and Dutta, B. N.,** Estimation of polyvinylpyrrolidone in plasma substitute preparations, *J. Proc. Inst. Chem. (Calcutta)*, 34, 9, 1962; *Chem. Abstr.*, 57, 4765e, 1962.
335. **Müller, K.,** Nachweis und Bestimmung von Polyvinylpyrrolidon (PVP) sowie Bestimmung von Wirkstoffen in PVP-haltigen Arzneimittelzubereitungen, *Pharm. Acta Helv.*, 43, 107, 1968; *Chem. Abstr.*, 68, 53312p, 1968.
336. **Frauenfelder, L. J.,** Universal chromatographic-colorimetric method for the determination of trace amounts of polyvinylpyrrolidone and its copolymers in foods, beverages, laundry products, and cosmetics, *J. Assoc. Off. Anal. Chem.*, 57, 796, 1974.
337. **Hagel, L. and Andersson, R.,** Automated nephelometric determination of polyvinylpyrrolidone in Salazopyrin, *Anal. Chim. Acta*, 86, 69, 1976.
338. **Larkin, R. L. and Kupel, R. E.,** Quantitative analysis of polyvinylpyrrolidone in atmospheric samples and biological tissues, *Am. Ind. Hyg. Assoc. J.*, 26, 558, 1965.
339. **Levy, G. B. and Fergus, D.,** Microdetermination of polyvinylpyrrolidone in aqueous solution and in body fluids, *Anal. Chem.*, 25, 1408, 1953.
340. **McFarlane, W. D., Grant, H. L., and Held, H. R.,** Studies on polyvinylpyrrolidone (PVP) as a tannin precipitant in wort and beer, *Proc. Am. Soc. Brewing Chem.*, 119, 1958; *Chem. Abstr.*, 53, 5584e, 1959.
341. **Riedhammer, T. M.,** Colorimetric determination of poly(N-vinyl-2-pyrrolidone) in contact lens solutions, *J. Assoc. Off. Anal. Chem.*, 62, 52, 1979.
342. **Von Frijtag Drabbe, C. A. J. and Reinhold, J. G.,** Measurement of polyvinylpyrrolidone concentrations in body fluids by means of a turbidity-producing reaction with phenol, *J. Lab. Clin. Med.*, 40, 616, 1952.
343. **Dwyer, R. F. and Lewandowski, R. J.,** Determination of polyvinylpyrrolidone by an iodine titration method, *Anal. Biochem.*, 9, 133, 1964.
344. **Shiraeff, D. A.,** Quantitative methods for trace amounts of polyvinylpyrrolidone, *J. Assoc. Off. Anal. Chem.*, 47, 724, 1964.
345. **Chinard, F. P.,** A colorimetric method for the estimation of polyvinylpyrrolidone (PVP) in whole blood, plasma, and urine, *J. Lab. Clin. Med.*, 39, 666, 1952.
346. **Meijer, A. E. F. H.,** The determination of polyvinylpyrrolidone in tissue extracts, *Clin. Chim. Acta*, 6, 713, 1961.

347. **Poullain, P. and Piette, M.,** Détermination de la masse sanguine par la polyvinylpyrrolidone, *Bull. Soc. Chim. Biol.*, 30, 496, 1948.
348. **Hecht, G., Joppich, G., and Weese, H.,** Peristonwirkung und nachweis bei exsiccotischen Säuglingen, *Monatsschr. Kinderheilk.*, 96, 145, 1948.
349. **Ammon, R. and Braunschmidt, G.,** Das Schicksal des Peristons in Organismus, *Biochem. Z.*, 319, 370, 1949; *Chem. Abstr.*, 43, 8539f, 1949.
350. **Ammon, R. and Nübling, W.,** Zur quantitativen Bestimmung der Kollidone, *Arzneim.-Forsch.*, 6, 565, 1956; *Chem. Abstr.*, 51, 1365f, 1957.
351. **Campbell, H. and Hunter G.,** Colorimetric estimation of polyvinylpyrrolidone by iodine, *Lancet*, 264(i), 197, 1953.
352. **Discombe, C. and Creig, H. B. W.,** Détermination de la polyvinylpyrrolidone dans l'eau et dans le plasma, *Ann. Biol. Clin. (Paris)*, 12, 415, 1954; *Chem. Abstr.*, 49, 4061b, 1955.
353. **Guttman, D. and Higuchi, T.,** Possible complex formation between macromolecules and certain pharmaceuticals. X. The interaction of some phenolic compounds with polyethylene glycols, polypropylene glycols, and polyvinylpyrrolidone, *J. Am. Pharm. Assoc. Sci. Ed.*, 45, 659, 1956.
354. **Tomita, S.,** Studies on the fluorimetric analysis of polyelectrolytes. I. Proteins, *Tokyo Kogyo Shikensho Hokoku*, 58, 99, 1963; *Chem. Abstr.*, 62, 8109f, 1965.
355. **Postel, W.,** Nachweis und Bestimmung von Polyvinylpyrrolidon im Bier, *Brauwissenschaft*, 26, 337, 1973; *Chem. Abstr.*, 80, 46456h, 1974.
356. **Verdier, E., Piro, J., and Montelongo, F. G.,** Dosage quantitatif des substances tensioactives par électroadsorption, *Talanta*, 18, 1237, 1971.
357. **Camozzo, S. and Dal Zotto, S.,** Behavior of polyvinylpyrrolidone towards the principal reagents for alkaloids and photocolorimetric determination, *Ann. Chim. (Rome)*, 43, 113, 1953; *Chem. Abstr.*, 47, 8588i, 1953.
358. **Dazzo, F. and Hubbell, D.,** A quantitative assay of insoluble polyvinylpyrrolidone, *Plant Soil*, 40, 435, 1974.
359. **Nelson, A. A. and Lusky, L. M.,** Pathological changes in rabbits from repeated intravenous injections of periston (polyvinyl pyrrolidone) or dextran, *Proc. Soc. Exp. Biol. Med.*, 76, 765, 1951.
360. **Freiman, D. G. and Gall, E. A.,** A staining method for the detection of polyvinylpyrrolidone (PVP) in tissue sections, *Am. J. Clin. Pathol.*, 25, 1427, 1955.
361. **Towers, R. P.,** Lymph-node changes due to polyvinyl pyrrolidone, *J. Clin. Pathol.*, 10, 175, 1957.
362. **Yamashita, M., Nakamoto, M., Matsumoto, N., and Hosokawa, S.,** New staining method for demonstration of PVP in tissue section, *Acta Pathol. Jpn.*, 18, 345, 1968; *Chem. Abstr.*, 71, 78031w, 1969.
363. **Pearse, A. G. E.,** Polyvinyl pyrrolidone (PVP), in *Histochemistry: Theoretical and Applied*, Vol. 2, 3rd ed., Churchill Livingstone, Edinburgh, 1972, 1162.
364. **Anon.,** Dow's "complexing resins" promise broad use, *Ind. Eng. Chem.*, 52(12), 34A and 37A, 1960.
365. **Tomalia, D. A.,** Vinyl oxazolidones, in *Functional Monomers — Their Preparation, Polymerization, and Application*, Yocum, R. H. and Nyquist, E. B., Eds., Marcel Dekker, New York, 1974, 58.
366. **Wolfson, B. B. and Banker, G. S.,** Solubilization and evaluation of poly-N-vinyl-5-methyl-2-oxazolidinone barbiturate systems, *J. Pharm. Sci.*, 54, 195, 1965.
367. **Blaug, S. M. and Rich, A. G.,** Interaction between poly-*N*-vinyl-5-methyl-2-oxazolidinone and certain pharmaceuticals in aqueous solution, *J. Pharm. Sci.*, 54, 30, 1965.
368. **Davidson, R. L., Ed.,** *Handbook of Water-Soluble Gums and Resins*, McGraw-Hill, New York, 1980.
369. **Modi, T. W.,** Polyvinyl alcohol, in *Handbook of Water-Soluble Gums and Resins*, Davidson, R. L., Ed., McGraw-Hill, New York, 1980, chap. 20.
370. **Kunieda, H. and Shinoda, K.,** Dissolution mechanism of water-soluble polymers (partially saponified polyvinyl acetates), in *Colloid Dispersions and Micellar Behavior (ACS Symposium Series*, Vol. 9), Mittal, K. L., Ed., American Chemical Society, Washington, D.C., 1975, chap. 20.
371. **Eagland, D., Wardlaw, G. C., and Thorn, I.,** Particle-macromolecule interactions in aqueous solutions. I. Solution and flow behavior of polyvinyl alcohol in an aqueous medium, *Colloid Polym. Sci.*, 256, 1073, 1978.
372. **Watanabe, T., Yahagi, T., and Fujiwara, S.,** EPR Studies of the molecular motion of nitroxide radicals in poly(vinyl alcohol) gels, *J. Am. Chem. Soc.*, 2, 5187, 1980.
373. **Ise, N. and Okubo, T.,** Mean activity coefficient of polyelectrolytes. II. Measurements of sodium salts of polyvinyl alcohols partially acetalized with glyoxylic acid, *J. Phys. Chem.*, 70, 1930, 1966.
374. **Imamura, S.-I., Tonomura, Y., Kawabata, N., and Kitao, T.,** Wet oxidation of water-soluble polymers, *Bull. Chem. Soc. Jpn.*, 54, 1548, 1981.
375. **Monte-Bovi, A. J.,** Preservatives and artificial sweeteners: interferences in the polyvinyl alcohol-borate-iodine complex, *J. Assoc. Off. Anal. Chem.*, 52, 891, 1969.
376. **Sidelkovskaya, F. P.,** *Khimie N-Vinilpirrolidona y ego Polimerov (Chemistry of N-Vinylpyrrolidone and its Polymers)*, Nauka, Moscow, 1970; *Chem. Abstr.*, 74, 100438v, 1971.

377. **Blecher, L., Lorenz, D. H., Lowd, H. L., Wood, A. S., and Wyman, D. P.,** Polyvinylpyrrolidone, in *Handbook of Water-Soluble Gums and Resins,* Davidson, R. L., Ed., McGraw-Hill, New York, 1980.
378. **Assarsson, P. and Eirich, F. R.,** Properties of amide in aqueous solution. I.A. Viscosity and density changes of amide-water systems. B. An analysis of volume deficiencies of mixtures based on molecular size differences (mixing of hard spheres), *J. Phys. Chem.,* 72, 2710, 1968.
379. **Assarsson, P. and Eirich, F. R.,** Interaction of water with alkyl-substituted amides, in *Molecular Association in Biological and Related Systems (Advances in Chemistry Series No. 84),* Goddard, E. D., Ed., American Chemical Society, Washington, D.C., 1968, chap. 1.
380. **Assarsson, P., Chen, N. Y., and Eirich, F. R.,** Thermodynamic aspects of the mixing of mono-amides and poly-peptides with water, in *Colloid Dispersions and Micellar Behavior (ACS Symposium Series,* Vol. 9), Mittal, K. L., Ed., American Chemical Society, Washington, D.C., 1975, chap. 21.
381. **Radding, W., Donzel, B., Ueyama, N., and Goodman, M.,** Conformation of six N-methylated diketopiperazines in solution, *J. Am. Chem. Soc.,* 102, 5999, 1980.
382. **Davis, J. E. and Senogles, E.,** γ-Irradiation of N-vinylpyrrolidin-2-one and its homopolymer, *Aust. J. Chem.,* 34, 1413, 1981.
383. **Davis, J. E., Sangster, D. F., and Senogles, E.,** Pulse radiolysis of aqueous solutions of N-vinylpyrrolidin-2-one and poly(N-vinylpyrrolidin-2-one), *Aust. J. Chem.,* 34, 1423, 1981.
384. **Tan, Y. Y. and Challa, G.,** The glass transition temperature of poly(N-vinyl pyrrolidone) and the effect of water, *Polymer (London),* 17, 739, 1976.

APPENDIX I

Format of the Chapters

The main material of these volumes is contained in the eight chapters which are arranged in the following manner. The three chapters following the Introduction (Chapter 1 of Volume I) are concerned with the nonionic polymers, first with the polyoxides, polyethers, and the related poly(ethylene imine) (Chapter 2); then with acrylic polymers (Chapter 3); and finally with vinyl polymers (Chapter 4). In each of these chapters, any polyelectrolyte character (i.e., for weak polyacids or polybases) is taken to be suppressed. Chapter 1 of Volume II is then concerned specifically with polyelectrolytes.

In these latter four chapters (Chapters 2 to 4 of Volume I and Chapter 1 of Volume II) the properties and behavior of the most important examples of the polymers are outlined; in each case a profile is given of certain practical aspects of handling the polymer — preparation, commercial sources and grades, purification, fractionation, characterization, properties of the bulk polymer,* and chemical and biological behavior — as a guide to the successful handling of the polymer in the laboratory. In addition, the published studies on the behavior of the polymer in aqueous and nonaqueous solution are summarized, and three important aspects — partial specific volume, specific refractive index, and dilute-solution viscosity — are chosen for more detailed discussion. It was originally intended, when these volumes were being planned, to survey critically the thermodynamic data that are available on these polymers in aqueous solution; however, there turned out to be marked discrepancies between the values from the various sources of, for example, the second virial coefficient A_2 (from which the Flory-Huggins interaction parameter χ and related thermodynamic quantities are mostly derived), and it was evident that a detailed examination of these discrepancies would not be justified within the scope of the present treatment. The literature sources compiled and data tabulated should nevertheless enable such a detailed critical examination to be carried out when required. In selecting literature on solution properties, the emphasis has been placed on the behavior at relatively low polymer concentration, but published work dealing with more concentrated systems, including gelation effects, has been included when the results are scientifically significant.

However, three aspects have been intentionally omitted from the above-mentioned chapters, and are given special treatment in Chapters 2 to 4 of Volume II. These are the interactions of the polymers with small-molecule cosolutes (Chapter 2); the interactions between two (or more) polymers (Chapter 3); and the interactions of these polymers in solution with interfaces (Chapter 4). These topics have been dealt with in this way partly because they involve special features which need to be considered together, and also because they are aspects that are relatively neglected in many discussions of these polymers.

The divisions in the contents of the chapters are not sharp, however, and some overlap must be expected. For Chapters 1 and 2 of Volume II, in particular, the consideration of polyelectrolytes in the first case almost invariably implies the interactions of the polyions with the small-molecule counterions (i.e., cosolutes) that accompany them to ensure electroneutrality of the system; by and large, only the simple salts of the polyelectrolyte (es-

* In view of the present emphasis on the behavior of these polymers in aqueous solution, the properties of the bulk polymer, i.e., solid state, may not seem as significant as with other polymers. It should be recognized, however, that the infrared spectrum of the solid may be used to characterize it, that the glass transition temperature (for amorphous polymers) or the melting point (for crystalline polymers) is important in relation to the removal of absorbed water (see Section III.E. of Chapter 1 in Volume I), and that the interactions in the solid state may be an important factor in the solubility (e.g., a highly crystalline polymer will usually be less soluble than an amorphous one). This state is also important in many practical applications (even for water-soluble materials), where the polymer may start out in this state (e.g., soluble containers for single-dose products) or end up in this state (e.g., coatings and adhesives).

pecially with alkali metal countercations and halide counteranions), where specific binding effects are not prominent, are considered in Chapter 1, while systems with transition metal ions, organic anions and cations, etc., as well as polyelectrolytes with added simple electrolytes, are left for consideration in Chapter 2.

Similarly, oligomeric cosolutes might logically be included either in Chapters 2 or 3 of Volume II, and indeed, certain borderline cases have, for completeness, been included in tabulated data in both chapters.

APPENDIX II

Nomenclature

The nomenclature employed here for the polymers follows in the main the common usage; thus, no attempt has been made to use the systematic nomenclature recommended by IUPAC,[1] although these forms are listed with the alternative names for the specific polymers. However, as discussed in Chapter 2 of Volume I, the names ''poly(ethylene oxide)'' and ''poly(propylene oxide)'' are used throughout not only for these true polymers, but also for the corresponding oligomers more commonly referred to as ''poly(ethylene glycol)'' and ''poly(propylene glycol)''. The nomenclature for polysalts is discussed in Chapter 1 of Volume II (Section I.C.1.). Copolymers are named simply ''poly(A-*co*-B)'' (where A and B are the monomers), followed if necessary by any further distinction as to comonomer distribution (random, alternating, block, graft, . . .).

REFERENCE

1. **Anon.,** Nomenclature of regular single-strand organic polymers, *Pure Appl. Chem.*, 48, 373, 1976.

APPENDIX III

Deka-Deci Notation

The frequent citation of molecular weights of polymers in the text and the tables in these volumes makes it convenient, from the viewpoint of brevity and clarity, to apply the ''deka-deci'' notation proposed by the present author,[1] and this notation is accordingly used throughout the volumes. On this notation, a value such as 124,000, i.e., 1.24×10^5, is represented as 1.24D5 (where ''D'' is to be pronounced ''deka''); correspondingly, fractional quantities such as 6.78×10^{-9} are represented as 6.78d9 (where ''d'' is to be pronounced ''deci'').

REFERENCE

1. **Molyneux, P.,** ''Deki-deci'' notation, *Nature (London)*, 224, 1237, 1969.

APPENDIX IV

Notes on the Tables

The tables, which make extensive use of the abbreviations and symbols in Appendix V, enable the extensive published work to be summarized and condensed into a compact form. The arrangement of each table is generally with the entries in chronological order for that subject, and then in order of first authors for that particular year. In individual entries, the designation "1,2" implies that only these two values of the specified variable were involved, whereas "1-4" implies three or more values within this range. The same applies to series of compounds, substituent groups or the like, e.g., "X = Li—Cs" implies that the whole series Li, Na, K, Rb, Cs was involved, and similarly "Me—Bu" implies the whole series Me, Et, Pr, Bu.

In the "Comments" column, entries before or minus the arrow (→) further specify the conditions used in the experiments, while those after this symbol summarize important results obtained; where the latter entry is given in brackets, this indicates that the quantity cited was evidently obtained in the course of the experiments but not reported, or that it may be deduced using this and other experimental data that were reported.

APPENDIX V

Abbreviations and Symbols

These are the main abbreviations and symbols that are used in these volumes. The abbreviations employed for polymers largely follow the format recommended by IUPAC[1,2] except that the restriction to capital letters is not adhered to in the case of polyacrylamide (PAAm) and polymethacrylamide (PMAAm), where the small letter m is used for distinction from the capital M used for methyl substituent groups. Copolymers (whose abbreviations are not in general listed here) are indicated by the form PA-B (where A and B are the two types of monomer unit). The symbols for physical quantities also largely follow the IUPAC recommendations[3] and common usage in the polymer field,[4] except that average values are represented by a bar over the symbol for the quantity (e.g., $\bar{M}$), and for distinction, partial specific quantities are indicated by a tilde (e.g., $\tilde{n}$, $\tilde{v}_2^0$) and root-mean-square values by a circumflex (e.g., $\hat{L}$, $\hat{R}$).

a	Index in the MHS equation (q.v.)
a_1	Water activity (relative vapor pressure, relative humidity)
a_3	Concentration of free small-molecule cosolute in binding systems (see Chapter 2 of Volume II)
A	Absorbance (of UV, VL, IR, . . . radiation)
A_2 (A_3,...)	Second (third,...) virial coefficient
AA	Acrylic acid
AAm	Acrylamide
Ac	Acetyl group (CH_3CO)—except in monomers (VAC, etc.) and polymers (PVAC, etc.)
AD	Adsorption studies of polymers at interfaces
ADI	Adsorption isotherms
ADK	Adsorption kinetics

AF	Adsorbed films of polymers at interfaces
ANS	1-Anilinonaphthalene-8-sulfonate(phenyl-1-naphthylamine-8-sulfonate)
at	Atactic
b_3	Concentration of bound small-molecule cosolute in binding systems (see Chapter 2 of Volume II)
B	Basemolar, i.e., basemole (monomoles)/ℓ
BI	Binding isotherm
BP	British Pharmacopeia
BPE	Boiling-point elevation method for relative molecular mass
Bu	Butyl
c_2	Polymer concentration—% w/v (i.e., g dl^{-1}) unless otherwise indicated
c_2^*	Critical overlap value of c_2
c_3	Cosolute concentration
C	Heat capacity (specific heat)
$\bar{C}_2^\circ$	Partial heat capacity of polymer
CAL	Calorimetry
CB	Cosolute binding
CC	Complex coacervation in multipolymer systems (see Chapter 3 of Volume II)
CD	Circular dichroism
CFT	Critical flocculation temperature (t_f)
CMC	Critical micelle concentration
CP	Cloud point (precipitation temperature, t_p)
CS	Cosolute solubility method (see Chapter 2 of Volume II)
CT	Conformational transition (especially, compact → expanded conformation for PMAA and other polymers)
cv	"Conventional" (with reference to the tacticity of PMAA, which is somewhat stereoregular when prepared by free radical means — see Chapter 3 of Volume I)
d	"deci" indicates negative power of 10 (— see "Deka-deci notation", Appendix III)
D	"Deka" indicates positive power of 10 (— see "Deka-deci notation", Appendix III)
D	Diffusion coefficient (translational) — of polymer, unless otherwise indicated; optical density (= absorbance, A)
Dd	Dodecyl
DD	Dynamic dialysis method for cosolute binding (see Chapter 2 of Volume II)
De	Decyl
DE	Donnan (membrane) equilibrium
DMAB	4-(N,N-Dimethylamino)azobenzene (Dimethyl Yellow)
DME	Dropping mercury electrode
DS	Dispersion stability (see Chapter 4 of Volume II, Table 3 for key to additional symbols in "DS>", etc.)
DTA	Differential thermal analysis
DVB	Divinylbenzene (cross-linking agent)
EA	Ethyl acrylate
EB	Electric birefringence techniques
EC	Electric conductance techniques
ED	Equilibrium dialysis

EDM	Electric dipole moment techniques
EL	Ellipsometry
EM	Electron microscopy
EMF	Electromotive force (potentiometry) methods
EO	Ethylene oxide unit, $-CH_2CH_2O-$
EP	Electric permittivity (dielectric constant) techniques — zero frequency unless otherwise indicated
EPH	Electrophoresis (including microelectrophoresis)
ESR	Electron spin (paramagnetic) resonance
Et	Ethyl
FB	Flow birefringence techniques
FL	Fluorescence
FPD	Freezing-point depression (cryoscopy)
FR	Filtration rate measurements (on water/solid dispersions)
g	Gauche conformation
g′	Antigauche conformation
GD	Gel diffusion techniques
GE	Gel equilibria
GLC	Gas-liquid chromatography
GPC	Gel permeation (exclusion, filtration) chromatography
Gu	Guanidine (e.g., GuHCl = guanidinium chloride)
H(ΔH)	Enthalpy (change)
HC	Heat capacity techniques
Hd	Hexadecyl (cetyl)
HD	Heat of dilution techniques (i.e., solution plus solvent — polymer concentration quoted is the mean of initial and final values)
HI	Heat of ionization
HM	Heat of mixing techniques (i.e., of pure components — polymer concentration quoted is the final value)
HME	Hanging-drop mercury electrode
HOAc	Acetic acid
HS	Heat of solution techniques
Hx	Hexyl
i (i_2)	Degree of ionization (neutralization, equivalence) of polyacid or polybase
IC	Incompatibility in multipolymer systems (see Chapter 3 of Volume II)
IEP	Interfacial electrical potential
IR	Infrared absorption spectra
it	Isotactic
IT	Interfacial tension (surface tension, interfacial pressure, surface pressure — see Chapter 3 of Volume II)
IVI	Interfacial viscosity
J	Spin-spin coupling constant (in NMR)
k	Boltzmann constant
k'	Huggins viscosity-slope parameter (see Section III. G. of Chapter 1)
K	Parameter in MHS equation (q.v.); Fikentscher viscosity number (see Chapter 4 of Volume I); binding constant in hyperbolic binding isotherm (see Chapter 2 of Volume II)
K_p	Liquid/liquid partition coefficient
KI	Kinetics of reactions
l	Length of chain unit or link in polymer chain
L	End-to-end distance for polymer chain

$\hat{L}$	Root-mean-square value of L
LS	Light scattering (including nephelometry and turbidimetry)
LT	Langmuir trough
m	Molal concentration [mol kg (solvent)$^{-1}$]
M	Molar concentration (mol dm^{-3})
M	Molecular weight (relative molecular mass) of polymer (unless otherwise specified) — subscripts n, sD, v, and w indicate number, sedimentation/diffusion, viscosity, and weight (mass) averages
M_o	Molecular weight of specified monomer unit (basemole, monomole)
MAA	Methacrylic acid
MAAm	Methacrylamide
MHS	Mark-Houwink-Sakurada equation (intrinsic viscosity vs. molecular weight relation) (see Section III. G. of Chapter 1) — parameters tabulated as: (a; K/cm^{3}g^{-1})
MHSM	Müh's extension of MHS equation for low M (see Section III. G. of Chapter 1) — parameters tabulated as: (a; K/cm^3 g^{-1}; μ)
MHSR	Rempp's extension of the MHS equation for low M (see Section III. G. of Chapter 1) — parameters tabulated as: (a; K/cm^3 g^{-1}; $[\eta]_o$)
MLA	Maleic acid
MMA	Methyl methacrylate
n	Amount of substance (moles); refractive index
$\tilde{n}$	Specific refractive index increment, dn/dc_2
NAS	2-Naphthylamine-6-sulfonate
NEP	N-Ethylpyrrolidone
NMP	N-Methylpyrrolidone
NMR	Nuclear magnetic resonance (and relaxation)
NOE	Nuclear Overhauser enhancement
NVP	N-Vinylpyrrolidone — except in polymers where VP is used
OA	Optical anisotropy
Oc	Octyl
Od	Octadecyl (stearyl)
OP	Osmotic pressure techniques
ORD	Optical rotary dispersion
p	Pressure
p_2	Fraction of chain units attached to an interface
P	Degree of polymerization
PAA	Poly(acrylic acid)
PAAm	Polyacrylamide
PAG	Polyacrylylglycinamide
PD	Phase diagram
PE	Polyethylene; partition (distribution) equilibrium method for cosolute binding (see Chapter 2 of Volume II)
PEA	Poly(ethyl acrylate)
PEI	Poly(ethylene imine)
PEO	Poly(ethylene oxide) and its oligomers
Ph	Phenyl, or phenylene (1,4 substituted unless otherwise indicated)
PIA	Poly(itaconic acid)
PIPAAm	Poly(N-isopropylacrylamide)
PMA	Poly(methyl acrylate)
PMAA	Poly(methacrylic acid)
PMAAm	Polymethacrylamide

PMAG	Polymethacrylylglycinamide
PMLA	Poly(maleic acid)
PMMA	Poly(methyl methacrylate)
Pn	Pentyl (amyl)
PO	Propylene oxide (unit)
POL	Polarography
POM	Poly(oxymethylene)
PPO	Poly(propylene oxide) and its oligomers
PS	Polymer solubility (in cosolute-binding studies - see Chapter 2 of Volume II); polystyrene; preferential solvation
PSSA	Poly (styrenesulfonic acid)
PT	Potentiometric titration
PVAA	Poly(vinyl acetate-*co*-vinyl alcohol), i.e., shortened form of PVAC-VAL
PVAC/VAL	Poly(vinyl acetate-*co*-vinyl alcohol)
PVAC	Poly(vinyl acetate)
PVAL	Poly(vinyl alcohol) — also PVAC-VAL with $<$ 2 mol % VAC
PVME	Poly(vinyl methyl ether)
PVMO	Polyvinylmethyloxazolidone
PVO	Polyvinyloxazolidone
PVP	Polyvinylpyrrolidone
P2VP	Poly(2-vinylpyridine)
P3MO	Poly(trimethylene oxide)
P4VP	Poly(4-vinylpyridine)
Py	Pyridyl (N-substituted, unless otherwise indicated)
Q	Degree of swelling (e.g., $Q_w = 1/w_2$ for gel)
r	Binding ratio ($\equiv b_3/c_2$) in polymer/cosolute binding (see Chapter 2 of Volume II)
R	Alkyl group
R	Radius of gyration
$\hat{R}$	Root-mean-square value of R
RA	Raman spectroscopy
RF	Retractile force
RG	Reversible gelation
RI	Refractive index measurement
r.t.	Room temperature
s^o	Sedimentation coefficient (at infinite dilution)
S	Degree of substitution
SA	Sarcosine anhydride
SAXS	Small-angle X-ray scattering
SE	Sedimentation equilibrium
SF	Spread films of polymers at interfaces
SR	Stress relaxation
SS	Sedimentation speed
st	Syndiotactic
SV	Specific volume (i.e., also density) measurements
t	Celsius temperature (°C)
t	Trans conformation
t_c	Critical precipitation temperature
t_f	Flocculation temperature for emulsions or dispersions of solids
t_g	Glass transition temperature

t_m	Melting point
t_p	Precipitation temperature (cloud point)
T	Absolute (thermodynamic) temperature (K)
T_1	Longitudinal (spin-lattice) relaxation time
T_2	Transverse (spin-spin) relaxation time
Td	Tetradecyl (myristyl)
TD	Translational diffusion
TNS	2-*p*-Toluidinylnaphthalene-6-sulfonate
TP	Thermodynamic parameters (especially for dilution of polymer solutions)
TRG	Thermally reversible gelation
u	Electrophoretic mobility
U	Urea
U_1	Equilibrium uptake of water by a polymer (g water/100 g dry polymer)
UC	Ultracentrifuge methods (especially SE,SS)
Ud	Undecyl
UF	Ultrafiltration methods for consolute binding (see Chapter 2 of Volume II)
US	Ultrasonic (absorption or velocity)
USA	Ultrasonic absorption measurements
USB	Uniform site-binding mode (see Chapter 2 of Volume II)
USNF	US National Formulary
USP	US Pharmacopeia
USV	Ultrasonic velocity measurements
UV	Ultraviolet absorption spectroscopy
v	Specific volume
$\tilde{v}_2^0$	Partial specific volume for polymer in solution at extreme dilution
V_i	Molar volume of component i
VAC	Vinyl acetate
VAL	Vinyl alcohol
VI	Viscosity (including general rheological) measurements
VL	Visible light absorption spectroscopy
VM	Volume change on mixing
VME	Vinyl methyl ether (methyl vinyl ether)
VMO	Vinylmethyloxazolidone
VO	Vinyloxazolidone
VP	Vapor pressure measurements; vinyl pyrrolidone — in polymers only
VPI	Vapor pressure (isopiestic method)
VPO	Vapor pressure osmometer
2VP	2-Vinylpyridine
4VP	4-Vinylpyridine
VU	Vinylurethane
w	Weight (mass) fraction
x	Mole fraction
X	Cation
Y	Anion
z	Electrovalence
ZP	Zeta potential
α	Thermal expansion coefficient

β	Linear charge density on polyelectrolyte chain
γ	Thermal pressure coefficient, $(dp/dT)_v$; activity coefficient
Γ	Donnan exclusion ratio
$\Gamma_2(\Gamma_{2m})$	Interfacial concentration of polymer (limiting "monolayer" value)
Γ_3	Reduced free cosolute concentration (see Chapter 2 of Volume II)
δ	NMR chemical shift (= O for tetramethylsilane)
δ_2	Interfacial film thickness
Δ	Change (finite)
ϵ	Extinction coefficient; electric permittivity (dielectric constant — zero frequency unless otherwise indicated)
ϵ', ϵ''	Real and imaginary parts of electric permittivity.
ξ	Reduced linear charge density parameter for a polyelectrolyte chain (see Chapter 1 of Volume II)
η	Viscosity
η_c	Viscosity number (see Section C.7 in Chapter 1 of Volume I)
η_{sp}	Specific viscosity (see Section C.7 in Chapter 1 of Volume I)
$[\eta]$	Intrinsic viscosity (limiting viscosity number)
$[\eta]_o$	Additive parameter in MHSR equation (q.v.)
θ	Fractional occupancy of binding sites on a polymer chain in solution (see Chapter 2 of Volume II); fractional occupancy of interfacial adsorption sites on a solid by adsorbed polymer (see Chapter 3 of Volume II)
Θ	Ideal (Flory, theta) temperature
$\Theta_+(\Theta_-)$	Value of Θ for system where phase separation occurs on heating (cooling)
κ	Enthalpy of dilution parameter; compressibility
$\tilde{\kappa}_2^o$	Partial specific compressibility of the polymer in solution at extreme dilution
λ	Wavelength
μ	Viscosity "steric hindrance parameter" of Müh in MHSM equation (see Section III. G of Chapter 1 in Volume I); electric dipole moment
μ_I	Interfacial electric dipole moment
υ	Frequency
Π	Osmotic pressure
ρ	Density
$\bar{\rho}_2$	Average density of polymer within the domain of the molecule
σ	Stereoregularity parameter (ranges from 0 for purely syndiotactic, through 0.5 for atactic, up to 1.0 for purely isotactic polymer); electrical conductivity
τ	Relaxation time; NMR chemical shift (= 10 for tetramethylsilane)
τ_R	Rotational relaxation time
ϕ	Volume fraction
ϕ_x	Volume fraction crystallinity
$\chi(\chi_{ij})$	Flory-Huggins interaction parameter (excess free energy of dilution parameter) for components i and j
ψ	Entropy of dilution parameter
ψ_2	Apparent specific volume of polymer in solution
1	Solvent component — water, unless otherwise specified
2	Polymer component (or first polymer, in multipolymer systems)
3	Small-molecule cosolute; second polymer; material forming the interface with the polymer solution

%	For polymer concentrations, weight in volume, i.e., g dl^{-1}; for copolymer compositions, mol %
+(−)	As second subscript on t_c and t_p (or t_f) indicates system where the precipitation (or flocculation) occurs on heating (cooling)
~	Noncovalent interactions (e.g., A ~ B); approximately

REFERENCES

1. **Anon.,** Recommendations for abbreviations of terms relating to plastics and elastomers, *Pure Appl. Chem.*, 18, 581, 1969.
2. **Anon.,** List of standard abbreviations (symbols) for synthetic polymers and polymer materials 1974, *Pure Appl Chem.*, 40, 475, 1974.
3. **Whiffen, D. H.,** *Manual of Symbols and Terminology for Physicochemical Quantities and Units,* IUPAC/Pergamon Press, Oxford, 1979.
4. **Elias, H.- G.,** Macromolecules, Vol. 1, Plenum Press, New York, 1977.

INDEX

A

B

D

K

L

M

N

O

P

R

S

T

U

V

W

X

Z

For Product Safety Concerns and Information please contact our EU
representative GPSR@taylorandfrancis.com
Taylor & Francis Verlag GmbH, Kaufingerstraße 24, 80331 München, Germany

www.ingramcontent.com/pod-product-compliance
Lightning Source LLC
LaVergne TN
LVHW081322300926
849161LV00008B/1852

* 9 7 8 1 3 1 5 8 9 8 5 7 5 *